Advances in Genome Science: Keeping up with Genome Sequence and Expression (Volume 3)

Editor

Christian Neri

Research Director at INSERM
Laboratory of Neuronal Cell Biology & Pathology
Psychiatry and Neuroscience Center U894
2 ter rue d'Alesia 75014 Paris
France

Bentham Science Publishers Ltd.
Executive Suite Y - 2
PO Box 7917, Saif Zone
Sharjah, U.A.E.
subscriptions@benthamscienc.org

CONTENTS

PREFACE

Genome science or genomics is essential to advancing knowledge in the fields of biology and medicine. Specifically, researchers learn about the molecular biology behind genetic expression in living organisms and related methods of treating human genetic diseases (including gene therapy). Advances in Genome Science is an eBook series which provides a multi-disciplinary view of some of the latest developments in genome research, allowing readers to capture the essentiality and diversity of genomics in contemporary science.

The third volume of this eBook series contains a selection of articles illustrating how genome sequencing and the study of genome structure and epigenetic regulation remain at the center of some of the most exciting developments in the fields of biology as well as human disease research.

Christian Neri

Research Director at INSERM

Laboratory of Neuronal Cell Biology & Pathology

Psychiatry and Neuroscience Center U894

2 ter rue d'Alesia 75014 Paris

France

List of Contributors

Allison B. Weckerle

Department of Biochemistry, Wake Forest School of Medicine, Medical Center Boulevard, Winston-Salem, NC 27157-1016, USA

Arnaud Duchon

Institute of Molecular Genetics and Cell Biology (IGBMC), INSERM U964, CNRS UMR7104, University of Strasbourg, Illkirch, France

Baltasar Mayo

Department of Microbiology and Biochemistry, Dairy Research Institute of Asturias Principality (IPLA-CSIC), Villaviciosa, Asturias, Spain

Carolyn M. Klinge

Department of Biochemistry & Molecular Biology, Center for Genetics and Molecular Medicine, University of Louisville School of Medicine, Louisville, KY. 40292, USA

Cláudia S. Baptista

Department of Veterinary Clinics, Institute of Biomedical Sciences Abel Salazar, University of Porto (ICBAS-UP), Porto, Portugal

Damien Maréchal

Institute of Molecular Genetics and Cell Biology (IGBMC), INSERM U964, CNRS UMR7104, University of Strasbourg, Illkirch, France

David Horvath

USDA-ARS, Bioscience Research Laboratory, 1605 Albrecht Blvd. Fargo ND 58105, USA

Diddier Prada

Unidad de Investigación Biomédica en Cáncer-Instituto Nacional de Cancerología-Instituto de Investigaciones Biomédicas, Universidad Nacional Autónoma de México (UNAM), Mexico City, Mexico

Douwe van Sinderen

Department of Microbiology, University College Cork, Cork, Ireland

Elias K. Michaelis

Higuchi Biosciences Center and Department of Pharmacology and Toxicology, The University of Kansas, Lawrence, KS 66047, USA

Estela Bastos

Institute for Biotechnology and Bioengineering, Centre of Genomics and Biotechnology, University of Trás-os-Montes and Alto Douro (IBB/CGB-UTAD), Vila Real, Portugal

Fátima Gärtner

Department of Pathology and Molecular Immunology, Institute of Biomedical Sciences Abel Salazar (ICBAS), University of Porto, Porto, Portugal and Institute of Pathology and Immunology (IPATIMUP), University of Porto, Porto, Portugal

Francesca Biagioni

Translational Oncogenomics Unit, National Cancer Institute Regina Elena Rome, Italy

Francesca Fausti

Translational Oncogenomics Unit, National Cancer Institute Regina Elena Rome, Italy

Giannina Descalzi

Department of Physiology, Faculty of Medicine, University of Toronto Centre for the Study of Pain, 1 King's College Circle, Toronto, Ontario, Canada

Giovanni Blandino

Translational Oncogenomics Unit, National Cancer Institute Regina Elena Rome, Italy

Giulia Fontemaggi

General Pathology Section, Department of Clinical and Experimental Medicine, Perugia University, Perugia, Italy

Henrique Guedes-Pinto

Institute for Biotechnology and Bioengineering, Centre of Genomics and Biotechnology, University of Trás-os-Montes and Alto Douro (IBB/CGB-UTAD), Vila Real, Portugal

Ivo G. Gut

CEA/DSV/IG-Centre National de Génotypage, Evry Cedex, France

Laura W. Dillon

Department of Biochemistry, Wake Forest School of Medicine, Medical Center Boulevard, Winston-Salem, NC 27157-1016, USA

Luis A. Herrera

Unidad de Investigación Biomédica en Cáncer-Instituto Nacional de Cancerología-Instituto de Investigaciones Biomédicas, Universidad Nacional Autónoma de México (UNAM), Mexico City, Mexico

Marco A. Andonegui

Unidad de Investigación Biomédica en Cáncer-Instituto Nacional de Cancerología-Instituto de Investigaciones Biomédicas, Universidad Nacional Autónoma de México (UNAM), Mexico City, Mexico

Marco Ventura

Laboratory of Probiogenomics, Department of Genetics, Biology of Microorganisms, Anthropology and Evolution, University of Parma, Italy

Mary L. Michaelis

Higuchi Biosciences Center and Department of Pharmacology and Toxicology, The University of Kansas, Lawrence, KS 66047, USA

Min Zhuo

Department of Physiology, Faculty of Medicine, University of Toronto Centre for the Study of Pain, 1 King's College Circle, Toronto, Ontario, Canada and

Department of Brain and Cognitive Sciences, Seoul National University, Seoul 151-746, Korea

Raquel Chaves

Institute for Biotechnology and Bioengineering, Centre of Genomics and Biotechnology, University of Trás-os-Montes and Alto Douro (IBB/CGB-UTAD), Vila Real, Portugal

Sabrina Strano

Molecular Chemoprevention Group, Scientific Direction, National Cancer Institute Regina Elena Rome, Italy

Sara Donzelli

Translational Oncogenomics Unit, National Cancer Institute Regina Elena Rome, Italy

Sara Santos

Institute for Biotechnology and Bioengineering, Centre of Genomics and Biotechnology, University of Trás-os-Montes and Alto Douro (IBB/CGB-UTAD), Vila Real, Portugal

Susan S. Kim

Department of Physiology, Faculty of Medicine, University of Toronto Centre for the Study of Pain, 1 King's College Circle, Toronto, Ontario, Canada

Tissa T. Manavalan

Department of Biochemistry & Molecular Biology, Center for Genetics and Molecular Medicine, University of Louisville School of Medicine, Louisville, KY. 40292, USA

Véronique Brault

Institute of Molecular Genetics and Cell Biology (IGBMC), INSERM U964, CNRS UMR7104, University of Strasbourg, Illkirch, France

Xinkun Wang

Higuchi Biosciences Center and Department of Pharmacology and Toxicology, The University of Kansas, Lawrence, KS 66047, USA

Yann Hérault

Institute of Molecular Genetics and Cell Biology (IGBMC), INSERM U964, CNRS UMR7104, University of Strasbourg, Illkirch, France and Institute Clinic Souris (ICS), IEFERM U964, CNRS UMR7104, University of Strasbourg, Illkirch, France

Yuh-Hwa Wang

Department of Biochemistry, Wake Forest School of Medicine, Medical Center Boulevard, Winston-Salem, NC 27157-1016, USA

Send Orders for Reprints to reprints@benthamscince.net
Advances in Genome Science, Vol. 3, 2014, 3-32

CHAPTER 1

Modeling Copy Number Variations in the Mouse

Yann Hérault[1,2,*], Arnaud Duchon[1], Damien Maréchal[1] and Véronique Brault[1]

[1]Institute of Molecular Genetics and Cell Biology (IGBMC), INSERM U964, CNRS UMR7104, University of Strasbourg, Illkirch, France and [2]Institute Clinic Souris (ICS), INSERM U964, CNRS UMR7104, University of Strasbourg, Illkirc, France

Abstract: Changes in the number of chromosomes and variations in the copy number of chromosomal regions are described in various pathological conditions, such as cancer and aneuploidy, but also in normal physiological condition. Our initial view of DNA replication and mitotic preservation of chromosomal integrity is now challenged as new technologies allow the observation of mosaic somatic changes in copy numbers of chromosome regions of different sizes. In the mouse various engineering strategies are available to better understand the significance of copy numbers in normal physiological condition. For example, inducing recombination during the G2 phase will allow the generation of deletions and duplications of regions of interest prior to mitosis. The effects of copy number variation and segmental aneuploidy can be observed now in daughter cells and explored in somatic mosaics for large chromosomal regions.

Keywords: Chromosomal engineering, Cre/loxP, deletion, duplication, mosaic genetics.

INTRODUCTION

Our knowledge about human genetic variation has considerably evolved in the past few years with the development of genome-wide technologies such as oligonucleotide microarray technologies and the next-generation sequencing with "paired-end" methods [1] that reveal structural variations and enable screening of the genome at a submicroscopic level. The discovery of frequent copy number variants (CNVs) between individuals, including deletions and duplications of

[*]**Address correspondence to Yann Hérault:** Institute of Molecular Genetics and Cell Biology (IGBMC), INSERM U964, CNRS UMR7104, University of Strasbourg, Illkirch, France; Tel: +33 388 65 5657; E-mail: herault@igbmc.fr

Christian Neri (Ed)

DNA sequences, has remodeled our view of genomic integrity. Thousands of different CNVs have been identified (http://projects.tcag.ca/variation) overlapping a large number of genes [2, 3] and accounting for about 13% of the human genome. Estimates suggest that CNVs account for up to 24 Mb [3, 4] of the genetic difference between individuals, exceeding single nucleotide polymorphism (SNP) (2.5 Mb) [5].

In addition of being a major source of genetic diversity in the human population, CNVs contribute to various pathological conditions such as cancer, aneuploidies and contiguous gene syndromes (CGS). The first observed CNVs were the result of large chromosomal rearrangements, visible under a light microscope and causing intellectual disabilities (ID) such as Down (trisomy 21), Turner (monosomy X), Fragile X (Xq27.3) and Cri du Chat (del 5p-) syndromes [6]. These genomic disorders are often sporadic as they are *de novo* changes. However, some CNVs can be inherited in a Mendelian way: Charcot-Marie-Tooth disease type 1A (CMT1A) is a dominant neuropathy associated with a duplication of the gene for peripheral myelin protein 22 (PMP22) [7], while a deletion of the b-globin gene cluster is responsible for the recessive anemia-thalassemia [8]. The number of genomic disorders has drastically increased since the 1980s with the constant discovery of submicroscopic (<5 Mb) genomic deletions and duplications associated with pathological phenotypes [9, 10]. Microdeletion of 15q11.2q12 was associated with Prader-Willi syndrome in 1981 [11] and the first recurrent microduplication was identified in 1991 in patients with CMT1A [7]. A 17p13.3 deletion was found in Miller-Dieker lissencephaly syndrome [12]. Recurrent genomic rearrangements of 17q12, encompassing HNF1 beta, were associated with diabetes, renal disease and ID [13]. 1.5 to 3.0-Mb hemizygous deletions of chromosome 22q11.2 causing haploinsufficiency of the *TBX1* gene were discovered in DiGeorge syndrome [14]. Changes in 17q21.31 or 16p11.2 copy numbers were associated with specific forms of cognitive disabilities [15-23]. The discovery of new CNVs has increased by 15-20% the diagnosis of recognized syndromes of mental retardation that affect 2-3% of the population [24]. Moreover, CNVs are more and more associated with complex traits such as autism [25-28], schizophrenia [29] or epilepsy [30-33] and susceptibility to infectious diseases [34] such as HIV [35-37], Crohn disease [38-41], psoriasis [42, 43] and malaria [44].

A major mechanism by which rearrangements can cause a phenotype is the alteration of the copy number of a gene (or genes) thus influencing its expression [45]. However, deletions and duplications can also cause gene expression disturbance due to a change in the position of a gene and hence its genetic environment [46]. This disturbance can be either the result of the physical dissociation between the transcription unit and its regulatory sequences that can sometimes lie as far as 1 Mb away from it [47, 48], or the result of the alteration of local or global regulation of chromatin structure [49-52]. These different mechanisms are observed in human disease states. For example, Aniridia (absence of the iris) is due to haploinsufficiency of *PAX6* at 11q13 and has been shown to result from a chromosomal rearrangement that disrupts the region downstream of the PAX6 transcription unit [53-56]. Human diseases resulting from aberrant gene expression through altered chromatin structures include FSHD (facioscapulohumeral dystrophy) (MIM 158900), a neuromuscular disorder affecting the facial and shoulder girdle muscles, and one case of α-Thalassemia (MIM 141800) in which the *HBA2* gene is silenced by methylation of its promoter region by an antisense RNA derived from the truncated neighboring *LUC7L* gene on the opposite strand [52]. Genomic rearrangements of the genome can also trigger phenotypes by unmasking recessive mutations or functional polymorphism of the remaining allele in the case of deletions. In the case of Sotos syndrome caused by the deletion 5q35 it was shown that the severity of the disease depends on the polymorphism of the remaining copy of the coagulation factor 12 gene [57].

CNVs often arise from meiotic recombination between highly similar (>97% sequence identity) duplicated sequences (termed low-copy repeats or LCRs) located less than 10 Mb apart. LCRs represent up to 5% of the haploid human genome [58] and can cause genomic instability as they may cause misalignment of chromosomes or chromatids and cause nonallelic homologous recombination (NAHR), resulting in unequal crossing-over [59]. But other mechanisms implicated in DNA replication and repair have also been proposed as responsible for CNVs formation: nonhomologous end joining (NHEJ) is a mechanism that repairs DNA double strand breaks and, when defective, can lead to translocations and telomere fusion, a hallmark of tumor cells [60]. The fork stalling and template switching (FoSTeS) model now superseded by the microhomology mediated break-induced repair (MMBIR) model, and DNA replication error mechanisms have recently been shown to play an important role in the origin of nonrecurrent rearrangements with a complex structure leading to genomic disorders [61-64]. Growing numbers of studies suggest those mechanisms are responsible for the generation of CNVs observed in many different pathologies such as Duchenne muscular dystrophy, Smith-Magenis syndrome, Pelizaeus-Merzbacher and Charcot-Marie-Tooth diseases, microdeletion syndromes or even more drastic chromosomal aberrations such as isochromosomes [65-69]. It was suggested that CNVs made of large inverted repeats at the end of a chromosome arise from the breakage-fusion-bridge cycle mechanism which occurs after replication of a chromosome that has lost its telomere due to a double strand break and which consists of rounds of fusion of sister chromatids forming a dicentric chromosome that is subsequently broken apart during anaphase [70, 71]. Finally, Carvalho and collaborators [72] proposed a new mechanism for chromosome rearrangements, involving inverted repeats present sometimes as far as more than 300 kb apart and leading to complex genomic rearrangements all consisting of the same intermixed duplications and triplications.

Mechanisms implicating DNA replication and repair also have the particularity of leading to CNVs mosaicism. Occurrence of somatic mosaicism has been mostly observed in rearrangements of the immunoglobulin and T-cell-receptor genes [73], in cancers [74-76] and for chromosomal aberrations connected with diseases [77] but leading to a milder phenotype of the disease [78, 79]. CNVs somatic mosaicism as a more general phenomenon occurring in healthy individuals was proposed only recently, with more and more data showing inter- and intra- tissues mosaicism for entire chromosome aneuploidies in germ cells, placenta, brain, skin, liver and blood [80, 81]. Their occurrence remains however unclear because the techniques used to study CNVs requires a large number of pooled cells. Recent studies have started pointing out the importance of CNVs somatic mosaicism: extensive chromosomal instability and *de novo* recurrent CNVs have been reported in human cleavage-stage [82] and in mouse embryonic stem cell [83], respectively, and somatic CNV mosaicism has been reported in different human tissues and organs [84] with studies on monozygotic twins revealing putative *de novo* somatic CNV events [85]. In 2010, the group of Anja Weise at the Institute of Human Genetics in Magdeburg (Germany) provided the first evidence of somatic mosaicism for CNVs between different cell types of one individual [86]. They used an approach that was defined to determine chromosomal parental origin in single cells based on fluorescent *in situ* hybridization ('parental-origin-determination fluorescence *in situ* hybridization' or pod-FISH) [87] to visualize CNVs on homologous chromosomes metaphase spreads from 10 healthy individuals and found CNV variation between different cell types but not within one cell type. They hence proposed that there is early embryonic chromosome instability resulting in stable mosaic patterns in human adult tissues.

While CNVs are now recognized as a major source of genetic variability between individuals, their biological effects are largely unexplored. Analysis of their pathological role and molecular mechanism are problematic to investigate in human and require an animal model which is genetically, morphologically and physiologically close. The mouse constitutes a model organism of choice with an anatomy, physiology and genetics highly similar to that of humans. 80% of mouse genes have an orthologous counterpart in the human genome [88] and 99% have a sequence match. Homologous genes are found in the same order and relative orientation in large blocks of conserved syntenic regions. In addition, the ability to manipulate the mouse genome has made the mouse the primary mammalian genetic model organism. CNVs can be artificially created using various chromosome engineering techniques that enable the precise manipulation of large genomic regions. The technology based on the Cre-*lox*P system is used to generate new chromosomes carrying deletions, duplications, inversions and translocations in targeted regions of interest and to study contiguous gene syndromes as well as normal developmental processes associated with CNVs.

I- GERM LINE GENETIC ENGINEERING USING EMBRYONIC STEM (ES) CELLS

Modelling chromosomal rearrangements such as deletions, duplications, inversions or translocations in the mouse genome has been made possible by combining gene targeting in embryonic stem (ES) cells [89-92] with site-specific recombinase (SSR) systems such as the Cre-*lox*P (Fig. **1**) [93]. Gene targeting enables the introduction of small DNA sequences known as *lox*P sites to predefined loci by homologous recombination. Subsequent expression of the bacteriophage Cre recombinase that catalyses recombination between *lox*P sites without any cofactor [94-97] allows the generation of large chromosomal rearrangements either in ES cells with subsequent transmission to the mouse by injecting the transformed ES cells to blastocysts, or directly in the mouse. The size of the *lox*P, small enough to be introduced very easily by genetic engineering and large enough to avoid problems associated with cryptic occurrence in eukaryotic genomes, has made the Cre-*lox*P system a very simple and powerful tool for mouse genomic engineering.

Cre-*lox*P recombination in ES cells can be easily generated for deletions of small genomic regions (<100 kb) with insertion of two *lox*P sites in a direct orientation and cis configuration at each end of the region [98, 99]. However, the frequency of recombination is too low to obtain larger deletions or duplications with *lox*P sites on different chromosomes and requires selection through the restoration of a positive selection marker [100-104] or the deletion of a negative selection marker [105-108]. To this end, targeting vectors were designed, containing the 5' or 3' part of the *HPRT* minigene together with one *lox*P site [100]. The use of this technique has been largely facilitated by the creation and the allocation of specific targeting vectors [109] that are now mapped on the mouse genome (Mutagenic Insertion and Chromosome Engineering Resource [MICER], http://www.sanger.ac.uk/Post Genomics/mousegenomics/) [110], avoiding the fastidious and time-consuming task of constructing the targeting vectors. In addition, it is now possible to easily generate a panel of deletions by using *lox*P sites within the genome of retroviral vectors to create nested insertions of the second *loxP* site extending from a few kb to several Mb around the locus of insertion of the first targeted *lox*P site [111].

Although Cre-*lox*P recombination is a very powerful and convenient technique, it has to be carefully planned as its efficiency depends on many factors. The obtention of deletion, duplication or inversion depends on the position (cis or trans) and the relative orientation of the two *lox*P sites, and on the cell cycle stage during which the Cre-mediated recombination occurs [112,113]. Cre-mediated recombination efficiency depends on the recombinase activity [114], the region targeted that could induced ES cell lethality after deletion [115], the genetic distance between the *loxP* sites in a cis configuration (10%-0.1%) and the configuration of the *lox*P sites [100, 113, 116]. The size of the deletions is limited by haploinsufficiency of genes within the deleted interval that can trigger ES cell lethality but this can be circumvented by making balancer deletion/duplication using *trans* recombination Fig. **1**.

The possibility of manipulating large chromosomal regions has provided the scientific community with the opportunity to develop mouse models of CGS. Models developed for del22q11.2 (DiGeorge syndrome) [117, 118] enabled the identification of *TBX1* as a candidate gene for the pathology [119]. Modelling Down syndrome (DS), the only viable autosomal aneuploidy in human, requires multiple chromosomal rearrangements as the around 300 human chromosome 21 orthologs are found in three syntenic regions on mouse chromosomes (Mmu) 10, 17 and 16. The phenotypic analysis of the Ts65Dn and Ts1Cje models obtained in a random manner and trisomic for part of the Mmu16 [120-122], the Ts1Rhr model trisomic for a ~5 Mb Down syndrome critical region (DCR) [101, 123] and the Ts1Yah model trisomic for the syntenic region on Mmu17 [124], both generated by chromosomal engineering, starts revealing the complexity of the genetic interactions at the origin of the disease (for review [125-127]). Mouse models have also been created for the 17q11.2 microdeletion and its reciprocal microduplication leading to Smith-Magenis and Potocki-Lupski syndromes, respectively [128-132], for the Prader Willi CSG [133], as well as for studying large genes or clusters of genes such as the *HoxD* genes cluster [99], the Duchenne muscular dystrophy gene [104], the amyloid precursor protein gene [105] or the *Nf1* gene [106]. Cre-*lox*P chromosome engineering in mouse ES cells was also developed to generate translocations or deletions similar to those found in cancer [115, 134-138]. Finally, in 2009, a mouse model was created for the first time to model a polygenic disease [139]. Autism, a common heterogeneous psychiatric disorder is a developmental brain disease characterized by deficit in social interaction and communication, and stereotyped repetitive behaviour. Autism has been shown to have a strong genetic basis with chromosomal anomalies accounting for 10 to 20% of the cases [140]. Among these anomalies, chromosome 15q11-13 duplication is the only recurrent cytogenetic aberration that could be associated with autism. Nakatani and collaborators [139] generated a 6.3 Mb duplication of the conserved linkage group on Mmu7 and showed that the mouse model generated recapitulated aspects of human autism such as social abnormalities and increased anxiety and fear, providing a new tool to decipher the physiological and molecular mechanism underlying this pathology.

The Zfn-nuclease technology also offers new possibilities. Joo Lee *et al.* [141] reported the engineering of deletions, duplications and inversions in cell culture of long genomic regions up to 835 Kb, without pre-insertion of *lox*P sites, just using pairs of dedicated Zfn-nucleases. While no selection is used, this approach requires the screening of a large number of treated cells and additional limiting dilution steps to isolate the clones carrying one specific rearrangement. The generation of deletions, duplications and inversions is based on single intra-chromosomal ligation of double strand break (DSB) endpoints, inter-chromosomal (or inter-chromatidal) ligations, and separated ligations respectively. Further investigations need to be done to validate the promising method for several loci of interest.

II- *IN VIVO* GERMLINE Cre-*lox*P RECOMBINATION

Creating chromosomal rearrangements in the whole organism offers an alternative to the *in vitro* protocol that is very labor-intensive and requires many rounds of ES cells genomic manipulations. It requires that the *lox*P sites recombination occurs either in germ line cells during meiosis or ubiquitously in all of the tissues or in early embryogenesis. The rearrangements can then be transmitted to the progeny. This can be achieved by using several Cre transgenic lines such as *Sycp1*-Cre [142], *Zp3-Cre* [143, 144], *Protamine*-Cre [145], *CMV-Cre* [146], *R26*-Cre [147], *PGK*-Cre [148] or *Hprt-Cre* [149]. Most of the large *in vivo* rearrangements are deletions that are obtained from two *lox*P sites in the same relative orientation inserted in *cis* in ES cells [99, 150-152]. The frequency of recombination in *cis* decreases with the increasing distance between the *lox*P sites (0.3%-1%), but still remains high enough to be workable up to distances reaching 28 Mb [153]. For *lox*P sites that are originating from different mouse founder lines and hence are not in *cis*, deletion can still be obtained by mating the two founder lines and selecting for recombination after classical crossing-over, a technique called STRING for Sequential Targeted Recombination Induced Genomic rearrangement [153]. Similarly, inversions have been obtained *in vivo* with two *lox*P sites located in the same chromosome but with a reverse orientation [150, 154]. The different ubiquitously expressing Cre transgenic lines are not all similarly efficient in making the recombination and can lead to mosaic individuals in the first generation [102, 155, 156]. An alternative to the ubiquitous Cre transgenic lines is the *Zp3*-Cre in which *cre* expression is controlled by regulatory sequences from the mouse zona pellucida gene, expressed exclusively in the growing oocyte prior to the completion of the first meiotic division. But this transgenic Cre line was also shown to produce some mosaicism in a mouse model for DiGeorge del22q1 syndrome [157].

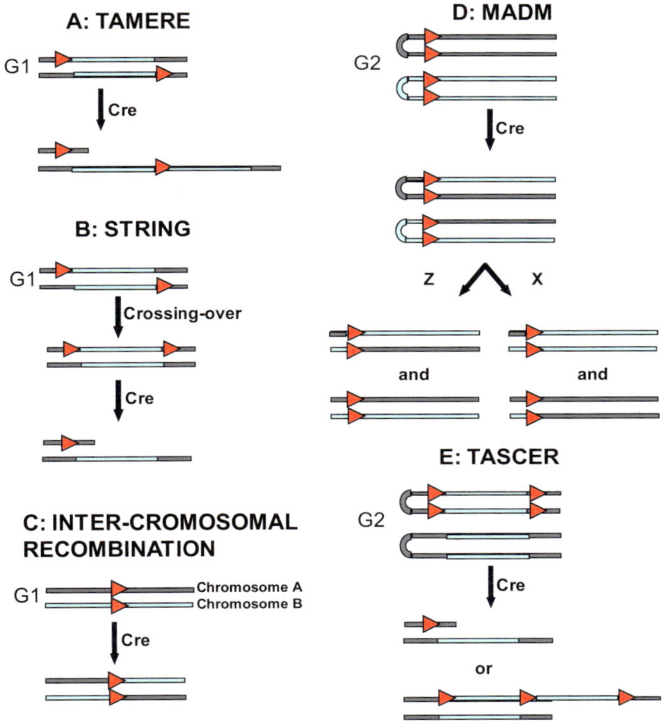

Figure 1: Schematic representation of the different *in vivo* strategies developed to model CNV in the mouse. (A) In the TAMERE strategy, *lox*P sites are inserted in a trans configuration and expression of the Cre recombinase during meiosis enables to generate cells containing a duplication with the reciprocal deletion. **(B)**The STRING approach takes advantage of the classical crossing-over to bring the two *lox*P sites into a *cis* configuration to generate a deletion. **(C)** Inter-chromosomal recombinations are obtained by introducing *lox*P sites at the site of the desired rearrangement and expressing the Cre recombinase. **(D)** The MADM strategy base on Cre-mediated inter-chromosomal recombination and G2-X segregation during mitosis, offers the possibility to introduce mosaicism *in vivo*. **(E)** The TASCER strategy takes advantages of the G2 stage of the S phase to recombine *lox*P sites in cis, leading to the deletion and the duplication of the sequence surrounded with the *lox*P.

Creating chromosomal duplications requires that the two *lox*P sites are on homologous chromosomes (*trans*) and is only possible when the two homologous chromosomes becomes physically close together. Herault *et al.* [142] designed a strategy called TAMERE for TArgeting MEiotic Recombination that makes use of a Cre under the control of the Synaptonemal Complex 1 promoter that is expressed during prophase of meiosis in male spermatocytes when chromatid pairs are closely aligned [158], enabling *trans* recombination [150, 159, 160]. However, this *trans* recombination strategy seems to be limited to short-distance recombination because no rearrangement was obtained for a distance of 3.9 Mb [161] compared to *in vivo* recombination for *cis* configurations which is still effective up to 28 Mb [153]. The TAMERE strategy Fig. **1** has been used extensively to rearrange and study the *HoxD* complex region [142, 154, 162, 163] and has also been used to address the functional relationship between *PrPc* and *Dp*l, two genes involved in Prion disease, by making a double deletion of these two genes that are separated by 16 kb [160].

Beside the Cre-*lox*P system described here, the Flp-*FRT* system is another tyrosine SSR system that has been used as tool for DNA and genome engineering. However, FLP recombinase was found to be less efficient than Cre, partly due to an unsuitable optimum temperature for mammalian cells [164]. Recently, a new Cre-like recombinase, called Dre, was identified from P1-related phage D6 by Sauer and McDermott [165]. Dre recombinase has DNA specificity for a 32 bp DNA site (rox) distinct from the *lox*P and shows no cross-recombination with the Cre-*lox*P system. Like the Cre, it catalyzes both integrative and excisive recombination and requires no co-factor. The Dre-*rox* system was validated in the mouse by Anastassiadis and collaborators [166] who tested the efficiency of Dre-rox recombination *in vivo* by developing ubiquitous mouse lines based on the CAGGs promoter [167] and the ROSA26 locus. This new SSR provides a new useful toolkit for mouse genome engineering. Finally, recent applications for large serine SRRs such as phiC31 and phiBT1 [168] in eukaryotic genome engineering have emerged as new additional tools.

Association of the Cre-*lox*P system with transposon-based approaches should considerably speed up the generation of large-scale rearrangements *in vivo*. Transposons are mobile genetic elements that function as a bipartite system with a transposase to move around to different positions within the genome, potentially perturbing gene function. They have been extensively used for random mutagenesis and have been shown to be efficient both *in vitro* in ES cells and *in vivo* in the germline or somatically [169-176]. The most widely studied transposon/transposase system for mammalian mutagenesis is the Sleeping Beauty transposon of the Tc1-like mariner family [177]. Transposons have the advantage to be active *in vivo* and only require breeding of transgenic mice, their insertion sites being defined thanks to the transposon serving as a tag. High frequency of Sleeping Beauty transposon remobilization was achieved to produce mice with new insertions around a selected starting point by mating the mice containing the transposon with a transgenic line expressing the hyperactive form of the Sleeping Beauty transposase (HSB16) [178] under the control of the mouse *Prm1* promoter fragment that triggers expression of the transposase only transiently in late spermatogenesis [179]. More recently, an alternative transposon called Piggybac that can carry larger payloads between its transposon repeats has been derived from the cabbage looper moth (*trichoplusia ni*) [173]. Transposons offer the possibility to easily distribute *lox*P sites throughout the mouse genome and, when associated with the Cre-*lox*P technology, permit to generate rapidly and easily genome-wide chromosome rearrangements [156]. Such a technique has already been successfully used to bring the entire megabase-scale *HoxD* regulatory landscape in front of the *HoxC* gene cluster *via* a targeted translocation [180].

III- SOMATIC GENETIC ENGINEERING

Mosaicism in copy number variation has been observed in human [181]. It is a direct consequence of the missegregation of a chromosome during mitosis: while the two copies of one homologous chromosome migrate in one daughter cell that will be trisomic, the second cell will be monosomic with only one copy.

The percentage of aneuploid mosaicism depends on the developmental stage, the lineage of the cell that is affected and its viability, and the type of chromosomal rearrangement produced. If the rearrangement occurs early during embryogenesis, the mosaicism can be generalized to all cell lineages and can result in major phenotypic changes that are often lethal. For example, in Turner syndrome more than 98% of conceptuses that are 45, X do not survive to term and only mosaic are thought to be compatible with life [182].

Mosaic aneuploidy can be registered in all somatic cell populations contributing to normal inter individual variability, with a frequency of 1.25–1.45% per chromosome and a percentage of aneuploid cells above 30% reported in blood cells [183] and brain [184-186]. Reasons for the presence of such aneuploid cells and its consequence remain for the moment speculative.

Consequences of somatic aneuploid mosaicism can be analysed in the mouse by generating segmental aneuploid mosaicism in mouse tissues using the Cre-*lox*P system during the G2 phase of mitosis. Two different strategies were developed using this system. Zong *et al.* [187] developed in 2005 a technique utilizing Cre-mediated inter-chromosomal recombination prior to cell division (mitotic recombination) and took advantage of the G2-X segregation during mitosis to induce recombination events between homologous chromosomes in somatic cells and to mark the resulting daughter cells with different genotypes. They called this technique Mosaic Analysis with Double Markers (MADM). In addition, this technique can also be used with a pre-introduced deletion or duplication on one chromosome to create CNV mosaics composed of wild-type daughter cells bearing the normal allele or mutant cells containing the allele with the deletion or the duplication in order to study sporadic loss of heterozygosity. Another strategy takes advantages of the property of the Cre recombinase to react in the G2 phase of mitosis to directly generate different types of cells with either the duplication or the deletion of the chromosomal region boxed with the *lox*P sites [188]. The TASCER (Targeted Asymmetric Sister Chromatid Recombination) takes advantage of the presence of *lox*P sites on sister chromatids after replication of DNA in G2 phase during mitosis in order to generate *in vivo* cells harbouring microdeletions and microduplications. *lox*P sites were inserted on the same chromosome in a *cis* configuration separated by up to 2.2 Mb. Upon mating *cis*/+ mice with a general Cre deleter line, mosaic animals were obtained with cells containing either the deletion or the duplication. Indeed, if *lox*P recombination occurs after DNA replication, the corresponding duplication can also be obtained in addition to the expected deletion, and corresponds to the intermediate reaction product containing three *lox*P sites. The duplication with presence of three *lox*P sites represents the intermediate product between the double *cis* configuration with four *lox*P sites of the G2 stage cells and the deletion end product. This is the first report of an *in vivo* generated duplication coming from a *cis* configuration, presumably because, in previous reports, *lox*P sites were not as distant, resulting in a better efficiency avoiding the presence of the intermediate genomic duplication configuration. Hence, the increased distance tends to decrease the Cre-mediated recombination frequency [116] somehow preserving the chromosome with a large duplication. Using TASCER on two chromosomal regions homologous to the telomeric part of human chromosome 21, cells with partial monosomy and trisomy were efficiently recovered at a relatively constant level in numerous different tissues. The TASCER mechanism necessitates that the Cre is particularly active during the G2 phase during cell proliferation. This phenomenon has been described *in vitro* in embryonic stem cells [113, 189] but with much less efficiency than *in vivo*. The reason for this discrepancy might be that expression of the Cre is more stable *in vivo* and that the *in vivo* environment is more favourable for the viability of the aneuploid cells. Change in the chromatin structure, with complexes such as the cohexin one inducing physical connection of sister chromatids, can also facilitate the mitotic recombination [190, 191].

Efficiency of the TASCER also depends on the type of Cre transgenic lines used, as these do not express the Cre recombinase with the same efficiency. Duchon *et al.* [188] were able to induce up to 2.2 Mb deletions and duplications with a set of three different Cre lines, but could not obtain a larger modification of 9.2 Mb that might require a Cre line with a stronger promoter such as the CAG promoter inserted in the *Hprt* locus of the $Hprt1^{tm1(Cre)Mnn}$ mouse line, which has been used successfully in a recent study [149]. The fact that cells bearing the duplication was found in many different organs despite the less stable presence of three *lox*P sites [192] indicates that either the event is quiet frequent or that the trisomic state might have a positive effect in some tissues and points out the necessity to be very careful when using the Cre-*lox*P system *in vivo* to generate deletions in somatic tissue.

Whereas chromosomal mosaicism has been observed in pathological conditions, such as Turner syndrome [193] or in cancer [194], Duchon *et al.* [188] did not report any dramatic changes in the survival of trisomic and monosomic cells produced. In some tissues such as the muscle, this can be explained by compensatory effects due to their organization in a plurinuclear syncitia. However, in other non-syncitial tissues, other mechanisms such as non-cell autonomous effect may be proposed to explain this phenomenon.

Although mosaic aneuploidy and CNVs have been reported in nonpathological conditions in different cell populations such as neurons [80, 81, 184-186, 195-197], their role and biological consequences to the organ structure and function in health and disease remain to be explored. In the brain, mosaic DNA rearrangements have been suggested to contribute to the mechanism for neuronal cell diversification [198]. However, these are only assumptions and there is to date little evidence supporting a functional consequence for mosaic CNVs in non-pathological conditions. To answer those questions, future studies need to examine the effect of somatic CNVs on cell physiology. TASCER offers the possibility to explore consequences of such a phenomenon in a model organism because it results into substantial segmental aneuploid mosaicism. It can also be used to induce partial aneuploid conditions such as those detected in cancer cells and to further study the consequence of loss or gain of copy number for chromosomal regions during tumorigenesis [194], avoiding the fastidious need of gene-by-gene inactivation or ES-cell MICER recombination system. TASCER hence offers new perspective for the generation and analysis of CNVs in health and diseases.

ACKNOWLEDGEMENTS

We thank members of the laboratory for helpful discussion, and members of the AnEUploidy consortium (www.aneuploidy.org). This work was funded by grants from the National Centre for Scientific Research (CNRS), the Fondation Jerome Lejeune, and the AnEUploidy project (LSHG-CT-2006-037627) supported by the European commission under FP6.

CONFLICT OF INTEREST

The authors confirm that this chapter contents have no conflict of interest.

DISCLOSURE

The chapter submitted for Advances in Genome Science, Vol. 3 entitled **"Endometrial cancer as a familial tumor: pathology and molecular carcinogenesis by K. Banno"** is an update of our article published in Current Genomics, 2009, 10: 127-132.

REFERENCES

[1] Tuzun E, Sharp AJ, Bailey JA, Kaul R, Morrison VA, Pertz LM, Haugen E, Hayden H, Albertson D, Pinkel D, Olson MV, Eichler EE. Fine-scale structural variation of the human genome. Nature Genetics, 2005; 37: 727-732.

[2] Redon R, Ishikawa S, Fitch KR, Feuk L, Perry GH, Andrews TD, Fiegler H, Shapero MH, Carson AR, Chen WW, Cho EK, Dallaire S, Freeman JL, Gonzalez JR, Gratacos M, Huang J, Kalaitzopoulos D, Komura D, MacDonald JR, Marshall CR, Mei R, Montgomery L, Nishimura K, Okamura K, Shen F, Somerville MJ, Tchinda J, Valsesia A, Woodwark C, Yang FT, Zhang JJ, Zerjal T, Zhang J, Armengol L, Conrad DF, Estivill X, Tyler-Smith C, Carter NP, Aburatani H, Lee C, Jones KW, Scherer SW, Hurles ME. Global variation in copy number in the human genome. Nature, 2006; 444: 444-454.

[3] Conrad DF, Pinto D, Redon R, Feuk L, Gokcumen O, Zhang YJ, Aerts J, Andrews TD, Barnes C, Campbell P, Fitzgerald T, Hu M, Ihm CH, Kristiansson K, MacArthur DG, MacDonald JR, Onyiah I, Pang AWC, Robson S, Stirrups K, Valsesia A, Walter K, Wei J, Tyler-Smith C, Carter NP, Lee C, Scherer SW, Hurles ME, Wellcome Trust Case C. Origins and functional impact of copy number variation in the human genome. Nature, 2010; 464: 704-712.

[4] Eichler EE, Nickerson DA, Altshuler D, Bowcock AM, Brooks LD, Carter NP, Church DM, Felsenfeld A, Guyer M, Lee C, Lupski JR, Mullikin JC, Pritchard JK, Sebat J, Sherry ST, Smith D, Valle D, Waterston RH, Human Genome Structural V. Completing the map of human genetic variation. Nature, 2007; 447: 161-165.

[5] Altshuler D, Pollara VJ, Cowles CR, Van Etten WJ, Baldwin J, Linton L, Lander ES. An SNP map of the human genome generated by reduced representation shotgun sequencing. Nature, 2000; 407: 513-516.

[6] Lupski JR. Genomic disorders: structural features of the genome can lead to DNA rearrangements and human disease traits. Trends in Genetics, 1998; 14: 417-422.

[7] Lupski J, de Oca-Luna R, Slaugenhaupt S, Pentao L, Guzzetta V, Trask B, Saucedo-Cardenas O, Barker D, Killian J, Garcia C, Chakravarti A, Patel P. DNA duplication associated with Charcot-Marie-Tooth disease type 1A. Cell, 1991; 66: 219-32.

[8] Higgs DR, Vickers MA, Wilkie AOM, Pretorius IM, Jarman AP, Weatherall DJ. A review of the molecular-genetics of the human alpha-globin gene-cluster. Blood, 1989; 73: 1081-1104.

[9] Schmickel RD. contiguous gene syndromes - a component of recognizable syndromes. Journal of Pediatrics, 1986; 109: 231-241.

[10] Schluth-Bolard C, Till M, Edery P, Sanlaville D. New chromosomal syndromes. Pathologie Biologie, 2008; 56: 380-387.

[11] Ledbetter DH, Riccardi VM, Airhart SD, Strobel RJ, Keenan BS, Crawford JD. deletions of chromosome-15 as a cause of the prader-willi syndrome. New England Journal of Medicine, 1981; 304: 325-329.

[12] Dobyns WB, Reiner O, Carrozzo R, Ledbetter DH. lissencephaly - a human brain malformation associated with deletion of the lis1 gene located at chromosome-17P13. Jama-Journal of the American Medical Association, 1993; 270: 2838-2842.

[13] Nagamani SC, Erez A, Shen J, Li C, Roeder E, Cox S, Karaviti L, Pearson M, Kang SH, Sahoo T, Lalani SR, Stankiewicz P, Sutton VR, Cheung SW. Clinical spectrum associated

with recurrent genomic rearrangements in chromosome 17q12. Eur J Hum Genet, 2010; 18: 278-84.

[14] Edelmann L, Pandita RK, Spiteri E, Funke B, Goldberg R, Palanisamy N, Chaganti RSK, Magenis E, Shprintzen RJ, Morrow BE. A common molecular basis for rearrangement disorders on chromosome 22q11. Human Molecular Genetics, 1999; 8: 1157-1167.

[15] Koolen DA, Sharp AJ, Hurst JA, Firth HV, Knight SJL, Goldenberg A, Saugier-Veber P, Pfundt R, Vissers L, Destree A, Grisart B, Rooms L, Van der Aa N, Field M, Hackett A, Bell K, Nowaczyk MJM, Mancini GMS, Poddighe PJ, Schwartz CE, Rossi E, De Gregori M, Antonacci-Fulton LL, McLellan MD, Garrett JM, Wiechert MA, Miner TL, Crosby S, Ciccone R, Willatt L, Rauch A, Zenker M, Aradhya S, Manning MA, Strom TM, Wagenstaller J, Krepischi-Santos AC, Vianna-Morgante AM, Rosenberg C, Price SM, Stewart H, Shaw-Smith C, Brunner HG, Wilkie AOM, Veltman JA, Zuffardi O, Eichler EE, de Vries BBA. Clinical and molecular delineation of the 17q21.31 microdeletion syndrome. Journal of Medical Genetics, 2008; 45: 710-720.

[16] Sharp AJ, Hansen S, Selzer RR, Cheng Z, Regan R, Hurst JA, Stewart H, Price SM, Blair E, Hennekam RC, Fitzpatrick CA, Segraves R, Richmond TA, Guiver C, Albertson DG, Pinkel D, Eis PS, Schwartz S, Knight SJL, Eichler EE. Discovery of previously unidentified genomic disorders from the duplication architecture of the human genome. Nature Genetics, 2006; 38: 1038-1042.

[17] Shaw-Smith C, Pittman AM, Willatt L, Martin H, Rickman L, Gribble S, Curley R, Cumming S, Dunn C, Kalaitzopoulos D, Porter K, Prigmore E, Krepischi-Santos ACV, Varela MC, Koiffmann CP, Lees AJ, Rosenberg C, Firth HV, de Silva R, Carter NP. Microdeletion encompassing MAPT at chromosome 17q21.3 is associated with developmental delay and learning disability. Nature Genetics, 2006; 38: 1032-1037.

[18] Tan TY, Aftimos S, Worgan L, Susman R, Wilson M, Ghedia S, Kirk EP, Love D, Ronan A, Darmanian A, Slavotinek A, Hogue J, Moeschler JB, Ozmore J, Widmer R, Savarirayan R, Peters G. Phenotypic expansion and further characterisation of the 17q21.31 microdeletion syndrome. Journal of Medical Genetics, 2009; 46: 480-489.

[19] Varela MC, Krepischi-Santos ACV, Paz JA, Knijnenburg J, Szuhai K, Rosenberg C, Koiffmann CP. A 17q21.31 microdeletion encompassing the MAPT gene in a mentally impaired patient. Cytogenetic and Genome Research, 2006; 114: 89-92.

[20] Ballif BC, Hornor SA, Jenkins E, Madan-Khetarpal S, Surti U, Jackson KE, Asamoah A, Brock PL, Gowans GC, Conway RL, Graham JM, Medne L, Zackai EH, Shaikh TH, Geoghegan J, Selzer RR, Eis PS, Bejjani BA, Shaffer LG. Discovery of a previously unrecognized microdeletion syndrome of 16p11.2-p12.2. Nature Genetics, 2007; 39: 1071-1073.

[21] Battaglia A, Novelli A, Bernardini L, Igliozzi R, Parrini B. Further Characterization of the New Microdeletion Syndrome of 16p11.2-p12.2. American Journal of Medical Genetics Part A, 2009; 149A: 1200-1204.

[22] Bijlsma EK, Gijsbers ACJ, Schuurs-Hoeijmakers JHM, van Haeringen A, van de Putte DEF, Anderlid BM, Lundin J, Lapunzina P, Jurado LAP, Delle Chiaie B, Loeys B, Menten B, Oostra A, Verhelst H, Amor DJ, Bruno DL, van Essen AJ, Hordijk R, Sikkema-Raddatz B, Verbruggen KT, Jongmans MCJ, Pfundt R, Reeser HM, Breuning MH, Ruivenkamp CAL. Extending the phenotype of recurrent rearrangements of 16p11.2: Deletions in mentally retarded patients without autism and in normal individuals. European Journal of Medical Genetics, 2009; 52: 77-87.

[23] Kumar RA, KaraMohamed S, Sudi J, Conrad DF, Brune C, Badner JA, Gilliam TC, Nowak NJ, Cook EH, Dobyns WB, Christian SL. Recurrent 16p11.2 microdeletions in autism. Human Molecular Genetics, 2008; 17: 628-638.

[24] Cooper GM, Coe BP, Girirajan S, Rosenfeld JA, Vu TH, Baker C, Williams C, Stalker H, Hamid R, Hannig V, Abdel-Hamid H, Bader P, McCracken E, Niyazov D, Leppig K, Thiese H, Hummel M, Alexander N, Gorski J, Kussmann J, Shashi V, Johnson K, Rehder C, Ballif BC, Shaffer LG, Eichler EE. A copy number variation morbidity map of developmental delay. Nature Genetics, 2011; 43: 838-U44.

[25] Sebat J, Lakshmi B, Malhotra D, Troge J, Lese-Martin C, Walsh T, Yamrom B, Yoon S, Krasnitz A, Kendall J, Leotta A, Pai D, Zhang R, Lee YH, Hicks J, Spence SJ, Lee AT, Puura K, Lehtimaki T, Ledbetter D, Gregersen PK, Bregman J, Sutcliffe JS, Jobanputra V, Chung W, Warburton D, King MC, Skuse D, Geschwind DH, Gilliam TC, Ye K, Wigler M. Strong association of *de novo* copy number mutations with autism. Science, 2007; 316: 445-449.

[26] Abrahams BS, Geschwind DH. Advances in autism genetics: on the threshold of a new neurobiology. Nature Reviews Genetics, 2008; 9: 341-355.

[27] Fernandez TV, Sanders SJ, Yurkiewicz IR, Ercan-Sencicek AG, Kim YS, Fishman DO, Raubeson MJ, Song Y, Yasuno K, Ho WSC, Bilguvar K, Glessner J, Chu SH, Leckman JF, King RA, Gilbert DL, Heiman GA, Tischfield JA, Hoekstra PJ, Devlin B, Hakonarson H, Mane SM, Gunel M, State MW. Rare Copy Number Variants in Tourette Syndrome Disrupt Genes in Histaminergic Pathways and Overlap with Autism. Biological Psychiatry, 2012; 71: 392-402.

[28] Chen YZ, Matsushita M, Girirajan S, Lisowski M, Sun E, Sul Y, Bernier R, Estes A, Dawson G, Minshew N, Shellenberg GD, Eichler EE, Rieder MJ, Nickerson DA, Tsuang DW, Tsuang MT, Wijsman EM, Raskind WH, Brkanac Z. Evidence for involvement of GNB1L in autism. American Journal of Medical Genetics Part B-Neuropsychiatric Genetics, 2012; 159B: 61-71.

[29] Stefansson H, Rujescu D, Cichon S, Pietilainen OPH, Ingason A, Steinberg S, Fossdal R, Sigurdsson E, Sigmundsson T, Buizer-Voskamp JE, Hansen T, Jakobsen KD, Muglia P, Francks C, Matthews PM, Gylfason A, Halldorsson BV, Gudbjartsson D, Thorgeirsson TE, Sigurdsson A, Jonasdottir A, Bjornsson A, Mattiasdottir S, Blondal T, Haraldsson M, Magnusdottir BB, Giegling I, Moller HJ, Hartmann A, Shianna KV, Ge DL, Need AC, Crombie C, Fraser G, Walker N, Lonnqvist J, Suvisaari J, Tuulio-Henriksson A, Paunio T, Toulopoulou T, Bramon E, Di Forti M, Murray R, Ruggeri M, Vassos E, Tosato S, Walshe M, Li T, Vasilescu C, Muhleisen TW, Wang AG, Ullum H, Djurovic S, Melle I, Olesen J, Kiemeney LA, Franke B, Sabatti C, Freimer NB, Gulcher JR, Thorsteinsdottir U, Kong A, Andreassen OA, Ophoff RA, Georgi A, Rietschel M, Werge T, Petursson H, Goldstein DB, Nothen MM, Peltonen L, Collier DA, St Clair D, Stefansson K, Group. Large recurrent microdeletions associated with schizophrenia. Nature, 2008; 455: 232-U61.

[30] Suls A, Claeys KG, Goossens D, Harding B, Van Luijk R, Scheers S, Deprez L, Audenaert D, Van Dyck T, Beeckmans S, Smouts I, Ceulemans B, Lagae L, Buyse G, Barisic N, Misson JP, Wauters J, Del-Favero J, De Jonghe P, Claes LRE. Microdeletions involving the SCN1A gene may be common in SCN1A-mutation-negative SMEI patients. Human Mutation, 2006; 27: 914-920.

[31] Erez A, Patel AJ, Wang XQ, Xia ZL, Bhatt SS, Craigen W, Cheung SW, Lewis RA, Fang P, Davenport SLH, Stankiewicz P, Lalani SR. Alu-specific microhomology-mediated

deletions in CDKL5 in females with early-onset seizure disorder. Neurogenetics, 2009; 10: 363-369.

[32] Paciorkowski AR, Thio LL, Rosenfeld JA, Gajecka M, Gurnett CA, Kulkarni S, Chung WK, Marsh ED, Gentile M, Reggin JD, Wheless JW, Balasubramanian S, Kumar R, Christian SL, Marini C, Guerrini R, Maltsev N, Shaffer LG, Dobyns WB. Copy number variants and infantile spasms: evidence for abnormalities in ventral forebrain development and pathways of synaptic function. European Journal of Human Genetics, 2011; 19: 1238-1245.

[33] Paciorkowski AR, Rosenfeld J, Thio LL, Marini C, Christian S, Guerrini R, Shaffer LG, Dobyns WB. Recurrent copy number variations in developmental epilepsies including infantile spasms. Annals of Neurology, 2011; 70: S160-S160.

[34] Barreiro LB, Quintana-Murci L. From evolutionary genetics to human immunology: how selection shapes host defence genes. Nature Reviews Genetics, 2010; 11: 17-30.

[35] Gonzalez E, Kulkarni H, Bolivar H, Mangano A, Sanchez R, Catano G, Nibbs RJ, Freedman BI, Quinones MP, Bamshad MJ, Murthy KK, Rovin BH, Bradley W, Clark RA, Anderson SA, O'Connell RJ, Agan BK, Ahuja SS, Bologna R, Sen L, Dolan MJ, Ahuja SK. The influence of CCL3L1 gene-containing segmental duplications on HIV-1/AIDS susceptibility. Science, 2005; 307: 1434-1440.

[36] Liu SJ, Yao L, Ding DL, Zhu HZ. CCL3L1 Copy Number Variation and Susceptibility to HIV-1 Infection: A Meta-Analysis. Plos One, 2010; 5.

[37] Pelak K, Need AC, Fellay J, Shianna KV, Feng S, Urban TJ, Ge DL, De Luca A, Martinez-Picado J, Wolinsky SM, Martinson JJ, Jamieson BD, Bream JH, Martin MP, Borrow P, Letvin NL, McMichael AJ, Haynes BF, Telenti A, Carrington M, Goldstein DB, Alter G, Vaccine NCHA. Copy Number Variation of KIR Genes Influences HIV-1 Control. Plos Biology, 2011; 9.

[38] Fellermann K, Stange DE, Schaeffeler E, Schmalzl H, Wehkamp J, Bevins CL, Reinisch W, Teml A, Schwab M, Lichter P, Radlwimmer B, Stange EF. A chromosome 8 gene-cluster polymorphism with low human beta-defensin 2 gene copy number predisposes to Crohn disease of the colon. American Journal of Human Genetics, 2006; 79: 439-448.

[39] McCarroll SA, Huett A, Kuballa P, Chilewski SD, Landry A, Goyette P, Zody MC, Hall JL, Brant SR, Cho JH, Duerr RH, Silverberg MS, Taylor KD, Rioux JD, Altshuler D, Daly MJ, Xavier RJ. Deletion polymorphism upstream of IRGM associated with altered IRGM expression and Crohn's disease. Nature Genetics, 2008; 40: 1107-1112.

[40] Jespersgaard C, Fode P, Dybdahl M, Vind I, Nielsen OH, Csillag C, Munkholm P, Vainer B, Riis L, Elkjaer M, Pedersen N, Knudsen E, Andersen PS. Alpha-Defensin DEFA1A3 Gene Copy Number Elevation in Danish Crohn's Disease Patients. Digestive Diseases and Sciences, 2011; 56: 3517-3524.

[41] Henckaerts L, Cleynen I, Brinar M, John JM, Van Steen K, Rutgeerts P, Vermeire S. Genetic Variation in the Autophagy Gene ULK1 and Risk of Crohn's Disease. Inflammatory Bowel Diseases, 2011; 17: 1392-1397.

[42] Hollox EJ, Huffmeier U, Zeeuwen P, Palla R, Lascorz J, Rodijk-Olthuis D, van de Kerkhof PCM, Traupe H, de Jongh G, den Heijer M, Reis A, Armour JAL, Schalkwijk J. Psoriasis is associated with increased beta-defensin genomic copy number. Nature Genetics, 2008; 40: 23-25.

[43] Pedrosa E, Carretero-Iglesia L, Boada A, Colobran R, Faner R, Pujol-Autonell I, Palou E, Esteve A, Pujol-Borrell R, Ferrandiz C, Juan M, Carrascosa JM. CCL4L Polymorphisms

and CCL4/CCL4L Serum Levels Are Associated with Psoriasis Severity. Journal of Investigative Dermatology, 2011; 131: 1830-1837.

[44] Flint J, Hill AVS, Bowden DK, Oppenheimer SJ, Sill PR, Serjeantson SW, Banakoiri J, Bhatia K, Alpers MP, Boyce AJ, Weatherall DJ, Clegg JB. HIGH-FREQUENCIES OF ALPHA-THALASSEMIA ARE THE RESULT OF NATURAL-SELECTION BY MALARIA. Nature, 1986; 321: 744-750.

[45] Lupski JR, Wise CA, Kuwano A, Pentao L, Parke JT, Glaze DG, Ledbetter DH, Greenberg F, Patel PI. GENE DOSAGE IS A MECHANISM FOR CHARCOT-MARIE-TOOTH DISEASE TYPE-1A. Nature Genetics, 1992; 1: 29-33.

[46] Kleinjan DA, van Heyningen V. Long-range control of gene expression: Emerging mechanisms and disruption in disease. American Journal of Human Genetics, 2005; 76: 8-32.

[47] Pfeifer D, Kist R, Dewar K, Devon K, Lander ES, Birren B, Korniszewski L, Back E, Scherer G. Campomelic dysplasia translocation breakpoints are scattered over 1 Mb proximal to SOX9: Evidence for an extended control region. American Journal of Human Genetics, 1999; 65: 111-124.

[48] Kimura-Yoshida C, Kitajima K, Oda-Ishii I, Tian E, Suzuki M, Yamamoto M, Suzuki T, Kobayashi M, Aizawa S, Matsuo I. Characterization of the pufferfish Otx2 cis-regulators reveals evolutionarily conserved genetic mechanisms for vertebrate head specification. Development, 2004; 131: 57-71.

[49] Kleinjan DJ, van Heyningen V. Position effect in human genetic disease. Human Molecular Genetics, 1998; 7: 1611-1618.

[50] Jiang GC, Yang F, van Overveld PGM, Vedanarayanan V, van der Maarel S, Ehrlich M. Testing the position-effect variegation hypothesis for facioscapulohumeral muscular dystrophy by analysis of histone modification and gene expression in subtelomeric 4q. Human Molecular Genetics, 2003; 12: 2909-2921.

[51] Saveliev A, Everett C, Sharpe T, Webster Z, Festenstein R. DNA triplet repeats mediate heterochromatin-protein-1-sensitive variegated gene silencing. Nature, 2003; 422: 909-913.

[52] Tufarelli C, Stanley JAS, Garrick D, Sharpe JA, Ayyub H, Wood WG, Higgs DR. Transcription of antisense RNA leading to gene silencing and methylation as a novel cause of human genetic disease. Nature Genetics, 2003; 34: 157-165.

[53] Fantes J, Redeker B, Breen M, Boyle S, Brown J, Fletcher J, Jones S, Bickmore W, Fukushima Y, Mannens M, Danes S, Vanheyningen V, Hanson I. ANIRIDIA-ASSOCIATED CYTOGENETIC REARRANGEMENTS SUGGEST THAT A POSITION EFFECT MAY CAUSE THE MUTANT PHENOTYPE. Human Molecular Genetics, 1995; 4: 415-422.

[54] Lauderdale J, Wilensky JS, Oliver ER, Walton DS, Glaser T. 3 ' deletions cause aniridia by preventing PAX6 gene expression. Proceedings of the National Academy of Sciences of the United States of America, 2000; 97: 13755-13759.

[55] Kleinjan DA, Seawright A, Schedl A, Quinlan RA, Danes S, van Heyningen V. Aniridia-associated translocations, DNase hypersensitivity, sequence comparison and transgenic analysis redefine the functional domain of PAX6. Human Molecular Genetics, 2001; 10: 2049-2059.

[56] Crolla JA, van Heyningen V. Frequent chromosome aberrations revealed by molecular cytogenetic studies in patients with aniridia. American Journal of Human Genetics, 2002; 71: 1138-1149.

[57] Kurotaki N, Shen JJ, Touyama M, Kondoh T, Visser R, Ozaki T, Nishimoto J, Shiihara T, Uetake K, Makita Y, Harada N, Raskin S, Brown CW, Hoglund P, Okamoto N, Lupski JR. Phenotypic consequences of genetic variation at hemizygous alleles: Sotos syndrome is a contiguous gene syndrome incorporating coagulation factor twelve (FXII) deficiency. Genetics in Medicine, 2005; 7: 479-483.

[58] Bailey JA, Yavor AM, Massa HF, Trask BJ, Eichler EE. Segmental duplications: Organization and impact within the current Human Genome Project assembly. Genome Research, 2001; 11: 1005-1017.

[59] Stankiewicz P, Lupski JR. Genome architecture, rearrangements and genomic disorders. Trends in Genetics, 2002; 18: 74-82.

[60] Espejel S, Franco S, Rodriguez-Perales S, Bouffler SD, Cigudosa JC, Blasco MA. Mammalian Ku86 mediates chromosomal fusions and apoptosis caused by critically short telomeres. Embo Journal, 2002; 21: 2207-2219.

[61] Lee JA, Carvalho CMB, Lupski JR. A DNA replication mechanism for generating nonrecurrent rearrangements associated with genomic disorders. Cell, 2007; 131: 1235-1247.

[62] Zhang F, Khajavi M, Connolly AM, Towne CF, Batish SD, Lupski JR. The DNA replication FoSTeS/MMBIR mechanism can generate genomic, genic and exonic complex rearrangements in humans. Nature Genetics, 2009; 41: 849-U115.

[63] Cocquempot O, Brault V, Babinet C, Herault Y. Fork Stalling and Template Switching As a Mechanism for Polyalanine Tract Expansion Affecting the DYC Mutant of HOXD13, a New Murine Model of Synpolydactyly. Genetics, 2009; 183: 23-30.

[64] Hastings PJ, Ira G, Lupski JR. A Microhomology-Mediated Break-Induced Replication Model for the Origin of Human Copy Number Variation. Plos Genetics, 2009; 5.

[65] Korbel JO, Urban AE, Affourtit JP, Godwin B, Grubert F, Simons JF, Kim PM, Palejev D, Carriero NJ, Du L, Taillon BE, Chen ZT, Tanzer A, Saunders ACE, Chi JX, Yang FT, Carter NP, Hurles ME, Weissman SM, Harkins TT, Gerstein MB, Egholm M, Snyder M. Paired-end mapping reveals extensive structural variation in the human genome. Science, 2007; 318: 420-426.

[66] Arlt MF, Mulle JG, Schaibley VM, Ragland RL, Durkin SG, Warren ST, Glover TW. Replication Stress Induces Genome-wide Copy Number Changes in Human Cells that Resemble Polymorphic and Pathogenic Variants. American Journal of Human Genetics, 2009; 84: 339-350.

[67] Choi BO, Kim NK, Park SW, Hyun YS, Jeon HJ, Hwang JH, Chung KW. Inheritance of Charcot-Marie-Tooth disease 1A with rare nonrecurrent genomic rearrangement. Neurogenetics, 2011; 12: 51-58.

[68] Bonaglia MC, Giorda R, Beri S, De Agostini C, Novara F, Fichera M, Grillo L, Galesi O, Vetro A, Ciccone R, Bonati MT, Giglio S, Guerrini R, Osimani S, Marelli S, Zucca C, Grasso R, Borgatti R, Mani E, Motta C, Molteni M, Romano C, Greco D, Reitano S, Baroncini A, Lapi E, Cecconi A, Arrigo G, Patricelli MG, Pantaleoni C, D'Arrigo S, Riva D, Sciacca F, Dalla Bernardina B, Zoccante L, Darra F, Termine C, Maserati E, Bigoni S, Priolo E, Bottani A, Gimelli S, Bena F, Brusco A, di Gregorio E, Bagnasco I, Giussani U, Nitsch L, Politi P, Martinez-Frias ML, Martinez-Fernandez ML, Guardia NM, Bremer A, Anderlid BM, Zuffardi O. Molecular Mechanisms Generating and Stabilizing Terminal 22q13 Deletions in 44 Subjects with Phelan/McDermid Syndrome. Plos Genetics, 2011; 7.

[69] Koumbaris G, Hatzisevastou-Loukidou H, Alexandrou A, Ioannides M, Christodoulou C, Fitzgerald T, Rajan D, Clayton S, Kitsiou-Tzeli S, Vermeesch JR, Skordis N, Antoniou P, Kurg A, Georgiou I, Carter NP, Patsalis PC. FoSTeS, MMBIR and NAHR at the human proximal Xp region and the mechanisms of human Xq isochromosome formation. Human Molecular Genetics, 2011; 20: 1925-1936.
[70] McClintock B. CHROMOSOME ORGANIZATION AND GENIC EXPRESSION. Cold Spring Harbor Symposia on Quantitative Biology, 1951; 16: 13-47.
[71] Pipiras E, Coquelle A, Bieth A, Debatisse M. Interstitial deletions and intrachromosomal amplification initiated from a double-strand break targeted to a mammalian chromosome. Embo Journal, 1998; 17: 325-333.
[72] Carvalho CMB, Ramocki MB, Pehlivan D, Franco LM, Gonzaga-Jauregui C, Fang P, McCall A, Pivnick EK, Hines-Dowell S, Seaver LH, Friehling L, Lee S, Smith R, del Gaudio D, Withers M, Liu PF, Cheung SW, Belmont JW, Zoghbi HY, Hastings PJ, Lupski JR. Inverted genomic segments and complex triplication rearrangements are mediated by inverted repeats in the human genome. Nature Genetics, 2011; 43: 1074-U59.
[73] Bergman Y. Allelic exclusion in B and T lymphopoiesis. Seminars in Immunology, 1999; 11: 319-328.
[74] Yang JJ, Bhojwani D, Yang WJ, Cai XJ, Stocco G, Crews K, Wang JH, Morrison D, Devidas M, Hunger SP, Willman CL, Raetz EA, Pui CH, Evans WE, Relling MV, Carroll WL. Genome-wide copy number profiling reveals molecular evolution from diagnosis to relapse in childhood acute lymphoblastic leukemia. Blood, 2008; 112: 4178-4183.
[75] Leary RJ, Lin JC, Cummins J, Boca S, Wood LD, Parsons DW, Jones S, Sjoblom T, Park BH, Parsons R, Willis J, Dawson D, Willson JKV, Nikolskaya T, Nikolsky Y, Kopelovich L, Papadopoulos N, Pennacchio LA, Wang TL, Markowitz SD, Parmigiani G, Kinzler KW, Vogelstein B, Velculescu VE. Integrated analysis of homozygous deletions, focal amplifications, and sequence alterations in breast and colorectal cancers. Proceedings of the National Academy of Sciences of the United States of America, 2008; 105: 16224-16229.
[76] Natrajan R, Warren W, Messahel B, Reis-Filho JS, Brundler MA, Dome JS, Grundy PE, Vujanic G, Pritchard-Jones K, Jones C. Complex patterns of chromosome 9 alterations including the p16(INK4a) locus in Wilms tumours. Journal of Clinical Pathology, 2008; 61: 95-102.
[77] Hassold TJ, Jacobs PA. TRISOMY IN MAN. Annual Review of Genetics, 1984; 18: 69-97.
[78] Gratacos M, Nadal M, Martin-Santos R, Pujana MA, Gago J, Peral B, Armengol L, Ponsa I, Miro R, Bulbena A, Estivill X. A polymorphic genomic duplication on human chromosome 15 is a susceptibility factor for panic and phobic disorders. Cell, 2001; 106: 367-379.
[79] Youssoufian H, Pyeritz RE. Mechanisms and consequences of somatic mosaicism in humans. Nature Reviews Genetics, 2002; 3: 748-758.
[80] Iourov IY, Vorsanova SG, Yurov YB. Chromosomal variation in mammalian neuronal cells: Known facts and attractive hypotheses. In: ed.^eds., International Review of Cytology - a Survey of Cell Biology, Vol 249, 2006; pp. 143-+.
[81] Yurov YB, Iourov IY, Vorsanova SG, Liehr T, Kolotii AD, Kutsev SI, Pellestor F, Beresheva AK, Demidova IA, Kravets VS, Monakhov VV, Soloviev IV. Aneuploidy and Confined Chromosomal Mosaicism in the Developing Human Brain. Plos One, 2007; 6.
[82] Vanneste E, Voet T, Le Caignec C, Ampe M, Konings P, Melotte C, Debrock S, Amyere M, Vikkula M, Schuit F, Fryns JP, Verbeke G, D'Hooghe T, Moreau Y, Vermeesch JR.

Chromosome instability is common in human cleavage-stage embryos. Nature Medicine, 2009; 15: 577-583.

[83] Liang Q, Conte N, Skarnes WC, Bradley A. Extensive genomic copy number variation in embryonic stem cells. Proceedings of the National Academy of Sciences of the United States of America, 2008; 105: 17453-17456.

[84] Piotrowski A, Bruder CEG, Andersson R, de Stahl TD, Menzel U, Sandgren J, Poplawski A, von Tell D, Crasto C, Bogdan A, Bartoszewski R, Bebok Z, Krzyzanowski M, Jankowski Z, Partridge EC, Komorowski J, Dumanski JP. Somatic mosaicism for copy number variation in differentiated human tissues. Human Mutation, 2008; 29: 1118-1124.

[85] Bruder CEG, Piotrowski A, Gijsbers A, Andersson R, Erickson S, de Stahl TD, Menzel U, Sandgren J, von Tell D, Poplawski A, Crowley M, Crasto C, Partridge EC, Tiwari H, Allison DB, Kornorowski J, van Ommen GJB, Boomsma DI, Pedersen NL, den Dunnen JT, Wirdefeldt K, Dumanski JP. Phenotypically concordant and discordant monozygotic twins display different DNA copy-number-variation profiles. American Journal of Human Genetics, 2008; 82: 763-771.

[86] Mkrtchyan H, Gross M, Hinreiner S, Polytiko A, Manvelyan M, Mrasek K, Kosyakova N, Ewers E, Nelle H, Liehr T, Volleth M, Weise A. Early Embryonic Chromosome Instability Results in Stable Mosaic Pattern in Human Tissues. Plos One, 2010; 5.

[87] Weise A, Gross M, Mrasek K, Mkrtchyan H, Horsthemke B, Jonsrud C, Von Eggeling F, Hinreiner S, Witthuhn V, Claussen U, Liehr T. Parental-origin-determination fluorescence *in situ* hybridization distinguishes homologous human chromosomes on a single-cell level. International Journal of Molecular Medicine, 2008; 21: 189-200.

[88] Waterston RH, Lindblad-Toh K, Birney E, Rogers J, Abril JF, Agarwal P, Agarwala R, Ainscough R, Alexandersson M, An P, Antonarakis SE, Attwood J, Baertsch R, Bailey J, Barlow K, Beck S, Berry E, Birren B, Bloom T, Bork P, Botcherby M, Bray N, Brent MR, Brown DG, Brown SD, Bult C, Burton J, Butler J, Campbell RD, Carninci P, Cawley S, Chiaromonte F, Chinwalla AT, Church DM, Clamp M, Clee C, Collins FS, Cook LL, Copley RR, Coulson A, Couronne O, Cuff J, Curwen V, Cutts T, Daly M, David R, Davies J, Delehaunty KD, Deri J, Dermitzakis ET, Dewey C, Dickens NJ, Diekhans M, Dodge S, Dubchak I, Dunn DM, Eddy SR, Elnitski L, Emes RD, Eswara P, Eyras E, Felsenfeld A, Fewell GA, Flicek P, Foley K, Frankel WN, Fulton LA, Fulton RS, Furey TS, Gage D, Gibbs RA, Glusman G, Gnerre S, Goldman N, Goodstadt L, Grafham D, Graves TA, Green ED, Gregory S, Guigo R, Guyer M, Hardison RC, Haussler D, Hayashizaki Y, Hillier LW, Hinrichs A, Hlavina W, Holzer T, Hsu F, Hua A, Hubbard T, Hunt A, Jackson I, Jaffe DB, Johnson LS, Jones M, Jones TA, Joy A, Kamal M, Karlsson EK, Karolchik D, Kasprzyk A, Kawai J, Keibler E, Kells C, Kent WJ, Kirby A, Kolbe DL, Korf I, Kucherlapati RS, Kulbokas EJ, Kulp D, Landers T, Leger JP, Leonard S, Letunic I, Levine R, Li J, Li M, Lloyd C, Lucas S, Ma B, Maglott DR, Mardis ER, Matthews L, Mauceli E, Mayer JH, McCarthy M, McCombie WR, McLaren S, McLay K, McPherson JD, Meldrim J, Meredith B, Mesirov JP, Miller W, Miner TL, Mongin E, Montgomery KT, Morgan M, Mott R, Mullikin JC, Muzny DM, Nash WE, Nelson JO, Nhan MN, Nicol R, Ning Z, Nusbaum C, O'Connor MJ, Okazaki Y, Oliver K, Larty EO, Pachter L, Parra G, Pepin KH, Peterson J, Pevzner P, Plumb R, Pohl CS, Poliakov A, Ponce TC, Ponting CP, Potter S, Quail M, Reymond A, Roe BA, Roskin KM, Rubin EM, Rust AG, Santos R, Sapojnikov V, Schultz B, Schultz J, Schwartz MS, Schwartz S, Scott C, Seaman S, Searle S, Sharpe T, Sheridan A, Shownkeen R, Sims S, Singer JB, Slater G, Smit A, Smith DR, Spencer B, Stabenau A,

Strange-Thomann NS, Sugnet C, Suyama M, Tesler G, Thompson J, Torrents D, Trevaskis E, Tromp J, Ucla C, Vidal AU, Vinson JP, von Niederhausern AC, Wade CM, Wall M, Weber RJ, Weiss RB, Wendl MC, West AP, Wetterstrand K, Wheeler R, Whelan S, Wierzbowski J, Willey D, Williams S, Wilson RK, Winter E, Worley KC, Wyman D, Yang S, Yang SP, Zdobnov EM, Zody MC, Lander ES, Mouse Genome Sequencing C. Initial sequencing and comparative analysis of the mouse genome. Nature, 2002; 420: 520-562.

[89] Evans MJ, Kaufman MH. ESTABLISHMENT IN CULTURE OF PLURIPOTENTIAL CELLS FROM MOUSE EMBRYOS. Nature, 1981; 292: 154-156.

[90] Gordon JW, Ruddle FH. INTEGRATION AND STABLE GERM LINE TRANSMISSION OF GENES INJECTED INTO MOUSE PRONUCLEI. Science, 1981; 214: 1244-1246.

[91] Doetschman T, Gregg RG, Maeda N, Hooper ML, Melton DW, Thompson S, Smithies O. TARGETED CORRECTION OF A MUTANT HPRT GENE IN MOUSE EMBRYONIC STEM-CELLS. Nature, 1987; 330: 576-578.

[92] Thomas KR, Capecchi MR. SITE-DIRECTED MUTAGENESIS BY GENE TARGETING IN MOUSE EMBRYO-DERIVED STEM-CELLS. Cell, 1987; 51: 503-512.

[93] Sauer B, Henderson N. SITE-SPECIFIC DNA RECOMBINATION IN MAMMALIAN-CELLS BY THE CRE RECOMBINASE OF BACTERIOPHAGE-P1. Proceedings of the National Academy of Sciences of the United States of America, 1988; 85: 5166-5170.

[94] Sauer B, Henderson N. CRE-STIMULATED RECOMBINATION AT LOXP-CONTAINING DNA-SEQUENCES PLACED INTO THE MAMMALIAN GENOME. Nucleic Acids Research, 1989; 17: 147-161.

[95] Sauer B. MANIPULATION OF TRANSGENES BY SITE-SPECIFIC RECOMBINATION - USE OF CRE RECOMBINASE. In: ed.^eds., Guide to Techniques in Mouse Development, 1993; pp. 890-900.

[96] Sternberg N, Hamilton D. BACTERIOPHAGE-P1 SITE-SPECIFIC RECOMBINATION .1. RECOMBINATION BETWEEN LOXP SITES. Journal of Molecular Biology, 1981; 150: 467-486.

[97] Gu H, Marth JD, Orban PC, Mossmann H, Rajewsky K. DELETION OF A DNA-POLYMERASE-BETA GENE SEGMENT IN T-CELLS USING CELL-TYPE-SPECIFIC GENE TARGETING. Science, 1994; 265: 103-106.

[98] Spitz F, Gonzalez F, Peichel C, Vogt TF, Duboule D, Zakany J. Large scale transgenic and cluster deletion analysis of the HoxD complex separate an ancestral regulatory module from evolutionary innovations. Genes & Development, 2001; 15: 2209-2214.

[99] Zakany J, Duboule D. Synpolydactyly in mice with a targeted deficiency in the HoxD complex. Nature, 1996; 384: 69-71.

[100] Ramirezsolis R, Liu PT, Bradley A. CHROMOSOME ENGINEERING IN MICE. Nature, 1995; 378: 720-724.

[101] Olson LE, Richtsmeier JT, Leszl J, Reeves RH. A chromosome 21 critical region does not cause specific down syndrome phenotypes. Science, 2004; 306: 687-690.

[102] Puech A, Saint-Jore B, Merscher S, Russell RG, Cherif D, Sirotkin H, Xu H, Factor S, Kucherlapati R, Skoultchi AI. Normal cardiovascular development in mice deficient for 16 genes in 550 kb of the velocardiofacial/DiGeorge syndrome region. Proceedings of the National Academy of Sciences of the United States of America, 2000; 97: 10090-10095.

[103] Suemori H, Noguchi S. Hox C cluster genes are dispensable for overall body plan of mouse embryonic development. Developmental Biology, 2000; 220: 333-342.

[104] Kudoh H, Ikeda H, Kakitani M, Ueda A, Hayasaka M, Tomizuka K, Hanaoka K. A new model mouse for Duchenne muscular dystrophy produced by 2.4 Mb deletion of dystrophin gene using Cre-loxP recombination system. Biochemical and Biophysical Research Communications, 2005; 328: 507-516.

[105] Li ZW, Stark G, Gotz J, Rulicke T, Gschwind M, Huber G, Muller U, Weissmann C. Generation of mice with a 200-kb amyloid precursor protein gene deletion by Cre recombinase-mediated site-specific recombination in embryonic stem cells (vol 93, pg 6158, 1996). Proceedings of the National Academy of Sciences of the United States of America, 1996; 93: 12052-12052.

[106] Schlake T, Schupp I, Kutsche K, Mincheva A, Lichter P, Boehm T. Predetermined chromosomal deletion encompassing the Nf-1 gene. Oncogene, 1999; 18: 6078-6082.

[107] Zhu YW, Jong MC, Frazer KA, Gong E, Krauss RM, Cheng JF, Boffelli D, Rubin EM. Genomic interval engineering of mice identifies a novel modulator of triglyceride production. Proceedings of the National Academy of Sciences of the United States of America, 2000; 97: 1137-1142.

[108] Nóbrega M, Zhu Y, Plajzer-Frick I, Afzal V, Rubin E. Megabase deletions of gene deserts result in viable mice. Nature, 2004; 431: 988-93.

[109] Zheng BH, Mills AA, Bradley A. A system for rapid generation of coat color-tagged knockouts and defined chromosomal rearrangements in mice. Nucleic Acids Research, 1999; 27: 2354-2360.

[110] Adams DJ, Biggs PJ, Cox T, Davies R, van der Weyden L, Jonkers J, Smith J, Plumb B, Taylor R, Nishijima I, Yu YJ, Rogers J, Bradley A. Mutagenic insertion and chromosome engineering resource (MICER). Nature Genetics, 2004; 36: 867-871.

[111] Su H, Wang XH, Bradley A. Nested chromosomal deletions induced with retroviral vectors in mice. Nature Genetics, 2000; 24: 92-95.

[112] Zheng BH, Mills AA, Bradley A. Introducing defined chromosomal rearrangements into the mouse genome. Methods, 2001; 24: 81-94.

[113] Yu YJ, Bradley A. Engineering chromosomal rearrangements in mice. Nature Reviews Genetics, 2001; 2: 780-790.

[114] Baubonis W, Sauer B. GENOMIC TARGETING WITH PURIFIED CRE RECOMBINASE. Nucleic Acids Research, 1993; 21: 2025-2029.

[115] Liu PT, Zhang HJ, McLellan A, Vogel H, Bradley A. Embryonic lethality and tumorigenesis caused by segmental aneuploidy on mouse chromosome 11. Genetics, 1998; 150: 1155-1168.

[116] Zheng BH, Sage M, Sheppeard EA, Jurecic V, Bradley A. Engineering mouse chromosomes with Cre-loxP: Range, efficiency, and somatic applications. Molecular and Cellular Biology, 2000; 20: 648-655.

[117] Lindsay EA, Botta A, Jurecic V, Carattini-Rivera S, Cheah YC, Rosenblatt HM, Bradley A, Baldini A. Congenital heart disease in mice deficient for the DiGeorge syndrome region. Nature, 1999; 401: 379-383.

[118] Lindsay EA, Vitelli F, Su H, Morishima M, Huynh T, Pramparo T, Jurecic V, Ogunrinu G, Sutherland HF, Scambler PJ, Bradley A, Baldini A. Tbx1 haploinsufficiency in the DiGeorge syndrome region causes aortic arch defects in mice. Nature, 2001; 410: 97-101.

[119] Jerome LA, Papaioannou VE. DiGeorge syndrome phenotype in mice mutant for the T-box gene, Tbx1. Nature Genetics, 2001; 27: 286-291.

[120] Davisson MT, Schmidt C, Akeson EC. SEGMENTAL TRISOMY OF MURINE CHROMOSOME-16 - A NEW MODEL SYSTEM FOR STUDYING DOWN-SYNDROME. Molecular Genetics of Chromosome 21 and Down Syndrome, 1990; 360: 263-280.

[121] Reeves RH, Irving NG, Moran TH, Wohn A, Kitt C, Sisodia SS, Schmidt C, Bronson RT, Davisson MT. A MOUSE MODEL FOR DOWN-SYNDROME EXHIBITS LEARNING AND BEHAVIOR DEFICITS. Nature Genetics, 1995; 11: 177-184.

[122] Sago H, Carlson EJ, Smith DJ, Kilbridge J, Rubin EM, Mobley WC, Epstein CJ, Huang TT. Ts1Cje, a partial trisomy 16 mouse model for Down syndrome, exhibits learning and behavioral abnormalities. Proceedings of the National Academy of Sciences of the United States of America, 1998; 95: 6256-6261.

[123] Olson LE, Roper RJ, Sengstaken CL, Peterson EA, Aquino V, Galdzicki Z, Siarey R, Pletnikov M, Moran TH, Reeves RH. Trisomy for the Down syndrome 'critical region' is necessary but not sufficient for brain phenotypes of trisomic mice. Human Molecular Genetics, 2007; 16: 774-782.

[124] Pereira PL, Magnol L, Sahun I, Brault V, Duchon A, Prandini P, Gruart A, Bizot JC, Chadefaux-Vekemans B, Deutsch S, Trovero F, Delgado-Garcia JM, Antonarakis SE, Dierssen M, Herault Y. A new mouse model for the trisomy of the Abcg1-U2af1 region reveals the complexity of the combinatorial genetic code of down syndrome. Human Molecular Genetics, 2009; 18: 4756-4769.

[125] Gotti S, Caricati E, Panzica G. Alterations of brain circuits in Down syndrome murine models. Journal of Chemical Neuroanatomy, 2011; 42: 317-326.

[126] Lana-Elola E, Watson-Scales SD, Fisher EMC, Tybulewicz VLJ. Down syndrome: searching for the genetic culprits. Disease Models & Mechanisms, 2011; 4: 586-595.

[127] Zhang L, Fu D, Belichenko PV, Liu C, Kleschevnikov AM, Pao A, Liang P, Clapcote SJ, Mobley WC, Yu YE. Genetic analysis of Down syndrome facilitated by mouse chromosome engineering. Bioeng Bugs, 2012; 3: 8-12.

[128] Walz K, Caratini-Rivera S, Bi WM, Fonseca P, Mansouri DL, Lynch J, Vogel H, Noebels JL, Bradley A, Lupski JR. Modeling del(17)(p11.2p11.2) and dup(17)(p11.2p11.2) contiguous gene syndromes by chromosome engineering in mice: Phenotypic consequences of gene dosage imbalance. Molecular and Cellular Biology, 2003; 23: 3646-3655.

[129] Walz K, Spencer C, Kaasik K, Lee CC, Lupski JR, Paylor R. Behavioral characterization of mouse models for Smith-Magenis syndrome and dup(17)(p11.2p11.2). Human Molecular Genetics, 2004; 13: 367-378.

[130] Yan J, Keener VW, Bi WM, Walz K, Bradley A, Justice MJ, Lupski JR. Reduced penetrance of craniofacial anomalies as a function of deletion size and genetic background in a chromosome engineered partial mouse model for Smith-Magenis syndrome. Human Molecular Genetics, 2004; 13: 2613-2624.

[131] Bi WM, Ohyama T, Nakamura H, Yan J, Visvanathan J, Justice MJ, Lupski JR. Inactivation of Rai1 in mice recapitulates phenotypes observed in chromosome engineered mouse models for Smith-Magenis syndrome. Human Molecular Genetics, 2005; 14: 983-995.

[132] Ricard G, Molina J, Chrast J, Gu WL, Gheldof N, Pradervand S, Schutz F, Young JI, Lupski JR, Reymond A, Walz K. Phenotypic Consequences of Copy Number Variation: Insights from Smith-Magenis and Potocki-Lupski Syndrome Mouse Models. Plos Biology, 2010; 8.

[133] Tsai TF, Jiang YH, Bressler J, Armstrong D, Beaudet AL . Paternal deletion from Snrpn to Ube3a in the mouse causes hypotonia, growth retardation and partial lethality and provides evidence for a gene contributing to Prader-Willi syndrome. Human Molecular Genetics, 1999; 8: 1357-1364.

[134] Smith AJH, Desousa MA, Kwabiaddo B, Heppellparton A, Impey H, Rabbitts P. A SITE-DIRECTED CHROMOSOMAL TRANSLOCATION INDUCED IN EMBRYONIC STEM-CELLS BY CRE-LOXP RECOMBINATION. Nature Genetics, 1995; 9: 376-385.

[135] Smith AJH, Xian J, Richardson M, Johnstone KA, Rabbitts PH. Cre-loxP chromosome engineering of a targeted deletion in the mouse corresponding to the 3p21.3 region of homozygous loss in human tumours. Oncogene, 2002; 21: 4521-4529.

[136] Vandeursen J, Fornerod M, Vanrees B, Grosveld G. CRE-MEDIATED SITE-SPECIFIC TRANSLOCATION BETWEEN NONHOMOLOGOUS MOUSE CHROMOSOMES. Proceedings of the National Academy of Sciences of the United States of America, 1995; 92: 7376-7380.

[137] Koike H, Horie K, Fukuyama H, Kondoh G, Nagata S, Takeda J. Efficient biallelic mutagenesis with Cre/loxP-mediated inter-chromosomal recombination. Embo Reports, 2002; 3: 433-437.

[138] Biggs PJ, Vogel H, Sage M, Martin LA, Donehower LA, Bradley A. Allelic phasing of a mouse chromosome 11 deficiency influences p53 tumorigenicity. Oncogene, 2003; 22: 3288-3296.

[139] Nakatani J, Tamada K, Hatanaka F, Ise S, Ohta H, Inoue K, Tomonaga S, Watanabe Y, Chung YJ, Banerjee R, Iwamoto K, Kato T, Okazawa M, Yamauchi K, Tanda K, Takao K, Miyakawa T, Bradley A, Takumi T. Abnormal Behavior in a Chromosome-Engineered Mouse Model for Human 15q11-13 Duplication Seen in Autism. Cell, 2009; 137: 1235-1246.

[140] Beaudet AL. Autism: highly heritable but not inherited. Nature Medicine, 2007; 13: 534-536.

[141] Lee HJ, Kweon J, Kim E, Kim S, Kim JS. Targeted chromosomal duplications and inversions in the human genome using zinc finger nucleases. Genome Res, 2012; 22: 539-48.

[142] Herault Y, Rassoulzadegan M, Cuzin F, Duboule D. Engineering chromosomes in mice through targeted meiotic recombination (TAMERE). Nature Genetics, 1998; 20: 381-384.

[143] Lewandoski M, Wassarman KM, Martin GR. Zp3-cre, a transgenic mouse line for the activation or inactivation of loxP-flanked target genes specifically in the female germ line. Current Biology, 1997; 7: 148-151.

[144] de Vries WN, Binns LT, Fancher KS, Dean J, Moore R, Kemler R, Knowles BB. Expression of Cre recombinase in mouse oocytes: A means to study maternal effect genes. Genesis, 2000; 26: 110-112.

[145] O'Gorman S, Dagenais NA, Qian M, Marchuk Y. Protamine-Cre recombinase transgenes efficiently recombine target sequences in the male germ line of mice, but not in embryonic stem cells. Proceedings of the National Academy of Sciences of the United States of America, 1997; 94: 14602-14607.

[146] Dupe V, Davenne M, Brocard J, Dolle P, Mark M, Dierich A, Chambon P, Rijli FM. *In vivo* functional analysis of the Hoxa-1 3' retinoic acid response element (3'RARE). Development, 1997; 124: 399-410.

[147] Soriano P. Generalized lacZ expression with the ROSA26 Cre reporter strain. Nature Genetics, 1999; 21: 70-71.

[148] Lallemand Y, Luria V, Haffner-Krausz R, Lonai P. Maternally expressed PGK-Cre transgene as a tool for early and uniform activation of the Cre site-specific recombinase. Transgenic Res, 1998; 7: 105-12.

[149] Tang SHE, Silva FJ, Tsark WMK, Mann JR. A Cre/loxP-deleter transgenic line in mouse strain 129S1/SvImJ. Genesis, 2002; 32: 199-202.
[150] Zakany J, Kmita M, Duboule D. A dual role for Hox genes in limb anterior-posterior asymmetry. Science, 2004; 304: 1669-1672.
[151] Kmita M, Kondo T, Duboule D. Targeted inversion of a polar silencer within the HoxD complex re-allocates domains of enhancer sharing. Nature Genetics, 2000; 26: 451-454.
[152] Stemmler MP, Hecht A, Kemler R. E-cadherin intron 2 contains cis-regulatory elements essential for gene expression. Development, 2005; 132: 965-976.
[153] Spitz F, Herkenne C, Morris MA, Duboule D. Inversion-induced disruption of the Hoxd cluster leads to the partition of regulatory landscapes. Nature Genetics, 2005; 37: 889-893.
[154] Kmita M, van der Hoeven F, Zakany J, Krumlauf R, Duboule D. Mechanisms of Hox gene colinearity: transposition of the anterior Hoxb1 gene into the posterior HoxD complex. Genes & Development, 2000; 14: 198-211.
[155] Besson V, Brault V, Duchon A, Togbe D, Bizot JC, Quesniaux VFJ, Ryffel B, Herault Y. Modeling the monosomy for the telomeric part of human chromosome 21 reveals haploinsufficient genes modulating the inflammatory and airway responses. Human Molecular Genetics, 2007; 16: 2040-2052.
[156] Wu S, Ying G, Wu Q, Capecchi MR. Toward simpler and faster genome-wide mutagenesis in mice. Nature Genetics, 2007; 39: 922-930.
[157] Merscher S, Funke B, Epstein JA, Heyer J, Puech A, Lu MM, Xavier RJ, Demay MB, Russell RG, Factor S, Tokooya K, Jore BS, Lopez M, Pandita RK, Lia M, Carrion D, Xu H, Schorle H, Kobler JB, Scambler P, Wynshaw-Boris A, Skoultchi AI, Morrow BE, Kucherlapati R. TBX1 is responsible for cardiovascular defects in Velo-Cardio-Facial/DiGeorge syndrome. Cell, 2001; 104: 619-629.
[158] Vidal F, Sage J, Cuzin F, Rassoulzadegan M. Cre expression in primary spermatocytes: A tool for genetic engineering of the germ line. Molecular Reproduction and Development, 1998; 51: 274-280.
[159] Herault Y, Kmita M, Sawaya CC, Duboule D. A nested deletion approach to generate Cre deleter mice with progressive Hox profiles. International Journal of Developmental Biology, 2002; 46: 185-191.
[160] Genoud N, Behrens A, Miele G, Robay D, Heppner FL, Freigang S, Aguzzi A. Disruption of Doppel prevents neurodegeneration in mice with extensive Prnp deletions. Proceedings of the National Academy of Sciences of the United States of America, 2004; 101: 4198-4203.
[161] Olson LE, Tien J, South S, Reeves RH. Long-range chromosomal engineering is more efficient *in vitro* than *in vitro*. Transgenic Research, 2005; 14: 325-332.
[162] Kmita M, Fraudeau N, Herault Y, Duboule D. Serial deletions and duplications suggest a mechanism for the collinearity of Hoxd genes in limbs. Nature, 2002; 420: 145-150.
[163] Kmita M, Tarchini B, Duboule D, Herault Y. Evolutionary conserved sequences are required for the insulation of the vertebrate Hoxd complex in neural cells. Development, 2002; 129: 5521-5528.
[164] Buchholz F, Ringrose L, Angrand PO, Rossi F, Stewart AF. Different thermostabilities of FLP and Cre recombinases: Implications for applied site-specific recombination. Nucleic Acids Research, 1996; 24: 4256-4262.
[165] Sauer B, McDermott J. DNA recombination with a heterospecific Cre homolog identified from comparison of the pac-c1 regions of P1-related phages. Nucleic Acids Research, 2004; 32: 6086-6095.

[166] Anastassiadis K, Fu J, Patsch C, Hu SB, Weidlich S, Duerschke K, Buchholz F, Edenhofer F, Stewart AF. Dre recombinase, like Cre, is a highly efficient site-specific recombinase in E-coli, mammalian cells and mice. Disease Models & Mechanisms, 2009; 2: 508-515.
[167] Schaft J, Ashery-Padan R, van der Hoeven F, Gruss P, Stewart AF. Efficient FLP recombination in mouse ES cells and oocytes. Genesis, 2001; 31: 6-10.
[168] Xu Z, Lee NCO, Dafhnis-Calas F, Malla S, Smith MCM, Brown WRA. Site-specific recombination in Schizosaccharomyces pombe and systematic assembly of a 400kb transgene array in mammalian cells using the integrase of Streptomyces phage phi BT1. Nucleic Acids Research, 2008; 36.
[169] Horie K, Kuroiwa A, Ikawa M, Okabe M, Kondoh G, Matsuda Y, Takeda J. Efficient chromosomal transposition of a Tc1/mariner-like transposon Sleeping Beauty in mice. Proceedings of the National Academy of Sciences of the United States of America, 2001; 98: 9191-9196.
[170] Horie K, Yusa K, Yae K, Odajima J, Fischer SEJ, Keng VW, Hayakawa T, Mizuno S, Kondoh G, Ijiri T, Matsuda Y, Plasterk RHA, Takeda J. Characterization of Sleeping Beauty transposition and its application to genetic screening in mice. Molecular and Cellular Biology, 2003; 23: 9189-9207.
[171] Carlson CM, Dupuy A, Fritz S, Roberg-Perez K, Fletcher CF, Largaespada DA. Transposon mutagenesis of the mouse germline. Genetics, 2003; 165: 243-256.
[172] Collier LS, Carlson CM, Ravimohan S, Dupuy AJ, Largaespada DA. Cancer gene discovery in solid tumours using transposon-based somatic mutagenesis in the mouse. Nature, 2005; 436: 272-276.
[173] Ding S, Wu XH, Li G, Han M, Zhuang Y, Xu T. Efficient transposition of the piggyBac resource (PB) transposon in mammalian cells and mice. Cell, 2005; 122: 473-483.
[174] Dupuy AJ, Fritz S, Largaespada DA. Transposition and gene disruption in the male germline of the mouse. Genesis, 2001; 30: 82-88.
[175] Keng VW, Yae K, Hayakawa T, Mizuno S, Uno Y, Yusa K, Kokubu C, Kinoshita T, Akagi K, Jenkins NA, Copeland NG, Horie K, Takeda J. Region-specific saturation germline mutagenesis in mice using the Sleeping Beauty transposon system. Nature Methods, 2005; 2: 763-769.
[176] Geurts AM, Collier LS, Geurts JL, Oseth LL, Bell ML, Mu D, Lucito R, Godbout SA, Green LE, Lowe SW, Hirsch BA, Leinwand LA, Largaespada DA. Gene mutations and genomic rearrangements in the mouse as a result of transposon mobilization from chromosomal concatemers. Plos Genetics, 2006; 2: 1413-1423.
[177] Ivics Z, Hackett PB, Plasterk RH, Izsvak Z. Molecular reconstruction of Sleeping beauty, a Tc1-like transposon from fish, and its transposition in human cells. Cell, 1997; 91: 501-510.
[178] Baus J, Liu L, Heggestad AD, Sanz S, Fletcher BS. Hyperactive transposase mutants of the Sleeping Beauty transposon. Molecular Therapy, 2005; 12: 1148-1156.
[179] Ruf S, Symmons O, Uslu VV, Dolle D, Hot C, Ettwiller L, Spitz F. Large-scale analysis of the regulatory architecture of the mouse genome with a transposon-associated sensor. Nature Genetics, 2011; 43: 379-U141.
[180] Tschopp P, Fraudeau N, Bena F, Duboule D. Reshuffling genomic landscapes to study the regulatory evolution of Hox gene clusters. Proceedings of the National Academy of Sciences of the United States of America, 2011; 108: 10632-10637.

[181] Hahnemann JM, Vejerslev LO. European collaborative research on mosaicism in CVS (EUCROMIC) - Fetal and extrafetal cell lineages in 192 gestations with CVS mosaicism involving single autosomal trisomy. American Journal of Medical Genetics, 1997; 70: 179-187.

[182] Allanson JE, Ohara P, Farkas LG, Nair RC. ANTHROPOMETRIC CRANIOFACIAL PATTERN PROFILES IN DOWN-SYNDROME. American Journal of Medical Genetics, 1993; 47: 748-752.

[183] Cheung SW, Shaw CA, Scott DA, Patel A, Sahoo T, Bacino CA, Pursley A, Li J, Erickson R, Gropman AL, Miller DT, Seashore MR, Summers AM, Stankiewicz P, Chinault AC, Lupski JR, Beaudet AL, Sutton VR. Microarray-based CGH detects chromosomal mosaicism not revealed by conventional cytogenetics. American Journal of Medical Genetics Part A, 2007; 143A: 1679-1686.

[184] Kingsbury MA, Yung YC, Peterson SE, Westra JW, Chun J. Aneuploidy in the normal and diseased brain. Cellular and Molecular Life Sciences, 2006; 63: 2626-2641.

[185] Rehen SK, Yung YC, McCreight MP, Kaushal D, Yang AH, Almeida BSV, Kingsbury MA, Cabral KMS, McConnell MJ, Anliker B, Fontanoz M, Chun J. Constitutional aneuploidy in the normal human brain. Journal of Neuroscience, 2005; 25: 2176-2180.

[186] Yurov YB, Iourov IY, Vorsanova SG, Demidova IA, Kravetz VS, Beresheva AK, Kolotii AD, Monakchov VV, Uranova NA, Vostrikov VA, Soloviev IV, Liehr T. The schizophrenia brain exhibits low-level aneuploidy involving chromosome 1. Schizophrenia Research, 2008; 98: 139-147.

[187] Zong H, Espinosa S, Su HH, Muzumdar MD, Luo LQ. Mosaic analysis with double markers in mice. Cell, 2005; 121: 479-492.

[188] Duchon A, Besson V, Pereira PL, Magnol L, Herault Y. Inducing segmental aneuploid mosaicism in the mouse through targeted asymmetric sister chromatid event of recombination. Genetics, 2008; 180: 51-59.

[189] Liu PT, Jenkins NA, Copeland NG. Efficient Cre-IoxP-induced mitotic recombination in mouse embryonic stem cells. Nature Genetics, 2002; 30: 66-72.

[190] Losada A, Hirano M, Hirano T. Cohesin release is required for sister chromatid resolution, but not for condensin-mediated compaction, at the onset of mitosis. Genes & Development, 2002; 16: 3004-3016.

[191] Takata H, Matsunaga S, Morimoto A, Ma N, Kurihara D, Ono-Maniwa R, Nakagawa M, Azuma T, Uchiyama S, Fukui K. PHB2 protects sister-chromatid cohesion in mitosis. Current Biology, 2007; 17: 1356-1361.

[192] Leneuve P, Colnot S, Hamard G, Francis F, Niwa-Kawakita M, Giovannini M, Holzenberger M. Cre-mediated germline mosaicism: a new transgenic mouse for the selective removal of residual markers from tri-lox conditional alleles. Nucleic Acids Research, 2003; 31.

[193] Modi D, Bhartiya D. Y chromosome mosaicism and occurrence of gonadoblastoma in cases of Turner syndrome and amenorrhoea. Reproductive Biomedicine Online, 2007; 15: 547-553.

[194] Weaver BAA, Cleveland DW. Aneuploidy: Instigator and inhibitor of tumorigenesis. Cancer Research, 2007; 67: 10103-10105.

[195] Arnoldus EPJ, Peters ACB, Bots G, Vanderploeg M. SOMATIC PAIRING OF CHROMOSOME-1 CENTROMERES IN INTERPHASE NUCLEI OF HUMAN CEREBELLUM. Human Genetics, 1989; 83: 231-234.

[196] Yurov YB, Lourov IY, Monakhov VV, Kolotii AD, Demidova EA, Beresheva AK, Kravets VS, Kutsev SI, Vorsanova SG. The variation of aneuploidy frequency and somatic chromosome pairing in fetal human brain. Chromosome Research, 2005; 13: 53-54.

[197] Westra JW, Peterson SE, Yung YC, Mutoh T, Barral S, Chun J. Aneuploid mosaicism in the developing and adult cerebellar cortex. Journal of Comparative Neurology, 2008; 507: 1944-1951.

[198] Muotri AR, Gage FH. Generation of neuronal variability and complexity. Nature, 2006; 441: 1087-1093.

Send Orders for Reprints to reprints@benthamscince.net
Advances in Genome Science, Vol. 3, **2014, 33-82** **33**

CHAPTER 2

Estrogen Regulation of microRNA Expression

Tissa T. Manavalan and Carolyn M. Klinge*

Department of Biochemistry & Molecular Biology, Center for Genetics and Molecular Medicine, University of Louisville School of Medicine, Louisville, KY. 40292, USA

Abstract: Estrogens have a wide variety of physiological functions as sex steroid hormones in women and in men. Estrogens have apparent anti-aging properties in brain, cardiovascular tissues, and bone. Estrogens regulate genes directly through binding to estrogen receptors alpha and beta (ERα and ERβ) that are ligand-activated transcription factors and indirectly by activating plasma membrane-associated ER which, in turns, activates intracellular signaling cascades leading to altered gene expression. Estrogens may also impact cellular signaling by binding to GPR30/GPER. MicroRNAs (miRNAs) are short (19-25 nucleotides), naturally-occurring, non-coding RNA molecules that base-pair with the 3' untranslated region of target mRNAs. This interaction either blocks translation of the mRNA or targets the mRNA transcript to be degraded. The human genome contains ~2,578 miRNAs. Aberrant patterns of miRNA expression are implicated in human diseases including estrogen-related diseases including breast cancer. miRNAs regulated by estrogens have been identified Zebrafish, mouse tissues, and in human cells and tissues including breast and myometrial tumors. The mechanism for estrogen regulation of miRNA expression and the role of estrogen-regulated miRNAs in normal homeostasis, reproduction, lactation, and in cancer is only beginning to be explored.

Keywords: Dicer, Drosha, endocrine-resistance, epithelial-mesenchymal transformation (EMT), estrogen, estrogen receptor, miRNA, mRNA stability, tamoxifen, transcription.

INTRODUCTION

Estrogens (estradiol, estrone, and estriol) are steroid hormones that regulate development and homeostasis in a wide variety of tissues including the brain, reproductive tract, vasculature, and breast. The ovary is the primary site of

***Address correspondence to Carolyn M. Klinge:** Department of Biochemistry & Molecular Biology, University of Louisville School of Medicine, Louisville, KY 40292, USA; Tel: 502-852-3668; Fax: 502-852-6222; Email: carolyn.klinge@louisville.edu

Christian Neri (Ed)

estradiol (E_2) synthesis and E_2 is the primary circulating estrogen in premenopausal women. Animal studies have shown that higher estrogen levels in females protect against aging by up-regulating the expression of antioxidant, longevity-related genes, *e.g.*, selenium-dependent glutathione peroxidase (GPx) and Mn-superoxide dismutase (Mn-SOD) [1], by protecting against stroke-related injury [2], by vasorelaxing effects [3], by direct myocardial protection [4], and by activating the insulin receptor substrate (IRS)-1 signaling pathway [5]. Women's life expectancy seems not to be influenced by hormone replacement therapy (HRT = estrogens plus a progestin, usually conjugated equine estrogens and medoxyprogesterone acetate (MPA)) in postmenopausal women, but atherosclerosis and bone loss are considerably delayed. In addition, HRT protects against Alzheimer's disease (AD) [6], perhaps by suppressing elevated gonadotropin levels, *i.e.*, luteinizing hormone (LH), in postmenopausal women since elevated LH is thought to play a key role in AD pathogenesis [7]. Other age-associated impairments are also reduced by estrogen. For example, premenopausal women have a reduced risk for cataracts compared with men of the same age group and women in the Farmingham study who used estrogen replacement therapy (ERT) showed reduced risk for cataracts [8]. miRNAs are a class of naturally-occurring, small, non-coding RNA molecules that are related to, but distinct from, small interfering RNAs (siRNAs) which regulate mRNA translation or stability [9-11]. There are very few studies on the hormonal regulation of miRNAs expression. Select changes in microRNA (miRNA) expression correlate with diagnostic markers used in breast cancer therapies, *e.g.*, estrogen receptor α (ERα) and tumor grade [12-22]. Although the number of publications on human miRNAs has increased exponentially since the first reports in 2001, there are fewer studies on estrogenic regulation of miRNAs or miRNAs that regulate estrogen receptor (ER) function (Fig. **1**). It is highly likely that estrogens regulate miRNA expression by both genomic (transcriptional) and non-genomic mechanisms of action. Identification and characterization of estrogen-regulated miRNAs may provide new biomarkers and therapeutic targets in aging as well as in diseases including breast cancer.

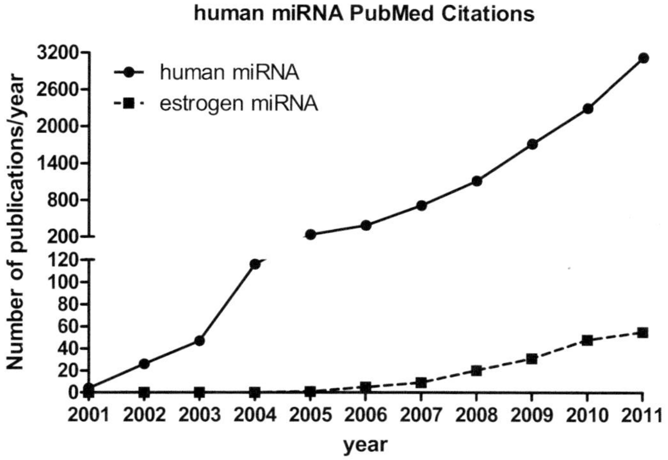

Figure 1: Increase in PubMed citations on miRNA and estrogen and miRNA. The search terms used were human AND miRNA (black closed circles) and human AND miRNA AND estrogen. Each point is the number of publications in the calendar year indicated.

GENOMIC ER ACTIVITIES

Initiation of transcription is a complex event occurring through the cooperative interaction of multiple factors at the target gene promoter. I will use the term ER to refer to either ERα or ERβ or to both subtypes. I will refer to each subtype individually as pertinent to the known functions of these two proteins. Estrogen action is primarily mediated through binding to ER. ERα and ERβ are members of the steroid/nuclear receptor superfamily of proteins of which there are 48 members in mammals [23]. ERα and ERβ are highly conserved within the DNA binding domain (DBD, C domain), consisting of two cys-cys zinc fingers (CI CII) [24], but differ in their N- and C- termini [25]. Structurally, ER is divided into 6 domains lettered A-F from N- to C-terminus. ER is believed to be the ancestral steroid receptor originating 600-1200 million years ago, presumably because of the critical role of estrogens in reproduction and maturation [26, 27].

In the simplest model, the binding of estradiol (E_2) *via* hydrogen bonding to residues within the ligand binding pocket of the ligand binding domain (LBD, E domain) results in conformational changes termed activation [28]. These conformational changes induced by E_2 binding result in loosening of contact between the N-terminus and the C-terminus of ERα and exposes nuclear localization signals within the DNA binding domain (DBD, C domain) and hinge region (D domain) as well as altering structure of the LBD [29]. Crystal structure studies of the LBD of ERα excluding the F domain shows that the LBD has 12 alpha helices and E_2 binding repositions helix 12 such that activation function-2 (AF-2) is exposed [30]. Helix 12 acts as a "switch" controlling accessibility of co-regulator interaction site. Ligand binding also facilitates ER dimerization (homodimerization or heterodimerization of ERα and ERβ).

E_2-liganded ER interacts directly with a specific DNA sequence called the estrogen response element (ERE = 5'-AGGTCAnnnTGACCT-3'), historically located in the promoter region and currently established, by techniques including ChIP-chip (chromatin immunoprecipitation for ERα followed by microarray analysis of bound DNA) and ChIP-seq (ChIP for ERα followed by sequencing of the bound DNA) has established that EREs are also located at great distances from the transcription start site including in the 3' flanking regions of target genes [31-38].

DNA binding increases ER interaction with co-regulators and basal transcription factors (reviewed in [39]). Fig. **2** models genomic ER action. EREs are enriched in genes upregulated by E_2-ERα, at least in MCF-7 cells [40]. Most of our knowledge about the molecular mechanisms of ERα in cells comes from studies in the MCF-7 human breast cancer cell line, although ChIP-chip and ChIP-seq were performed in ERα-transfected U2OS cells [41] and ERα-negative, basal-like MDA-MB-231 breast cancer cells transfected with wildtype (wt) ERα or a DNA binding mutant of ERα to distinguish between direct ERE-binding sites and other ERα binding sites [38]. These studies revealed cell-specific ERα cistromes.

ERα binds thousands of sites within the MCF-7 genome and FoxA1 [42-45] and PBX1 (Pre-B-cell leukemia homeobox 1) [46] appear to act as "pioneer factors" that remodel condensed chromatin to facilitate ERα binding [47]. As expected, EREs are the most common binding sites for genes upregulated by E_2-ERα [40] and E_2-ERβ [48] in MCF-7 cells, and in the wtERα-transfected MDA-MB-231 cells [38]. ER-ERE interactions enhance the recruitment of co-activator/chromatin remodeling complexes resulting in histone modifications, nucleosomal repositioning, increased accessibility to the DNA template for RNA polymerase II interaction, and increased target gene transcription (reviewed in [49-55]).

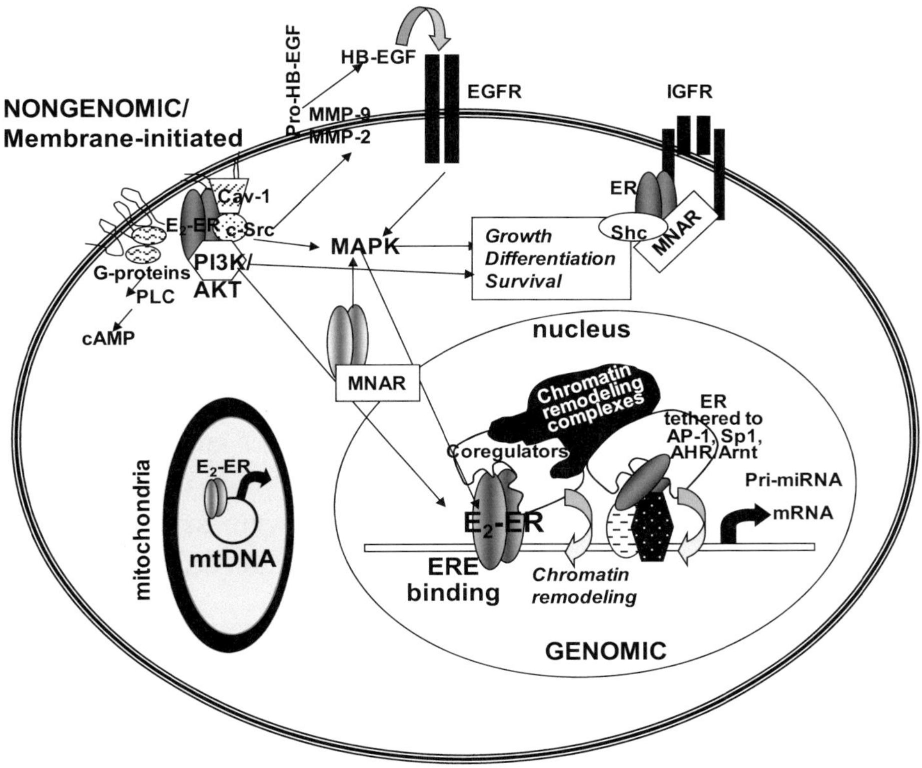

Figure 2: Cell model showing genomic and nongenomic activities of ERα and ERβ. ERα and ERβ are located in the cytoplasm, nucleus, and mitochondria. For genomic (nuclear) ER activity, E_2 binds and activates ER resulting in dimerization, ERE binding or interaction with other transcription factors, *e.g.* AP-1 bound to DNA, co-regulator and chromatin remodeling complex recruitment, chromatin remodeling, and increased transcription of target genes. For nongenomic/membrane-initiated estrogen signaling, E_2 binds ERα in caveolae in the plasma membrane [150, 334]. ERα interacts with G-proteins, the p85 subunit of PI3K, c-Src, and Cav-1 to initiate PI3K/AKT and MAPK signaling cascades [89, 335]. ERα interacts with MNAR [165]and Shc [133]in the cytoplasm. ERα interacts with the EGF- and IGF-1 receptors in plasma membranes. In mitochondria, ERβ interact with the D-loop of mtDNA [336, 337].

ERα can also be activated by phosphorylation of ser118 in the N-terminal A/B domain that activates activation function-1 (AF-1) in the absence of ligand binding [56]. ERα interacts directly with ERK2 and ChIP-on-chip analysis identified overlapping binding sites for ERα and ERK2 in E_2-treated MCF-7 cells

[57]. ERα is also phosphorylated and stabilized by LMTK3 on ser-167 which stabilizes ERα and inhibits proteasomal degradation, and, in a negative feedback loop, E_2 represses LMTK3 transcription in MCF-7 cells [58].

In HeLa cells transfected with fluorescent fluorescent-tagged ERα, E_2 causes rapid intracellular and intranuclear movement of ERα to form punctuate nuclear speckles that appear to indicate ERα-nuclear matrix interaction [59-61]. In addition to direct ER-ERE binding, ER also activates transcription *via* a "tethering mechanism" in which ER interacts directly with transcription factors, *e.g.* Sp1 [62] and AP-1 [63], bound to their response elements (Fig. **2**). E_2-ERα can suppress NF-κB target gene expression by direct interaction with p65 [64, 65]. Notably ChIP-seq, ChIP-PET (ChIP for ERα followed by paired-end tag sequencing) and ChIP-chip experiments have identified a number of transcription factor binding sites with which ERα interacts in MCF-7 cells including: AP-1, CEBP, FOXA1, PAX6, RORA, PITX2, and GATA2 [66]. There are also cell-specific ERα binding sites. For example, ERα overexpression in HeLa cells identified only 9% of common promoter binding sites with MCF-7 [67].

Specific effects of ERβ on gene expression are largely distinct from ERα [68]. ERβ-regulated genes have been studied in cell lines devoid of endogenous ERα and ERβ expression, *i.e.*, U2OS [69-71], HEK293 [72], and Hs578T [73]; or in ERβ-stable overexpressing MCF-7 cells [48, 74]. Notably, E_2-ERβ suppresses the transcription of growth-promoting genes [73]. ChIP-seq for ERβ in MCF-7 cells transfected with ERβ revealed that most ERβ binding sites were EREs and the other sites were those known to be involved in ERβ-tethering including AP1, E2F, and SP1 [48].

These DNA-protein and protein-protein interactions recruit co-activator/chromatin remodeling complexes resulting in histone modifications that lead to nucleosomal remodeling, increased accessibility to the DNA template for RNA polymerase II interaction, and increased target gene transcription (reviewed in [49-51]).

By definition, co-activators are proteins that interact directly with transcription factors to enhance transcription [75]. It is important to note that the term co-activator or co-repressor is used when referring to an ER (or other NR) co-regulator is gene-, cell-type, and context- specific [55]. This indicates that proteins classified as co-activators can also repress transcription and corepressors such as SMRT are gene- and cell- specific co-activators for ER [76]. Co-activators promote the assembly of the transcription initiation complex in part by altering chromatin structure and 'loosening' DNA-histone interactions, facilitated by increased histone lysine residue acetylation, methylation, ubiquitination, or sumoylation [77]. Once the transcription initiation complex is complete, RNA polymerase II (RNA pol II) is recruited to the transcription start site and begins transcription. By my count, at least 64 different ER co-activators and 24 corepressors have been functionally identified as interacting with ERα and/or ERβ (reviewed in [49, 52, 54, 78, 79], see also http://www.nursa.org/index.cfm). The current model predicts that ERE-bound, agonist-liganded ER recruits co-activator proteins to enhance gene expression [80]. In contrast for those genes at which tamoxifen (TAM) is an antagonist of ER transactivation, the LBD of TAM-occupied ERα does not interact with co-activators due to key conformational differences between agonist and antagonist – occupied ERα in AF-2 [81]. TAM-occupied ERα interacts with corepressors, *e.g.* NCoR or SMRT [82-86], that recruit histone deacetylase (HDAC) complexes thus keeping chromatin condensed and blocking transcriptional activation. Antiestrogen ICI 182,780 (Fulvestrant)-occupied ERα is targeted to the 26S proteasome for degradation [61, 87]. In contrast, although ICI 182,780 inhibits ERβ-mediated transcription, it stabilizes the ERβ protein [88].

RAPID, MEMBRANE-INITIATED, NONGENOMIC ESTROGEN ACTION

In addition to its classical genomic/transcriptional effects mediated by ER-DNA interaction, described above, E_2 has rapid "nongenomic, extra-nuclear, or membrane-initiated" effects that occur very rapidly, *i.e.*, within seconds-minutes after E_2 administration [89-98]. These effects are independent and distinct from the genomic, *i.e.*, ER-mediated transcription, activities reviewed in the preceding section. Fig. **2** highlights some of the established nongenomic activities of ER. Rapid estrogen-stimulated intracellular activities are mediated by plasma membrane (PM)-associated ER and/or by an 'orphan' G-protein coupled receptor GPR30/ GPER [93, 99-109]. Evidence of a PM-associated ER population is supported by experiments in which a cell-impermeable E_2–bovine serum albumin or other E_2-conjugate was shown to rapidly initiate intracellular kinase cascade activities including MAPK/ERK (p42/p44 MAPKs), endothelial nitric oxide synthase (eNOS), and PI3K/AKT [110-119]. In MCF-7 cells, E_2 rapidly increased PIP2-phospholipase C activity [120], mobilized intracellular Ca^{2+}, activated MAPK [90, 121-140], and PI3K/AKT [131, 141-146]. Immunohistochemical techniques have visualized membrane ERα in a variety of cell types including endometrial and breast cancer cells [147], pituitary cells [148], and lung cancer cells/tumors [135]. ERβ was shown to be located in the PM of primary cortical neurons in recent confocal imaging studies [149]. The amount of PM localized ER is cell type-specific and is generally ~ 5-10% of total ER [98].

The mechanism by which ERα and its splice variant ERα46 localizes to the PM involves palmitoylation [150-153]. In addition, ER interaction with intracellular proteins is important in PM association. In MCF-7 cells, the adaptor proteins Shc shuttles ERα from the nucleus to the PM where ERα interacts with the IGF-1 receptor (IGF-1R) [133, 140]. Similarly, another scaffold protein, called MNAR (modulator of nongenomic activity of ER), has been implicated in ERα-cSrc interaction and MAPK signaling [129, 139, 142, 154-166]. A role for a membrane caveolae-localized ERα46 in rapid NO release *via* PI3K/Akt activated eNOS has been reported in endothelial cells [150, 161, 167, 168].

In addition to PM-associated ER, GPR30/GPER30 has reported to serve as a membrane estrogen receptor because it binds E_2 with high affinity (Kd = 2.7nM) and activates adenylate cyclase, thus increasing cAMP levels [114, 117, 169-172]. GPR30 is distinct from ERα and ERβ in that ICI 182,780 and tamoxifen bind GPR30 with high affinity and mimic the effects of E_2 [171]. Although the role of GPR30 in MCF-7 and SKBr3 breast cancer cells has been questioned [90], it appears likely that GPR30 is a *bone fide* membrane estrogen receptor in some cell types [173-179].

ERα36

ERα36 was identified as a target of GPR30 regulation [180]. ERα36 is produced from an alternative site the *ESR1* gene that lacks AF-1 (A and B domains), AF-2, and the last 138 aa of ERα66 which are replaced with a novel 22 aa sequence, but contains the DBD and hinge regions (C-D domains) and most of the LBD (E domain) [181]. ERα36 predominantly localizes to the PM where it activates MAPK and AKT [182]. ERα36 acts as a dominant negative effector for ERα66 and ERβ. ERα36 is expressed in ERα66-negative cell lines and breast tumors and was associated with shorter disease-free survival [183]. Interestingly, this group reported that GPR30 regulates ERα36 expression by stimulating activation of AP-1 and that the activity of the G1-GPR30-specific agonist was blocked by an anti-ERα36-specific antibody [180]. Regulation of ERα36 by miRNA will be described later.

microRNAs (miRNAs)

Evidence from the Encyclopedia of DNA project (ENCODE) has revealed that 80-93 % of the genome is "expressed" [184-186]. However, the function of this large relative mass of 'dark matter RNAs' (excluding rRNA, mRNA, tRNA, and mtRNA) is not yet fully understood [187]. The transcribed RNAs are largely conserved between humans and mice suggesting that these noncoding RNAs (ncRNAs) have important functions. Evidence of the importance of small cellular noncoding RNAs (ncRNAs), which includes short interfering RNAs (siRNAs), microRNAs (miRNAs), and PIWI-interacting RNA (piRNAs), was recently reviewed [188] and includes roles in cancer, cardiovascular diseases and diabetes [189]. Among the small ncRNAs are microRNAs (miRNAs). The importance of miRNAs is reflected by the 2008 Albert Lasker Award for Basic Medical research award to Drs. Victor Ambros [190] and Gary Ruvkun [191] who discovered and characterized the first miRNAs in *C. elegans* and Dr. David Baulcombe who discovered let-7 miRNA in plants [192] (see also http://www.laskerfoundation.org/).

miRNAs are distinct from small interfering RNAs (siRNAs) [9-11]. About half of miRNAs are expressed from introns of protein-coding transcripts and miRNAs have 5' and 3' sequence features that form boundaries including transcription start sites, CpG islands, and transcription factor binding recognition elements [193]. miRNAs may be differentially processed from the sense and antisense strands of the same hairpin RNA or transcripts from the same locus, thus expanding the number of miRNAs from a single genomic locus [194].

The pathway of mature miRNA biogenesis is depicted in Fig. **3**. miRNA genes are mostly transcribed by RNA polymerase II into primary-micro-RNAs (pri-miRNAs) that are capped and polyadenylated [195]. Pri-miRNAs contain self-base-pairing stem-loop structure that is necessary for critical processing within the nucleus by Drosha, an endonuclease of the RNAse III family, and its cofactor DGCR8 into short (60 to 70 nt) imperfect hairpin structure precursor-miRNAs (pre-miRNAs) [196]. E_2 was shown to inhibit the processing of pri-miRNAs into pre-miRNAs Drosha through ER-Drosha complex interactions [197]. These pre-miRNAs are exported from the nucleus by exportin and Ran-GTP. Pre-miRNAs are processed by the cytoplasmic RNAse II enzyme Dicer to form mature ~22 nt transiently double-stranded miRNA duplexes that are transferred to Argonaute proteins (Ago1, Ago2, Ago3, and Ago4 [198]) in the RNA-induced silencing complex (RISC), leading to unwinding of the duplexes to form single stranded microRNAs. RISC guides RNA silencing with the miRNA binding either to the 3' untranslated region (3' UTR) or to the open reading frame (ORF) of its target mRNA [199-202]. Most commonly, because of imperfect complimentarity of the base pairing between the miRNA and the 3'UTR, the RISC complex causes translational repression by RISC interaction with eIF6 which prevents assembly of 80S ribosomal assembly [203] or by inhibition of translation [16]. Thus, miRNA-mRNA 3'UTR interaction results in a decrease in target protein, not mRNA. The 7 to 8 nucleotide region of basepairing between the 5' end of the mature miRNA and the mRNA is called the 'seed sequence'. Base pairing of the miRNA-RISC complex within the ORF requires almost perfect complimentarity for its mRNA target and the mRNA is either degraded or translation is blocked [195]. The miRNA-containing ribonucleoprotein particle (miRNP)-silenced mRNA is directed to the P-bodies, where the mRNA is either released from its inhibition upon a cellular signal and/or actively degraded [204]. Comparative genomics analyses have revealed that over 45,000 miRNA binding sites within human 3'UTRs that are conserved above background levels [205]. This number indicates that > 60% of human protein-coding genes have been under selective pressure to maintain pairing to miRNAs [205]. Some miRNAs may also increase translation of select mRNAs in a cell cycle-dependent manner [206].

miRNAs have important roles in regulating various cellular processes including replication, differentiation, and apoptosis [11, 201, 207-222]. The human genome contains ~ 2,578 miRNAs (June 2013, http://www.mirbase.org/) [223]. MiRNAs are expressed in a tissue-specific manner [224]. Each miRNA targets ~ 200 transcripts directly or indirectly [225], but the *bone fide* physiological targets of the vast majority of miRNAs is virtually unknown.

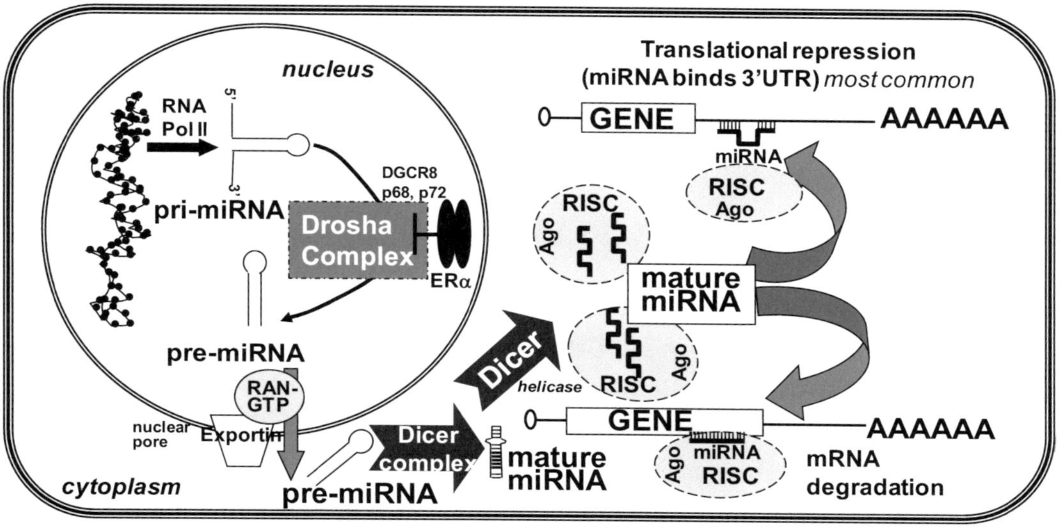

Figure 3: Model of miRNA biogenesis and function. Primary transcripts of microRNAs (pri-miRNAs) are transcribed by RNA polymerase II, processed by the RNAse III enzyme, Drosha and its cofactor DGCR8, to precursor microRNAs (pre-miRNAs) and are then exported from the nucleus by Exportin/RAN-GTP [195]. In the cytoplasm, pre-miRNAs are processed by the RNAse III enzyme, Dicer to form mature ~22 nt transiently double-stranded miRNA duplexes that are transferred to Argonaute proteins (Ago1, Ago2, Ago3, and Ago4 [198]) in the RNA-induced silencing complex (RISC), leading to unwinding of the duplexes to form single stranded miRNAs. The mature miRNAs bind either to the 3' untranslated region (3' UTR) or to the open reading frame (ORF) of its target mRNA [199-202]. Binding of miRNA/RISC complex with the 3'UTR causes translational repression [16]. Thus, miRNA-mRNA 3'UTR interaction results in a decrease in target protein, not mRNA.

NOMENCLATURE OF miRNA

miRNAs are given a three lettered prefix depending on the species that they originate in *e.g.*, hsa for *homo sapiens* and mmu for mouse [226]. If miRNAs originate from different genomic loci, they are assigned a numerical suffix, *i.e,* hsa-miR-miR-29b-1 and hsa-miR-29b-2. miR-21 and miR-21* refers to miRNAs that arise from the same hairpin precursor. The * indicates that it is a less predominant form of the miRNA in RISC [227]. If miRNAs differ by a few bases they are given a lettered suffix of the form miR-125a and miR-125b. Those miRNAs that originate from opposite arms of a chromosome are assigned suffixes of the type miR-142-5p and miR-142-3p. A miRNA family refers to miRNAs that arise a common ancestor and whose sequences are similar *e.g.* miR-221 and miR-222 family. miRNA cluster arises due to gene duplication, *e.g.*, the miR-200 cluster of miRNAs are located in two chromosomes, *i.e.*, miR-200a, miR-200b and miR-429 are located on chromosome 1 and miR-200c and miR-141 are located on chromosome 12 [228]. Each cluster is transcribed into a common precursor RNA.

REGULATION OF miRNA EXPRESSION

miRNA expression is regulated at different levels. The ratio of the ribonulcease Drosha and its binding partner DGCR8 is tightly regulated to ensure proper pri-miRNA processing [229]. DGCR8 stabilizes Drosha. Drosha in turn regulates DGCR8 levels by cleaving and thus inactivating it. Thus a tight feedback loop maintains the cellular Drosha/DGCR8 ratio [230]. A number of co-activators and co-repressors can alter Drosha activity. Transforming growth factor beta (TGFβ) signaling and bone morphogenetic protein (BMP) and SMAD proteins stimulate, while the nuclear factor NF90-NF45 heterodimer suppresses Drosha activity [231, 232]. ERα along with helicases p68 and p72, and Drosha have been shown to affect Drosha complex formation, thus repressing pri-miRNA processing [233].

Defects in Exportin 5 results in accumulation of pre-miRNAs in the nucleus [234]. Another ribonuclease that is a key point of regulation in the miRNA biogenesis pathway is Dicer. Altered Dicer expression can affect processing of pre-miRNA to mature miRNA. Dicer cofactors TRBP and PACT are critical in maintaining the stability of Dicer [235, 236]. Increased phosphorylation of TRBP by MAPK signaling enhances Dicer activity and promotes miRNA processing [237]. miR-221, miR-222, and miR-29a target DICER1 and miR-200a overexpression increases Dicer protein expression [238].

After cleavage of the pre-miRNA by the ribonuclease Dicer, one of the strands of the mature miRNA is incorporated into the RISC by Ago2 protein. Although the human genome encodes about 8 Ago proteins, Ago2 is the major protein involved in RNA cleavage and silencing. Within RISC, Ago2 unwinds the miRNA duplexes to form single-stranded miRNA-5p and -3p products. The passenger strand is usually, but not always, degraded, and the mature miRNA (guide strand) is incorporated into RISC. Which miRNA strand is in the functional RISC varies between cell types. Some pre-miRNAs are processed by Ago2 slicer catalytic activity instead of Dicer [239].

The level of Ago2 determines the amount of mature miRNA synthesized. It is subject to regulation at the transcriptional and post-transcriptional level. For *e.g.* in MCF-7 breast cancer cells Ago2 expression is inhibited by E_2 and EGF-MAPK signaling [240].

ALTERED miRNA EXPRESSION IN BREAST CANCER

The spectrum of miRNAs expressed in solid tumors, *i.e.*, prostate, colon, stomach, pancreas, lung, and breast, is different from normal tissues [224]. Aberrant patterns of miRNA expression have been reported in human breast cancer [12, 13, 18, 20, 21, 196, 208, 210-216, 222, 224, 225, 241-248] and recently reviewed [195]. The first miRNA study in breast cancer indicated differential expression of miRNAs in concordance with other well-established markers of breast cancer stage and patient prognosis including ERα and PR expression, tumor stage, number of positive lymph nodes, and vascular invasion [20]. Different miRNA expression profiles were also associated with ErbB2+ *versus* ER+ tumors [22]. More recently, patients whose breast tumors showed reduced miR-126, miR-206, or miR-335 were found to have reduced survival, regardless of ERα or ErbB2 status [18].

A number of genes involved in breast cancer progression have been identified by *in silico* analysis to be targets of miRNAs that are deregulated in breast cancer [249] and some have been experimentally proven. Breast tumors express higher miR-21 than normal breast and that antisense to miR-21 suppressed MCF-7 breast cancer cell growth *in vitro* and as tumor xenografts in mice by regulating Bcl-2 [13]. We reported that overexpression of miR-21 in MCF-7 cells increased soft agar colony formation, reflecting increased tumorigenicity of these cells [250]. We demonstrated that miR-21 binds to a seed element in the 3'-UTR of the programmed cell death 4 (PDCD4) gene and reduces Pdcd4 protein expression

[250]. In addition to PDCD4, miR-21 has also been shown to promote tumor formation by targeting TPM1, Maspin and BCL2 [251-254].

miRNAs promote metastasis and invasive properties of breast cancer cells. Expression profiling of miRNAs in solid tumors and breast tissues have identified miRNAs associated with breast cancer subtypes [12]. Let-7a, miR-21, miR-141 and miR-214 were expressed in luminal cell type while miR-145 and miR-205 were associated with the myoepithelial cell type [248].

Other examples of the miRNAs that are deregulated in breast cancer include tumor suppressors, *e.g.*, Let-7, miR-125, miR-200, and oncogenic miRs, *e.g.*, miR-21 and miR-155 [255]. Overexpression of miR-125a and miR-125b decreased ERBB2 and ERBB3 mRNA and protein levels, inhibited phosphorylation of ERK1/2 and AKT, and inhibited the anchorage-independent growth of ERα-negative/ERBB2-overexpressing SKBR3 breast cancer cells [256]. These data indicate that miR-125a and miR-125b act as tumor suppressors and their induction in HER2-overexpressing breast tumors might be beneficial as a treatment strategy. Down-regulation of the tumor suppressor Let-7a alters the expression of its targets RAS and HMGA2 [257]. Down-regulation of Let-7a, miR-200 and miR-205 promotes epithelial-to-mesenchymal transition (EMT) in breast cancer [258-260]. The role of miRNAs in EMT was recently reviewed [261]. The targets of miR-200 include transcription factors ZEB1, ZEB2; PLCG1, TUBB3 and BM1 [262-264]. Other metastatic promoting miRNAs that promote migration and invasion of cancer cells include miR-10b, miR-520c and miR-373 [265, 266]. miR-10b targets HOXD10 protein expression [267]. Overexpression of miR-10b suppresses HOXD10 leading to an invasive phenotype [265].

The miR-17-20 cluster located on chromosome 13q31 encodes six miRNAs: miR-17-5p, 18a, 19a, 20a, 19b, and 92, and acts as a tumor suppressor [19]. Viral transduction of the miR-17-20 cluster in MCF-7 breast cancer cells inhibited cell proliferation by suppressing cyclin D1 expression [19]. We observed that expression of members of the miR-17/20 cluster were higher in LY2 endocrine resistant cells compared to MCF-7 cells and that 4-OHT increased the expression of miR-17, 18a, and 19a (Fig. **4**).

The breast cancer oncogene and nuclear receptor co-activator AIB1/SRC-3/NCOA3 is regulated by miR-17-5p and there is a reciprocal relationship between reduced miR-17-5p and increased AIB1 in breast cancer cells [268]. Overexpression of miR-17-5p reduced E_2-stimulated proliferation of MCF-7 breast cancer cells, indicating a role for deregulation of miR-17-5p in breast cancer [268].

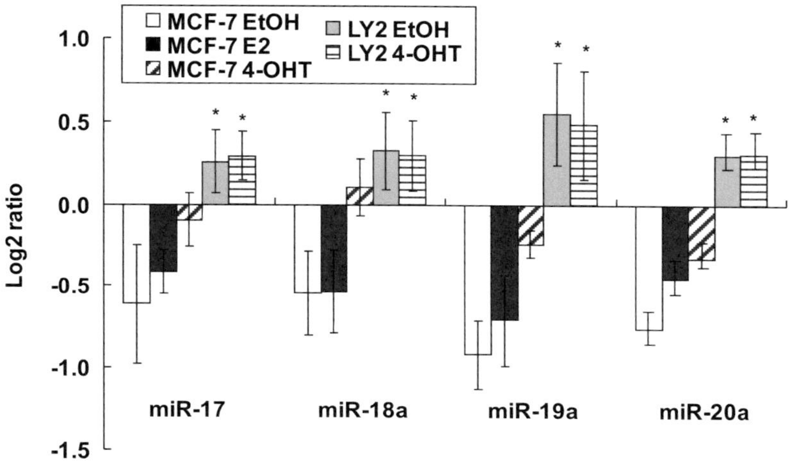

Figure 4: Expression of members of the miR-17/20 cluster are higher in LY2 endocrine-resistant compared to MCF-7 cells. MCF-7 and LY2 human breast cancer cells were serum-starved (5% DCC serum in phenol red-free medium) for 48 h and then treated with ethanol (EtOH, vehicle control), 10 nM E_2, or 100 nM 4-OHT for 6 h. miRNA expression was examined by Exiqon microarray. Each bar represents the mean ± SEM of 5 separate experimental determinations (biological replicates). * Significantly different from EtOH control in MCF-7 cells, $P < 0.05$.

The miR-17-92 cluster is amplified in solid cancers [269] and has a dual role as a tumor suppressor and an oncogenic miRNA in breast and other cancers [270]. miR-17/92 promotes breast cancer by suppressing HBP1 [271] and ZBTB4 [272].

Analysis of miRNA expression in breast tumors using tissue microarray identified miRNAs differentially expressed in breast cancer tissue when compared to normal tissue, *e.g.,* miR-10b, miR-21, miR-145 were upregulated in breast cancer tissue [273]. There was also a correlation between miRNA expression, ER and PR status, lymph node status in these samples. In another study, miR-342 and miR-520g were found to be overexpressed in ER-positive and HER2 positive tumors when compared to normal tissue [274].

Circulating miRNAs have been detected in the blood, plasma and serum of breast cancer patients [275]. The level of plasma miRNAs was higher in breast cancer patients when compared to control groups [276]. For example miR-425*, Let 7c were higher in the plasma of women with early stage breast cancer compared to healthy controls. Higher expression of miR-10b and miR-34 in the serum of breast cancer patients correlated with metastasis [275]. Systemic miR-195 and Let-7a were higher in breast cancer patients when compared to normal subjects [277]. The expression of these miRNAs also correlated with ER status. These reports indicate that miRNAs could be valuable diagnostic markers in the prognosis and treatment of breast cancer patients.

ESTROGENIC REGULATION OF miRNA EXPRESSION

Regulation of miRNA expression by estrogens has been reviewed [278, 279]. A PubMed search for estrogen AND miRNA revealed 198 papers (Fig. **1**). However, in that list and in total there are, to our knowledge, there are only ~ 20 studies in which miRNA regulation by E_2 has been directly examined (see below). Indeed, although a software application that will retrieve all miRNA:mRNA functional pairs in an experimentally derived set of genes was recently developed and used to identify E_2-regulated mRNA genes in breast cancer [280], this paper does not experimentally address miRNA changes regulated by E_2.

E_2 REGULATION OF miRNAs IN ANIMAL STUDIES

The effect of E_2 in miRNA expression was reported zebrafish [281], August Copenhagen Irish (ACI) rats [282], and mouse splenocytes [283]. miRNA expression in female mammalian reproductive tissues in transgenic and knockout mouse models was reviewed and findings related to changes in miRNAs in the ovary and uterus in response to deletion of Dicer [284], LH, and during development [285]. Changes in miRNA expression in mouse uterus during implantation have been cataloged [286]. Importantly, the authors of this review concluded that the expression, regulation, and function of miRNAs within specific tissues and cells "still needs to be determined" [285].

A study of the effect of E_2 on miRNA expression in the adult (3 mos) zebrafish male (*Danio rerio*) identified altered expression of 38 miRNAs in the whole body homogenates [281]. E_2 was added to the aquariums at a final concentration of 5 μg/liter (18 nM) and although various times of treatment were analyzed, most miRNA changes in response to E_2 were observed after 12 h. miRNAs were regulated by E_2 in a tissue-specific manner with E_2 downregulating miRNAs in the liver and increasing miRNA expression in the skin of the zebrafish. For example, miR-122 was decreased by E_2 in skin, but increased in gills, intestine. and liver. Among the most up-regulated miRNAs were miR-196b and let-7h, and miR-130c and miR-101a were the most down-regulated. The authors identified *Hoxb8a* as a target of miR-196b and showed that E_2, by increasing miR-196b, decreased Hoxb8a [281]. The authors concluded that miR-196b may serve as "a biomarker of exposure to environmental estrogens and endocrine-disrupting chemicals that fish may encounter in their aquatic environment" [281].

In another study, miRNA expression was analyzed after 6, 12, and 18 weeks of E_2-induced mammary carcinogenesis in female ACI rats [282]. After 6 and 12 weeks of E_2 exposure, 15 miRNAs were down-regulated *e.g.*, miR-22, miR-99a, miR-106a, miR-127, miR-499, and 19 miRNAs were-upregulated., *e.g.,* miR-17-5p, miR-20a, miR-21, miR-129-3p, miR-106a, miR-22, and miR-127. By 18 weeks of E_2 treatment, the mammary glands were characterized by lobular involution and hyperplasia, and only 1 miRNA was down-regulated (miR-139) and 5 miRNAs were up-regulated (miR-20b, miR-21, miR-103, mir-107, miR-129-3p, and miR-148a). Genes targeted by three of the altered miRNAs were examined: miR-20a regulates E2F1, miR-106a regulates RBI, and miR-127 regulates BCL6. Western blot of mammary gland lysates after 12 wks of E_2 showed that levels of RBI and E2F1 were decreased and BCL6 protein was increased, data that are in agreement with the increase miR-20a and miR-106a and the decrease in miR-127 detected [282].

A study by Nothnick *et al.* identified miRNAs that are altered by E_2 in mouse uterus [287]. E_2 significantly increased the expression of miR-155, miR-429 and miR-451 and decreased the expression of miR-181b and miR-204. ICI 182, 780 blocked the increase in miR-204 expression by E_2, indicating that the expression of this miR is ER-mediated.

E_2 decreased miR-146a, miR 125a, miR-125b, let-7e, miR-126, miR-145, and miR-143 and increased miR-223, miR-451, miR-486, miR-148a, miR-18a, and miR-708 expression in mouse splenic lymphocytes [283]. Notably, transfection of cells with miR-146a decreased LPS-induced IFNγ. AS to miR-223 blocked LPS-induced IFNγ secretion in splenocytes from E_2 treated mice. This is the first report on E_2 regulation of miRNA expression in immune cells [283], but provided no mechanism by which E_2 regulated these changes.

E_2 REGULATION OF miRNAs IN HUMAN CELL LINES

Although the precise sequence of events leading to breast tumors are not understood, lifetime exposure to estrogens is widely accepted as a major risk factor for the development of breast cancer [288]. Some investigators have documented that E_2 is carcinogenic in human breast epithelial cells [289-291]. However, epidemiological evidence disputing the carcinogenicity of E_2 in humans has been published [292]. Hence, studying E_2 regulation of miRNA expression in breast cancer cell lines may offer insight into mechanisms of estrogen action in breast carcinogenesis and progression.

Recently, GRO-seq (global nuclear run-on and sequencing) identified all the genomic binding sites for RNA polymerases (RNA pol I, II, and III) and all RNA transcripts in E_2-treateed MCF-7 cells [293]. The authors identified 119 miRNA transcripts as E_2 regulated by E_2 one of the three time points examined. Another genome wide analysis of miRNA expression was performed in MCF-7 and ZR-75-1 luminal-like breast cancer cells to study to the effect of E_2 on miRNA expression [294]. In that study E_2 increased miR-760 and miR-424 and decreased miR-618, miR-570 and mir-107 expression. It will be of interest to correlate binding events, transcriptional regulation, and functional outcome in these large-scale studies.

E_2 and the ERα-selective agonist 4,4',4"-(4-Propyl-[1H]-pyrazole-1,3,5-triyl)trisphenol (PPT) [295] decreased miR-206 expression in MCF-7 cells whereas 2,3-bis(4-hydroxyphenyl)propionitrile (DPN), an ERβ-selective agonist [296], increased miR-206, pointing to a regulatory loop [208]. Interestingly, miR-206 also reduced β-actin. The authors of this report called miR-206 a tumor suppressor and found that miR-206 was higher in ERα-negative MDA-MB-231 cells [208], offering a mechanism in addition to ERα promoter methylation [297-301] for reducing ERα expression in MDA-MB-231 cells.

Results from miRNA profiling studies in endometrial cancer were recently reviewed [302]. As for estrogen regulation of miRNAs in the uterus, one study identifying miRNAs expressed in myometrial and leiomyoma smooth muscle cells (MSMC and LSMC) using microarray and real time PCR reported that E_2 inhibited the expression of miR-21 in MSMC and LSMC, whereas E_2 increased and inhibited miR-26a in MSMC and LSMC, respectively [303]. In contrast, ICI 182,780 increased the expression of miR-20a and miR-21 in MSMC and LSMC, and miR-26a in MSMC, while inhibiting the expression of miR-26a in LSMC [303]. No mechanistic studies or miRNA mRNA target gene studies were performed to identify the mechanism(s) involved in these cell-specific differences in miRNA regulation by E_2 and ICI or their downstream targets.

To identify E_2 regulated miRNAs in a classical estrogen-responsive human breast cancer cell line, we treated ERα-positive MCF-7 cells with 10nM E_2 or EtOH (vehicle control) for 6 h to identify primary E_2 target miRNAs [278]. RNA was harvested, labeled either with Cy3 or Cy5, and hybridized with two identical, dual-color miRNA microarrays from LC Sciences. This array contained probes to detect mature miRNA sequences as well as precursor (pre)-miRNAs in the Sanger miRNA registry 7.0 (http://microrna.sanger.ac.uk/sequences/). The differentially expressed transcripts that were consistent on both chips are summarized in Table **1**. 38 miRNA genes were regulated by E_2: 9 were reduced and 29 were increased. A summary about what is known about each of these E_2-responsive miRNAs in terms of breast cancer and estrogen-responsiveness is included in Table **1** [278].

Table 1: Comparison of miRNA expression in MCF-7 cells treated for 6 h with E_2 *versus* EtOH. MCF-7 cells were grown in phenol red free IMEM medium containing 5% dextran-coated charcoal-stripped fetal bovine serum for 48 h prior to a 6 h treatment with ethanol (EtOH, vehicle control) or 10 nM E_2. RNA was isolated using miRVana and sent to LC Sciences for miRNA microarray analysis. All miRNA gene changes included in this table are statistically significant as analyzed by LC Sciences. Negative values indicate decreased expression and positive values indicate increased expression with E_2. NS = no significant change in miR gene expression with E_2-treatment.

miRNA	log2 (E_2/EtOH)	Comments regarding the possible connection of the identified miRNA gene with breast cancer and/or estrogenic responses. *Bone fide* targets of miRNAs are indicated.
let-7a	-0.3	Expressed in ERα positive human breast tumors [22]. Expression was higher in ERα+ than ERα- tumors [12]. Expressed in ZR-75, MCF-7, BT474, SK-BR-3, and MDA-MB-231 breast cancer cells [338]. Expression of Let-7 family members is reduced in 'breast cancer stem cells'(CD44+CD24−/low) [339].

miRNA	**log2 (E_2/EtOH)**	**Comments regarding the possible connection of the identified miRNA gene with breast cancer and/or estrogenic responses. *Bone fide* targets of miRNAs are indicated.**
		Targets of let 7 are Ras [216] and Caspase-3 [340]. Targets ERα36 [341]
let-7c	0.3	Expression was higher in PR+ *versus* PR- human breast tumors [22] and in ERα+ than ERα- tumors [12]. let-7c increased TRAIL-induced caspase 3 activation in MDA-MB-453 breast cancer cells and let-7c was predicted to target CD95L [342]. Downregulated in breast tumors [343].
let-7d	0.25	Let-7d was increased in acute promyelocytic leukemia patients [344]
let-7f	-0.25	Let-7f was > in node negative *versus* positive human breast tumors [22] and higher in ERα+ tumors [12]. let-7f in mammary gland was reduced by E_2 treatment of female ACI rats [282]. Targets ERα36 [341]
let-7g	1.66	Let-7g was expressed in ErbB2 positive human breast tumors [22]. Targets ERα36 [341]
let-7i	1.34	Targets ERα36 [341]
miR-106b	1.14	Higher in PR+ than PR– tumors, but higher in ERα- than ERα+ breast tumors [12]. Overexpressed in breast tumors compared to normal breast and reduced p21 mRNA and protein and thus stimulates G1-S cell cycle progression in human mammary epithelial cells [345]. miR-106b in mammary gland was increased by 6 wks of E_2 treatment of female ACI rats [282].
miR-149	-3.17	miR-149 in mammary gland was increased by 6 wks of E_2 treatment of female ACI rats [282].
miR-15a	2.32	Higher in low *versus* high tumor stage in human breast tumors [20] and greater in ERα+ than ERα- tumors [22]. Higher in PR+ than PR- tumors [22]. miR-15a is a tumor suppressor [346]. miR-15a negatively regulates Bcl2 at a posttranscriptional level [347].
miR-15b	1.13	Upregulated in hepatocellular, cervical, and colon carcinomas & malignant melanoma [348]. Regulates chemotherapy-induced EMT in human tongue cancer cells by targeting BMI1 [349]
miR-151	0.27	Higher in ERα+/lymph node negative breast tumors from patients with a short time to distant metastasis (TDM) *versus* those with a long TDM [15].
miR-16	0.79	Higher in ERα+ than ERα- tumors [22]. miR-16-1 negatively regulates Bcl2 at a posttranscriptional level [347]. The miR-16 family negatively regulates cell cycle progression by inducing G0/G1-cell accumulation [350] by reducing CCND1 (cyclin D1), CCND3, CCNE1, and CDK6 [351].
miR-182	0.83	Higher in ERα+ than ERα- human breast tumors and not significantly higher in ErbB2 –*vs.* positive tumors [22] significantly higher in PR+ *vs.* PR– tumors [22]. miR-182 inhibited TRAIL-induced caspase 3 activation in MDA-MB-453 breast cancer cells and miR-182 was predicted to target caspase 3 and FADD [342].
miR-183	0.98	miR-183-96-182 cluster is overexpressed in prostate tissue [352]
miR-195	2	Higher in ErbB2- *vs.* ErbB2 positive tumors [22], but not significantly > in ERα+ than ERα- or PR+ *vs.* PR– tumors [22]. miR-195 expression was increased by hypoxia in MCF-7 cells [353]. miR-195 inhibits CCND1,CCND3, CCNE1, and CDK6 protein expression [351].
miR-200a	2.58	Correlated with ERα status in human breast tumors [22] and was significantly > in ERα+ than ERα- and PR+ than PR- breast tumors [22]. miR-200a is expressed in MCF-7 and other epithelial breast cancer cell lines [259]. miR-200 expression was reduced in tamoxifen-resistant MCF-7 cells [17].
miR-200b	0.7	miR-200b is expressed in MCF-7 and other epithelial breast cancer cell lines [259]. The miR-200b/200a/429 cluster of miR-200 family is the only tissue-specific cluster that displays substantial variability between individual human mammary epithelial genotypes [354]
miR-200c	-0.42	miR-200c is expressed in MCF-7 cells [212] and is higher than miR-200a or -200b in MCF-7 [259]. Targets ZEB1 and ZEB2 [263] Loss of miR-200c causes EMT [355]
miR-203	1.84	Expression is increased in ovarian, breast and melanoma cancers [225]. Higher expression in high *versus* low tumor stage in human breast tumors [20] and increased in

miRNA	log2 (E_2/EtOH)	Comments regarding the possible connection of the identified miRNA gene with breast cancer and/or estrogenic responses. *Bone fide* targets of miRNAs are indicated.
		tamoxifen-resistant MCF-7 cells [17].
miR-20a	0.83	Increased in lung, breast, stomach, prostate, colon, and pancreatic tumors [224]. miR-20a expression was low in MCF-7 and other breast cancer cell lines and overexpression of the miR-17/20 locus in MCF-7 inhibited cell proliferation and cyclin D1 expression [19]. miR-20a was increased in mammary gland after 6 or 12 wks of E_2 treatment of female ACI rats [282]. Both E_2 and ICI decreased miR-20a in human endometrial stromal cells [356]. Targets of miR-20a are PCAF, RUNX1, and TGFBR2 [357].
miR-21	-0.14	Significantly up-regulated in tissues or cell lines of breast cancer [20] [358]. overexpressed in all solid tumors (lung, breast, stomach, prostate, colon, and pancreatic) [224]. Significantly higher in ERα+ than ERα-, in ErbB2 - *vs*. ErbB2 +, and in PR+ *vs*. PR– breast tumors [22]. Expression was higher in breast tumor compared to adjacent normal breast tissue [359]. miR-21 expression was increased by hypoxia in MCF-7 [353]. miR-21 in mammary gland was increased after 18 wks of E_2 treatment of female ACI rats [282]. Both E_2 and ICI decreased miR-21 in human endometrial stromal and glandular epithelial cells, but when combined, miR-21 expression returned to basal [356]. E_2 suppressed and ICI increased miR-21 in human myometrial smooth muscle cells [303]. E_2 inhibited the ICI-induced increase in miR-21 in these cells [303]. E_2 (75% decrease) and Progesterone (41% decrease) reduced miR-21 expression on the uterus of ovex mice [286]. miR-21 expression was significantly reduced in tamoxifen-resistant MCF-7 cells [17].
miR-23a	0.31	Increased by hypoxia in MCF-7 cells [353]; reduced in tamoxifen-resistant MCF-7 cells [17].
miR-23b	0.32	increased by hypoxia in MCF-7 cells [353]
miR-25	1.6	Higher in ERα+ than ERα- breast tumors [12] and PR+ *vs*. PR– breast tumors [22], but significantly higher in ErbB2-negative *vs*. positive breast tumors [22]. increased in ovarian, breast and melanoma cancers [225]. miR-25 was increased in mammary gland after 6 wks of E_2 treatment of female ACI rats [282].
miR-26a	0.87	Significantly > in ERα+ than ERα- breast tumors [20, 22]. significantly higher in ErbB2 – *vs*. ErbB2-positive breast tumors [22] and PR+ than PR- tumors [20]. miR-26a expression was increased by hypoxia in MCF-7 cells [353]. Both E_2 and ICI increased miR-26a in human endometrial glandular epithelial cells, but when combined, miR-26a expression was suppressed below basal [356]. E_2 and ICI increased miR-26a in human myometrial smooth muscle cells, but each inhibited miR-26a in human leiosarcoma cells [303].
miR-26b	2.07	Higher in ERα+ than ERα- breast tumors [22]; significantly higher PR+ *vs*. PR–tumors [20]. Higher in ErbB2+ *vs*. positive breast tumors [22]. miR-26b expression was increased by hypoxia in MCF-7 cells [353]
miR-27a	1.73	Higher in ErbB2- *vs* ErbB2+ and PR+ *vs*. PR- breast tumors [22], but not significantly higher in ER+ *vs*. ER– tumors [20]. miR-27a expression was increased by hypoxia in MCF-7 cells[353]
miR-27b	1.94	Higher in ERα+ than ERα- breast tumors [22], in ErbB2-*vs* ErbB2+ tumors [22], and in PR+ *vs*. PR– tumors [22].
miR-30b	2.17	Higher in node negative *versus* positive human breast tumors [20], higher in ERα+ than ERα-, in ErbB2 - *vs* ErbB+ and in PR+ *vs*. PR– breast tumors [22]. miR-30b expression was increased by hypoxia in MCF-7 cells [353]
miR-320	-0.84	Downregulated in human breast cancer [225]. Expressed in stromal cells and directly targets ETS2, MMP, and EMILIN2 [360]. Downregulated in squamous lung carcinoma [361]
miR-328	-3.92	Overexpression of miR-328 in A431 human epithelial carcinoma cells reduced cell adhesion, aggregation, and migration by repressing CD44 expression [362]. Downregulates breast cancer resistance protein BCRP/ABCG2[363].
miR-342	-0.26	Higher in ERα+ than ERα-,in ErbB2 – *vs*. ErbB2+, and in PR+ *vs*. PR– tumors [22]. miR-342 expression was reduced in tamoxifen-resistant MCF-7 cells [17].

miRNA	log2 (E_2/EtOH)	**Comments regarding the possible connection of the identified miRNA gene with breast cancer and/or estrogenic responses. *Bone fide* targets of miRNAs are indicated.**
miR-365	1.47	Decreased by E_2 treatment in female ACI rat mammary gland [282].
miR-423	-1.49	Promotes cell growth and regulates G1/S transition by targeting p21Cip1/Waf1 in hepatocellular carcinoma [364] Expression was found to be higher in tumors of patients who developed metastatic breast cancer [358].
miR-489	0.59	Metastasis suppressor: decreased expression in metastatic sublines of MDA-MB-231[18]. miR-489 expression was reduced in tamoxifen-resistant MCF-7 cells [17].
miR-7	1.84	Tumor suppressor in gastric cancer development and IL-1□ and TNFα repress miR-7[365]. Higher in ERα+ than ERα- breast cancer cell lines and increased by E2 in MCF-8 cells [238]. Represses EGFR and IGFR1 in triple negative breast cancer cells [238].
miR-92	0.45	Higher in ERα+ than ERα- and in PR+ *vs*. PR– breast tumors, but not higher in ErbB2- *vs*. ErbB2+ tumors [22]. miR-92 was increased in mammary gland after 6 wks of E_2 treatment of female ACI rats [282]. miR-92 is in the miR-17/20 cluster and overexpression of the miR-17/20 cluster in MCF-7 cells inhibited basal and E_2-stimulated cell proliferation and cyclin D1 transcription [19].
miR-98	1.55	Overexpressed in breast tumor compared to adjacent normal breast tissue [359].

E_2 upregulated the expression of miR-17-92 and the paralgous miR-106a-363 clusters in MCF-7 cells [304]. Interestingly, rather than ER binding the promoters of these miRNA clusters, E_2 indirectly upregulates the miR-17-92 cluster by increasing the transcription of cMYC which binds to the miR-17-92 promoter and increases transcription. Further, microarrays identified 23 miRNAs to be downregulated by E_2 in MCF-7 cells [305]. Of these 23, the expression of 8 were confirmed by quantitative real time PCR (QPCR) in MCF-7, BT474, T47D and ZR-75-1 breast cancer cells. The expression of pri-miR-21 and pri-miR-181a was also found to be downregulated at 3 h of E_2 treatment in MCF-7 cells. Work done in our laboratory has also shown that E_2 regulates miRNA expression in MCF-7 breast cancer cells [306, 307]. E_2 decreased miR-21 expression which in turn increased the expression of its target genes, PDCD4 and BCL2 [253]. Contrary to our results, Nakshatri *et al.* reported an increase in miR-21 expression in MCF-7 cells after E_2 treatment [308]. However, others have likewise reported that E_2 reduces miR-21 expression [304, 305, 309].

miRNAs REGULATING ER EXPRESSION

miRNAs can influence estrogen-regulated gene expression by directly reducing ERα mRNA stability or translation. Eight miRNAs have been reported to reduce ERα protein levels: miR-18a, miR 18b, miR-193b, miR-302c, miR-22, miR-201, miR-

221, and miR-222 [279] (Fig. **5**). Two miR-206 recognition sites were identified in the 3'UTR of ERα and transfection of an expression vector for miR-206 in MCF-7 cells reduced both mRNA and protein levels of ERα [208]. Treatment of MCF-7 cells with 1 nM E_2 or the ERα agonist PPT (10nM) reduced miR-206 levels by ~ 80%. In contrast, the ERβ agonist DPN (10nM) increased miR-206 expression by ~ 60%. Interestingly, the investigators found that miR-206 levels were significantly higher in ERα-negative MDA-MB-231 cells than in MCF-7 cells, suggesting a mechanism for miR-206 in repressing ERα protein levels in MDA-MB-231 cells. The authors suggested that miR-206 may function in a mutually negative feedback loop to temporally regulate ERα expression and ductal/lobuloalveolar proliferation [208]. MiR-206 is inversely correlated with ERα expression, but not ERβ, in human breast tumors [310]. Further, transfection of MCF-7 human breast cancer cells with an expression plasmid for pre-miR-206 reduced ERα mRNA expression ~ 25%, reduced the basal expression levels of PR, cyclin D1, and pS2 (all well-established ERα-regulated genes), and inhibited cell proliferation with or without E_2 [310].

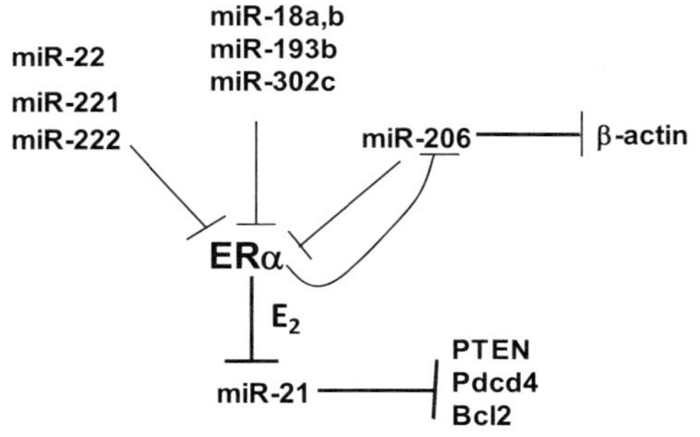

Figure 5: MiRNA regulation of ERα and ERα regulation of miRNA expression. MiRNAs that inhibit ERα protein expression are shown. E_2-ER represses miR-206 and miR-21 expression. Target genes of miR-21 are PTEN, PDCD4, and BCL2. MiR-206 also represses β-actin expression.

miR-221/222 is higher in ERα negative than ERα positive breast cancer cell lines and human breast tumors [311, 312]. Two miR-221 and miR-222 seed elements were identified in the 3'UTR of ERα and transfection of miR-221 and miR-222 suppressed ERα protein, but not mRNA in ERα positive MCF-7 and T47D cells. Conversely, knockdown of miR-221 and miR-222 in ERα-negative MDA-MB-468 partially restored ERα protein expression and increased tamoxifen-induced apoptosis [311]. A comparison of miRNA expression in ERα-positive and ERα-negative cells show that the expression of miR-193 and miR-221 is higher in ERα-positive cells compared to ERα-negative cells [313]. Targeting miR-221/222 and other miRNAs that downregulate ERα may be effective in restoring endocrine sensitivity. Anti-miR-221 suppressed the growth of TAM-resistant breast cancer cells as xenografts in nude mice [314].

The expression of miR-22 was found to be significantly lower in ERα-positive cell lines, MCF-7, T-47D and BT474 when compared to ERα-negative MDA-MB-231 and SK-BR-3 [315]. miR-22 targets ERα by binding its 3'UTR and thus inhibits E_2 signaling in breast cancer cells [316]. Further, miR-22 regulates ERα protein expression in a pancreatic cancer cell line. Curcumin increased miR-22 by 65% in BxPC-3 human pancreatic carcinoma cells and reduced ERα protein [317]. Follow-up studies showed that transfection of an antisense RNA oligonucleotide of miRNA-22 into BxPC-3 cells increased ERα protein by ~ 1.9-fold. Thus, miR-22 regulates ERα protein levels and the authors suggest a role for ERα as anti-tumorigenic in pancreatic cancer.

A protein lysate microarray (LMA)-based strategy in which a library of pre-miRs was transiently transfected into MCF-7 and BT-474 wells in 384-well plates and ERα protein was subsequently analyzed in protein lysates that were printed on nitrocellulose-coated slides [318] miR-18a, -18b. -193, -206 and -302c were reported to reduce ERα by directly binding sites in the 3'UTR of ERα. Further, the authors reported an inverse correlation between the expression of miR-18a, -18b and ERα-negative breast tumor samples [318].

miRNAs THAT REGULATE ER CO-REGULATORS

miRNAs can affect estrogen-regulated gene expression by reducing the expression of the co-activator SRC-3/AIB1/NCOA3 or other co-regulators. miR-17-5p inhibited translation of SRC-3/AIB1/NCOA3 [268]. Transfection of CHO-K1 cells with ERα and miR-17-5p inhibited E_2-stimulated ERE-driven luciferase reporter activity by 50%. This report also demonstrated that transfection of MCF-7 cells (which do not express miR-17-5p) with miR-17-5p reduced E_2-induced proliferation and E_2-induced endogenous cyclin D1 transcription [268].

There are 3 reports on miRNA regulation of corepressors that target ERα. miR-10a and -10b repress SMRT/NCOR2 [319]. miR-184 represses SMRT/NCOR2 translation [320], but how miR-184 affects ER expression in breast cancer cells is unknown. MTA1 (metastatic tumor antigen 1) repressed miR-661, but the effect on ERα transcription was not evaluated [321]. miR-615-3p repressed LCoR expression [322], but whether this affects ERα was not studied. Clearly, little is known about regulation of ER co-activators and corepressors by miRNAs.

E_2 REGULATION OF Ago2 AND ERα IN HUMAN BREAST CANCER CELL LINES:

Argonaut-2 (Ago2), the catalytic subunit of the RISC complex that mediates miRNA-dependent cleavage/degradation in mammals [200, 216, 323], expression is higher in ERα-negative, HER2-positive (basal) than ERα-positive/HER2 negative (luminal) human breast cancer cell lines and tumors [14]. E_2 and the ERα-agonist PPT, but not the ERβ-agonist DPN, increased Ago2 protein expression in MCF-7 cells [14]. Further studies showed that EGF acts through the MAPK pathway to increase Ago2 protein stability, but there were no studies examining the mechanism by which E_2 and PPT, presumably through ERα, increase Ago2 protein levels. Surprisingly, Ago2 overexpression in MCF-7 cells increased ERα protein levels by 3-fold, despite also increasing miR-206 that reduces ERα. The authors concluded that this "discordant" finding indicates that there is a greater concentration of miRNAs than target proteins involved in ERα suppression than those that target ERα itself [14]. Microarray profiling shows that the expression of Ago1 and Ago2 proteins is higher while Dicer and TRBP1 is lower in ER-negative *versus* ER-positive breast cancer cells [324].

microRNA AND ENDOCRINE-RESISTANT BREAST CANCER

Altered miRNA expression is likely to play a role in endocrine-resistance in breast cancer. A PubMed search for 'MicroRNA and endocrine resistance in breast cancer' generated 10 publications as of March 2012. miR-221/222 was reported to promote TAM-resistance by targeting ERα and the cell cycle regulator p27 (also known as Kip1) [325]. Indeed, miR-221/222 is overexpressed in ERα-negative cells lines and tumors, as well as HER2-positive breast tumors [326]. Overexpression of miR-221/222 also associates with Fulvestrant-resistance [326]. Some of the signaling pathways involved in promoting miR-221/222-mediated

Fulvestrant resistance include β-catenin, TGF-β and p53 [326]. miR-221/222 is also increased in $CD44^{+}CD24^{-/low}lineage^{-}$ human breast cancer stem cells, indicating a role for these stem cells in endocrine resistance [327].

In MCF-7 cells stably overexpressing the oncogenic HER2Δ16 mutant, miR-15a/16 downregulated the expression of anti-apoptotic/pro-survival BCL2 and promoted TAM resistance [328]. miR-342 was downregulated in two TAM-resistant cell lines derived from MCF-7 cells called LCC2 and TAMR1 [329]. Overexpression of miR-342 conferred TAM-sensitivity and increased apoptosis. miR-451, a tumor suppressor miRNA, was suppressed in, TAM-resistant breast cancer cells [330]. miR-451 targets 14-3-3ζ an anti-apoptotic gene that is overexpressed in TAM-resistant tumors and is also associated with poor clinical outcome [330]. Thus the loss of miR-451 in ERα-positive breast cancer upregulates 14-3-3ζ and contributes to TAM-resistance.

Increased expression of ERα-36, a truncated form of the full length ERα-66, that blocks ERα-66 genomic activity while activating MAPK signaling, has been reported in tamoxifen-resistant breast tumors [331]. Let-7a targets ERα-36 and loss of Let-7 family members conferred TAM-resistance by activating non-genomic estrogen signaling mediated by ERα-36 [332].

Our laboratory identified miRNAs that are differentially regulated by TAM in endocrine-sensitive MCF-7 and endocrine-resistant LY2 human breast cancer cells [306]. We identified 97 miRNAs regulated in the opposite direction in MCF-7 and LY2 cells. QPCR was performed to confirm the expression 12 miRNAs that showed significant opposite expression between the two cell lines. Examples include higher miR-200a,b,c in MCF-7 than LY2 and higher miR-10a, miR-22, miR-29a, miR-125b, and miR-222 in LY2 than in MCF-7 [306]. Bioinformatic prediction identified 36 putative mRNA targets of the 12 miRNAs whose expression was validated by QPCR. Some of the mRNA targets include *PDCD4*, *BCL2*, *CYP1B1*, *ERBB3*. *ZEB1*, a target of miR-200 family of miRNAs and a promoter of EMT, was found to be overexpressed in LY2 cells when compared to MCF-7 cells. This was a significant finding as it indicated that LY2 cells have undergone EMT.

Transient overexpression of miR-200b or mR-200c in LY2 cells sensitized the cells to growth inhibition by TAM and fulvestrant [333]. Knockdown of ZEB1 in LY2 cells recapitulated the effect of miR-200b and miR-200c overexpression resulting in inhibition of LY2 cell proliferation by TAM and fulvestrant, but not the aromatase inhibitor exemestane. Demethylating agent 5-aza-2′-deoxycytidine (5-aza-dC) in combination with histone deacetylase inhibitor trichostatin A (TSA) increased miR-200b and miR-200c in LY2 cells, suggesting methylation and chromatin modification (deacetylation) as part of the mechanism for the reduced expression. Concomitant with the increase in miR-200b and miR-200c, ZEB1 expression was decreased and cells appeared more epithelial in morphology and were sensitized to TAM and fulvestrant inhibition. Likewise, knockdown of ZEB1 increased antiestrogen sensitivity of LY2 cells resulting in inhibition of cell proliferation. These data suggest that reduced miRNA-200b and miR-200c expression contributes to endocrine resistance in breast cancer cells and that the reduced expression of these miR-200 family members in endocrine-resistant cells can be reversed by 5-aza-dC+TSA.

CONCLUSIONS

Estrogen signaling plays a critical role in regulating reproduction, lactation, bone density, cardiovascular function, neuronal signaling, immune function, and homeostasis in a wide variety of tissues. Research on the mechanisms by which E_2 and other estrogens regulate diverse physiological effects has established both genomic and nongenomic mechanisms involving ERα, ERβ, and GPR30 in signal transduction (Fig. **2**). miRNAs are small, non-coding RNAs that bind to the 3’ UTR of target mRNAs and either block the translation of the message and/or target the mRNA transcript to be degraded. Although many studies have described miRNA changes in breast tumors and compared ERα-positive *versus* ERα-negative miRNA signatures as potential biomarkers, there are few studies examining E_2- or antiestrogen- responsive miRNAs in cell or tissue systems. In those few studies that have identified E_2-induced alterations in miRNA expression, the mechanisms elaborated for the E_2 effect (s) on miRNA expression remain largely undefined. Further, E_2 regulates miRNA expression in a cell-type-dependent manner. Thus, identification of E_2-regulated miRNAs and the function of miRNAs within specific tissues and cells remains a field in need of further exploration.

ACKNOWLEDGEMENTS

This work was supported by NIH CA138410 to C.M.K.

CONFLICT OF INTEREST

The authors confirm that this chapter contents have no conflict of interest.

DISCLOSURE

The chapter submitted for Advances in Genome Science, Vol. 3 entitled **"Estrogen Regulation of MicroRNA Expression by C.M. Klinge"**is an update of our article published in Current Genomics 10:169-183, 2009 and C.M. Klinge "miRNAs and estrogen action. Trends in Endocrinology & Metabolism" 23:223-233, 2012.

REFERENCES

[1] Vina, J., Sastre, J., Pallardo, F.V., Gambini, J., Borras, C. Modulation of longevity-associated genes by estrogens or phytoestrogens. Biol. Chem. **2008**, *389*: 273-277.

[2] Wise, P.M. Estrogen therapy: Does it help or hurt the adult and aging brain? Insights derived from animal models. Neuroscience **2006**, *138*: 831-835.

[3] Wynne, F.L., Payne, J.A., Cain, A.E., Reckelhoff, J.F., Khalil, R.A. Age-related reduction in estrogen receptor-mediated mechanisms of vascular relaxation in female spontaneously hypertensive rats. Hypertension **2004**, *43*: 405-412.

[4] Moolman, J.A. Unravelling the cardioprotective mechanism of action of estrogens. Cardiovasc. Res. **2006**, *69*: 777-780.

[5] Alonso, A., Moreno, M., Ordonez, P., Fernandez, R., Perez, C., Diaz, F., *et al.* Chronic estradiol treatment improves brain homeostasis during aging in female rats. Endocrinology **2008**, *149*: 57-72.

[6] Lamberts, S.W., van den Beld, A.W., van der Lely, A.J. The endocrinology of aging. Science **1997**, *278*: 419-424.

[7] Casadesus, G., Rolston, R.K., Webber, K.M., Atwood, C.S., Bowen, R.L., Perry, G., *et al.* Menopause, estrogen, and gonadotropins in alzheimer's disease. Adv. Clin. Chem. **2008**, *45*: 139-153.

[8] Worzala, K., Hiller, R., Sperduto, R.D., Mutalik, K., Murabito, J.M., Moskowitz, M., *et al.* Postmenopausal estrogen use, type of menopause, and lens opacities: The framingham studies. Arch. Intern. Med. **2001**, *161*: 1448-1454.

[9] Zamore, P.D., Haley, B. Ribo-gnome: The big world of small rnas. Science **2005**, *309*: 1519-1524.

[10] Zeng, Y. Principles of micro-rna production and maturation. Oncogene **2006**, *25*: 6156-6162.

[11] Couzin, J. Genetics. Erasing micrornas reveals their powerful punch. Science **2007**, *316*: 530.

[12] Blenkiron, C., Goldstein, L.D., Thorne, N.P., Spiteri, I., Chin, S.F., Dunning, M.J., *et al.* Microrna expression profiling of human breast cancer identifies new markers of tumor subtype. Genome Biol **2007**, *8*: R214.

[13] Si, M.L., Zhu, S., Wu, H., Lu, Z., Wu, F., Mo, Y.Y. Mir-21-mediated tumor growth. Oncogene **2007**, *26*: 2799-2803.

[14] Adams, B.D., Claffey, K.P., White, B.A. Argonaute-2 expression is regulated by egfr/mapk signaling and correlates with a transformed phenotype in breast cancer cells. Endocrinology **2008**: en.2008-0984.

[15] Foekens, J.A., Sieuwerts, A.M., Smid, M., Look, M.P., de Weerd, V., Boersma, A.W.M., *et al.* Four mirnas associated with aggressiveness of lymph node-negative, estrogen receptor-positive human breast cancer. Proceedings of the National Academy of Sciences **2008**, *105*: 13021-13026-.

[16] Lowery, A.J., Miller, N., McNeill, R.E., Kerin, M.J. Micrornas as prognostic indicators and therapeutic targets: Potential effect on breast cancer management. Clin. Cancer Res. **2008**, *14*: 360-365.

[17] Miller, T.E., Ghoshal, K., Ramaswamy, B., Roy, S., Datta, J., Shapiro, C.L., *et al.* Microrna-221/222 confers tamoxifen resistance in breast cancer by targeting p27(kip1). J. Biol. Chem. **2008**, *283*: 29897-29903.

[18] Tavazoie, S.F., Alarcon, C., Oskarsson, T., Padua, D., Wang, Q., Bos, P.D., *et al.* Endogenous human micrornas that suppress breast cancer metastasis. Nature **2008**, *451*: 147-152.

[19] Yu, Z., Wang, C., Wang, M., Li, Z., Casimiro, M.C., Liu, M., *et al.* A cyclin d1/microrna 17/20 regulatory feedback loop in control of breast cancer cell proliferation. J. Cell Biol. **2008**, *182*: 509-517.

[20] Iorio, M.V., Ferracin, M., Liu, C.-G., Veronese, A., Spizzo, R., Sabbioni, S., *et al.* Microrna gene expression deregulation in human breast cancer. Cancer Res. **2005**, *65*: 7065-7070.

[21] Jiang, J., Lee, E.J., Gusev, Y., Schmittgen, T.D. Real-time expression profiling of microrna precursors in human cancer cell lines. Nucl. Acids Res. **2005**, *33*: 5394-5403.

[22] Mattie, M.D., Benz, C.C., Bowers, J., Sensinger, K., Wong, L., Scott, G.K., *et al.* Optimized high-throughput microrna expression profiling provides novel biomarker assessment of clinical prostate and breast cancer biopsies. Mol Cancer **2006**, *5*: 24.

[23] Maglich, J.M., Sluder, A., Guan, X., Shi, Y., McKee, D.D., Carrick, K., *et al.* Comparison of complete nuclear receptor sets from the human, caenorhabditis elegans and drosophila genomes. Genome Biol **2001**, *2*: RESEARCH0029.

[24] Green, S., Kumar, V., Theulaz, I., Wahli, W., Chambon, P. The n-terminal DNA-binding 'zinc finger' of the oestrogen and glucocorticoid receptors determines target gene specificity. EMBO J. **1988**, *7*: 3037-3044.

[25] Klinge, C.M. Estrogen receptor interaction with estrogen response elements. Nucleic Acids Res **2001**, *29*: 2905-2919.

[26] Thornton, J.W. Evolution of vertebrate steroid receptors from an ancestral estrogen receptor by ligand exploitation and serial genome expansions. Proc. Natl. Acad. Sci. U. S. A. **2001**, *98*: 5671-5676.

[27] Thornton, J.W., Need, E., Crews, D. Resurrecting the ancestral steroid receptor: Ancient origin of estrogen signaling. Science **2003**, *301*: 1714-1717.

[28] Tanenbaum, D.M., Wang, Y., Williams, S.P., Sigler, P.B. Crystallographic comparison of the estrogen and progesterone receptor's ligand binding domains. Proc. Natl. Acad. Sci. USA **1998**, *95*: 5998-6003.

[29] Metivier, R., Stark, A., Flouriot, G., Hubner, M.R., Brand, H., Penot, G., *et al.* A dynamic structural model for estrogen receptor-alpha activation by ligands, emphasizing the role of interactions between distant a and e domains. Mol. Cell **2002**, *10*: 1019-1032.

[30] Ruff, M., Gangloff, M., Wurtz, J.M., Moras, D. Estrogen receptor transcription and transactivation: Structure-function relationship in DNA- and ligand-binding domains of estrogen receptors. Breast Cancer Res **2000**, *2*: 353-359.

[31] Carroll, J.S., Meyer, C.A., Song, J., Li, W., Geistlinger, T.R., Eeckhoute, J., *et al.* Genome-wide analysis of estrogen receptor binding sites. Nat. Genet. **2006**, *38*: 1289-1297.

[32] Carroll, J.S., Brown, M. Estrogen receptor target gene: An evolving concept. Mol. Endocrinol. **2006**, *20*: 1707-1714.

[33] Liu, Y., Gao, H., Marstrand, T.T., Strom, A., Valen, E., Sandelin, A., *et al.* The genome landscape of er{alpha}- and er{beta}-binding DNA regions. Proceedings of the National Academy of Sciences **2008**: 0712085105.

[34] Lin, C.Y., Vega, V.B., Thomsen, J.S., Zhang, T., Kong, S.L., Xie, M., *et al.* Whole-genome cartography of estrogen receptor alpha binding sites. PLoS Genet **2007**, *3*: e87.

[35] Kwon, Y.-S., Garcia-Bassets, I., Hutt, K.R., Cheng, C.S., Jin, M., Liu, D., *et al.* Sensitive chip-dsl technology reveals an extensive estrogen receptor {alpha}-binding program on human gene promoters. PNAS **2007**, *104*: 4852-4857.

[36] Welboren, W.-J., Sweep, F.C.G.J., Span, P.N., Stunnenberg, H.G. Genomic actions of estrogen receptor {alpha}: What are the targets and how are they regulated? Endocr Relat Cancer **2009**, *16*: 1073-1089.

[37] Welboren, W.-J., van Driel, M.A., Janssen-Megens, E.M., van Heeringen, S.J., Sweep, F.C.G.J., Span, P.N., *et al.* Chip-seq of er[alpha] and rna polymerase ii defines genes differentially responding to ligands. EMBO J. **2009**, *advanced online publication*.

[38] Stender, J.D., Kim, K., Charn, T.H., Komm, B., Chang, K.C.N., Kraus, W.L., *et al.* Genome-wide analysis of estrogen receptor {alpha} DNA binding and tethering mechanisms identifies runx1 as a novel tethering factor in receptor-mediated transcriptional activation. Mol. Cell. Biol. **2010**, *30*: 3943-3955.

[39] McKenna, N.J., Lanz, R.B., O'Malley, B.W. Nuclear receptor co-regulators: Cellular and molecular biology. Endocr. Rev. **1999**, *20*: 321-344.

[40] Bourdeau, V., Deschenes, J., Laperriere, D., Aid, M., White, J.H., Mader, S. Mechanisms of primary and secondary estrogen target gene regulation in breast cancer cells. Nucl. Acids Res. **2008**, *36*: 76-93.

[41] Krum, S.A., Miranda-Carboni, G.A., Lupien, M., Eeckhoute, J., Carroll, J.S., Brown, M. Unique er{alpha} cistromes control cell type-specific gene regulation. Mol. Endocrinol. **2008**, *22*: 2393-2406.

[42] Carroll, J.S., Liu, X.S., Brodsky, A.S., Li, W., Meyer, C.A., Szary, A.J., *et al.* Chromosome-wide mapping of estrogen receptor binding reveals long-range regulation requiring the forkhead protein foxa1. Cell **2005**, *122*: 33-43.

[43] Eeckhoute, J., Carroll, J.S., Geistlinger, T.R., Torres-Arzayus, M.I., Brown, M. A cell-type-specific transcriptional network required for estrogen regulation of cyclin d1 and cell cycle progression in breast cancer. Genes Dev. **2006**, *20*: 2513-2526.

[44] Eeckhoute, J., Keeton, E.K., Lupien, M., Krum, S.A., Carroll, J.S., Brown, M. Positive cross-regulatory loop ties gata-3 to estrogen receptor {alpha} expression in breast cancer. Cancer Res. **2007**, *67*: 6477-6483.

[45] Lupien, M., Eeckhoute, J., Meyer, C.A., Wang, Q., Zhang, Y., Li, W., *et al.* Foxa1 translates epigenetic signatures into enhancer-driven lineage-specific transcription. Cell **2008**, *132*: 958-970.

[46] Magnani, L., Ballantyne, E.B., Zhang, X., Lupien, M. Pbx1 genomic pioneer function drives erα signaling underlying progression in breast cancer. PLoS Genet **2011**, *7*: e1002368.

[47] Tang, Q., Chen, Y., Meyer, C., Geistlinger, T., Lupien, M., Wang, Q., *et al.* A comprehensive view of nuclear receptor cancer cistromes. Cancer Res. **2011**, *71*: 6940-6947.

[48] Grober, O., Mutarelli, M., Giurato, G., Ravo, M., Cicatiello, L., De Filippo, M., *et al.* Global analysis of estrogen receptor beta binding to breast cancer cell genome reveals an extensive interplay with estrogen receptor alpha for target gene regulation. BMC Genomics **2011**, *12*: 36.

[49] Klinge, C.M. Estrogen receptor interaction with co-activators and co-repressors. Steroids **2000**, *65*: 227-251.

[50] Lonard, D.M., O'Malley, B.W. Nuclear receptor co-regulators: Judges, juries, and executioners of cellular regulation. Mol. Cell **2007**, *27*: 691-700.

[51] Rosenfeld, M.G., Lunyak, V.V., Glass, C.K. Sensors and signals: A co-activator/corepressor/epigenetic code for integrating signal-dependent programs of transcriptional response. Genes Dev. **2006**, *20*: 1405-1428.

[52] Lonard, D.M., Kumar, R., O'Malley, B.W. Minireview: The src family of co-activators: An entree to understanding a subset of polygenic diseases? Mol. Endocrinol. **2010**, *24*: 279-285.

[53] Lonard, D.M., Nawaz, Z. Co-activators and corepressors. *Nuclear receptors and genetic disease*: Academic Press 2001.

[54] O'Malley, B.W., Kumar, R. Nuclear receptor co-regulators in cancer biology. Cancer Res. **2009**, *69*: 8217-8222.

[55] O'Malley, B.W., McKenna, N.J. Editorial: Co-activators and corepressors: What's in a name? Mol. Endocrinol. **2008**, *22*: 2213-2214.

[56] Lannigan, D.A. Estrogen receptor phosphorylation. Steroids **2003**, *68*: 1-9.

[57] Madak-Erdogan, Z., Lupien, M., Stossi, F., Brown, M., Katzenellenbogen, B.S. Genomic collaboration of estrogen receptor {alpha} and extracellular signal-regulated kinase 2 in regulating gene and proliferation programs. Mol. Cell. Biol. **2011**, *31*: 226-236.

[58] Giamas, G., Filipovic, A., Jacob, J., Messier, W., Zhang, H., Yang, D., *et al.* Kinome screening for regulators of the estrogen receptor identifies lmtk3 as a new therapeutic target in breast cancer. Nat. Med. **2011**, *17*: 715-719.

[59] Stenoien, D.L., Mancini, M.G., Patel, K., Allegretto, E.A., Smith, C.L., Mancini, M.A. Subnuclear trafficking of estrogen receptor-alpha and steroid receptor co-activator-1. Mol. Endocrinol. **2000**, *14*: 518-534.

[60] Stenoien, D.L., Nye, A.C., Mancini, M.G., Patel, K., Dutertre, M., O'Malley, B.W., *et al.* Ligand-mediated assembly and real-time cellular dynamics of estrogen receptor alpha-co-activator complexes in living cells. Mol. Cell. Biol. **2001**, *21*: 4404-4412.

[61] Stenoien, D.L., Patel, K., Mancini, M.G., Dutertre, M., Smith, C.L., O'Malley, B.W., *et al.* Frap reveals that mobility of oestrogen receptor-alpha is ligand- and proteasome-dependent. Nat Cell Biol **2001**, *3*: 15-23.

[62] Porter, W., Wang, F., Wang, W., Duan, R., Safe, S. Role of estrogen receptor/sp1 complexes in estrogen-induced heat shock protein 27 gene expression. Mol. Endocrinol. **1996**, *10*: 1371-1378.

[63] Paech, K., Webb, P., Kuiper, G.G., Nilsson, S., Gustafsson, J., Kushner, P.J., *et al.* Differential ligand activation of estrogen receptors eralpha and erbeta at ap1 sites. Science **1997**, *277*: 1508-1510.

[64] Quaedackers, M.E., van den Brink, C.E., van der Saag, P.T., Tertoolen, L.G.J. Direct interaction between estrogen receptor [alpha] and nf-[kappa]b in the nucleus of living cells. Mol. Cell. Endocrinol. **2007**, *273*: 42-50.

[65] Kalaitzidis, D., Gilmore, T.D. Transcription factor cross-talk: The estrogen receptor and nf-[kappa]b. Trends Endocrinol. Metab. **2005**, *16*: 46-52.

[66] Gu, F., Hsu, H.-K., Hsu, P.-Y., Wu, J., Ma, Y., Parvin, J., *et al.* Inference of hierarchical regulatory network of estrogen-dependent breast cancer through chip-based data. BMC Systems Biology **2010**, *4*: 170.

[67] Heldring, N., Isaacs, G.D., Diehl, A.G., Sun, M., Cheung, E., Ranish, J.A., *et al.* Multiple sequence-specific DNA-binding proteins mediate estrogen receptor signaling through a tethering pathway. Mol. Endocrinol. **2011**, *25*: 564-574.

[68] Zhao, C., Dahlman-Wright, K., Gustafsson, J.-Å. Estrogen signaling *via* estrogen receptor β. J. Biol. Chem. **2010**, *285*: 39575-39579.

[69] Monroe, D.G., Getz, B.J., Johnsen, S.A., Riggs, B.L., Khosla, S., Spelsberg, T.C. Estrogen receptor isoform-specific regulation of endogenous gene expression in human osteoblastic cell lines expressing either eralpha or erbeta. J. Cell. Biochem. **2003**, *90*: 315-326.

[70] Vivar, O.I., Zhao, X., Saunier, E.F., Griffin, C., Mayba, O.S., Tagliaferri, M., *et al.* Estrogen receptor [beta] binds to and regulates three distinct classes of target genes. J. Biol. Chem. **2010**, *285*: 22059-22066.

[71] Stossi, F., Barnett, D.H., Frasor, J., Komm, B., Lyttle, C.R., Katzenellenbogen, B.S. Transcriptional profiling of estrogen-regulated gene expression *via* estrogen receptor (er) {alpha} or er{beta} in human osteosarcoma cells: Distinct and common target genes for these receptors. Endocrinology **2004**, *145*: 3473-3486.

[72] Zhao, C., Putnik, M., Gustafsson, J.-Å., Dahlman-Wright, K. Microarray analysis of altered gene expression in erβ-overexpressing hek293 cells. Endocrine **2009**, *36*: 224-232.

[73] Secreto, F., G, Monroe, D.G., Dutta, S., Ingle, J.N., Spelsberg, T.C. Estrogen receptor alpha/beta isoforms, but not betacx, modulate unique patterns of gene expression and cell proliferation in hs578t cells. J. Cell. Biochem. **2007**, *9999*: n/a.

[74] Zhao, C., Gao, H., Liu, Y., Papoutsi, Z., Jaffrey, S., Gustafsson, J.-Å., *et al.* Genome-wide mapping of estrogen receptor-β–binding regions reveals extensive cross-talk with transcription factor activator protein-1. Cancer Res. **2010**, *70*: 5174-5183.

[75] Horwitz, K., Jackson, T., Bain, D., Richer, J., Takimoto, G., Tung, L. Nuclear receptor co-activators and corepressors. Mol. Endocrinol. **1996**, *10*: 1167-1177.

[76] Peterson, T.J., Karmakar, S., Pace, M.C., Gao, T., Smith, C.L. The silencing mediator of retinoic acid and thyroid hormone receptor (smrt) corepressor is required for full estrogen receptor {alpha} transcriptional activity. Mol. Cell. Biol. **2007**, *27*: 5933-5948.

[77] Yang, X.J., Seto, E. Hats and hdacs: From structure, function and regulation to novel strategies for therapy and prevention. Oncogene **2007**, *26*: 5310-5318.

[78] Privalsky, M.L. The role of corepressors in transcriptional regulation by nuclear hormone receptors. Annu. Rev. Physiol. **2004**, *66*: 315-360.

[79] Smith, C.L., O'Malley, B.W. Co-regulator function: A key to understanding tissue specificity of selective receptor modulators. Endocr. Rev. **2004**, *25*: 45-71.

[80] Beato, M., Candau, R., Chavez, S., Mows, C., Truss, M. Interaction of steroid hormone receptors with transcription factors involves chromatin remodelling. J. Steroid Biochem. Molec. Biol. **1996**, *56*: 47-59.

[81] Shiau, A.K., Barstad, D., Loria, P.M., Cheng, L., Kushner, P.J., Agard, D.A., *et al.* The structural basis of estrogen receptor/co-activator recognition and the antagonism of this interaction by tamoxifen. Cell **1998**, *95*: 927-937.

[82] Smith, C.L., Nawaz, Z., O'Malley, B.W. Co-activator and corepressor regulation of the agonist/antagonist activity of the mixed antiestrogen, 4-hydroxytamoxifen. Mol. Endocrinol. **1997**, *11*: 657-666.

[83] Shang, Y., Hu, X., DiRenzo, J., Lazar, M.A., Brown, M. Cofactor dynamics and sufficiency in estrogen receptor-regulated transcription. Cell **2000**, *103*: 843-852.

[84] Jung, D.J., Lee, S.K., Lee, J.W. Agonist-dependent repression mediated by mutant estrogen receptor alpha that lacks the activation function 2 core domain. J. Biol. Chem. **2001**, *276*: 37280-37283.

[85] Shang, Y., Brown, M. Molecular determinants for the tissue specificity of serms. Science **2002**, *295*: 2465-2468.

[86] Huang, H.J., Norris, J.D., McDonnell, D.P. Identification of a negative regulatory surface within estrogen receptor alpha provides evidence in support of a role for corepressors in regulating cellular responses to agonists and antagonists. Mol. Endocrinol. **2002**, *16*: 1778-1792.

[87] Wijayaratne, A.L., Nagel, S.C., Paige, L.A., Christensen, D.J., Norris, J.D., Fowlkes, D.M., *et al.* Comparative analyses of mechanistic differences among antiestrogens. Endocrinology **1999**, *140*: 5828-5840.

[88] Peekhaus, N., Chang, T., Hayes, E., Wilkinson, H., Mitra, S., Schaeffer, J., *et al.* Distinct effects of the antiestrogen faslodex on the stability of estrogen receptors-alpha and -beta in the breast cancer cell line mcf-7. J. Mol. Endocrinol. **2004**, *32*: 987-995.

[89] Chambliss, K.L., Shaul, P.W. Estrogen modulation of endothelial nitric oxide synthase. Endocr. Rev. **2002**, *23*: 665-686.

[90] Pedram, A., Razandi, M., Levin, E.R. Nature of functional estrogen receptors at the plasma membrane. Mol. Endocrinol. **2006**, *20*: 1996-2009.

[91] Kim, J.K., Levin, E.R. Estrogen signaling in the cardiovascular system. Nucl Recept Signal **2006**, *4*: e013.

[92] Levin, E.R. Integration of the extranuclear and nuclear actions of estrogen. Mol. Endocrinol. **2005**, *19*: 1951-1959.

[93] Levin, E.R. Minireview: Extranuclear steroid receptors: Roles in modulation of cell functions. Mol. Endocrinol. **2011**, *25*: 377-384.

[94] Pedram, A., Razandi, M., Kim, J.K., O'Mahony, F., Lee, E.Y., Luderer, U., *et al.* Developmental phenotype of a membrane only estrogen receptor {alpha} (moer) mouse. J. Biol. Chem. **2009**, *284*: 3488-3495.

[95] Jeng, Y.-J., Kochukov, M., Watson, C. Membrane estrogen receptor-alpha-mediated nongenomic actions of phytoestrogens in gh3/b6/f10 pituitary tumor cells. Journal of Molecular Signaling **2009**, *4*: 2.

[96] Catalano, S., Barone, I., Giordano, C., Rizza, P., Qi, H., Gu, G., *et al.* Rapid estradiol/er{alpha} signaling enhances aromatase enzymatic activity in breast cancer cells. Mol. Endocrinol. **2009**, *23*: 1634-1645.

[97] Watson, C.S., Jeng, Y.-J., Kochukov, M.Y. Nongenomic actions of estradiol compared with estrone and estriol in pituitary tumor cell signaling and proliferation. FASEB J. **2008**, *22*: 3328-3336.

[98] Ellis R, L. Plasma membrane estrogen receptors. Trends in Endocrinology & Metabolism **2009**, *20*: 477-482.

[99] Watson, C.S., Alyea, R.A., Jeng, Y.J., Kochukov, M.Y. Nongenomic actions of low concentration estrogens and xenoestrogens on multiple tissues. Mol. Cell. Endocrinol. **2007**, *274*: 1-7.

[100] Sandén, C., Broselid, S., Cornmark, L., Andersson, K., Daszkiewicz-Nilsson, J., Mårtensson, U.E.A., *et al.* G protein-coupled estrogen receptor 1/g protein-coupled receptor 30 localizes in the plasma membrane and traffics intracellularly on cytokeratin intermediate filaments. Mol. Pharmacol. **2011**, *79*: 400-410.

[101] Recchia, A.G., De Francesco, E.M., Vivacqua, A., Sisci, D., Panno, M.L., Andò, S., *et al.* The g protein-coupled receptor 30 is up-regulated by hypoxia-inducible factor-1α (hif-1α) in breast cancer cells and cardiomyocytes. J. Biol. Chem. **2011**, *286*: 10773-10782.

[102] Cheng, S.-B., Quinn, J.A., Graeber, C.T., Filardo, E.J. Downmodulation of the g-protein-coupled estrogen receptor, gper, from the cell surface occurs *via* a transgolgi-proteasome pathway. Journal of Biological Chemistry **2011**.

[103] Wang, D., Hu, L., Zhang, G., Zhang, L., Chen, C. G protein-coupled receptor 30 in tumor development. Endocrine **2010**, *38*: 29-37.

[104] Stratton, R.C., Squires, P.E., Green, A.K. 17β-estradiol elevates cgmp and, *via* plasma membrane recruitment of protein kinase giα, stimulates ca2+ efflux from rat hepatocytes. J. Biol. Chem. **2010**, *285*: 27201-27212.

[105] Madeo, A., Maggiolini, M. Nuclear alternate estrogen receptor gpr30 mediates 17β-estradiol–induced gene expression and migration in breast cancer–associated fibroblasts. Cancer Res. **2010**, *70*: 6036-6046.

[106] Kolkova, Z., Noskova, V., Ehinger, A., Hansson, S., Casslén, B. G protein-coupled estrogen receptor 1 (gper, gpr 30) in normal human endometrium and early pregnancy decidua. Mol. Hum. Reprod. **2010**, *16*: 743-751.

[107] Ignatov, A., Ignatov, T., Roessner, A., Costa, S., Kalinski, T. Role of gpr30 in the mechanisms of tamoxifen resistance in breast cancer mcf-7 cells. Breast Cancer Res. Treat. **2010**, *123*: 87-96.

[108] Prossnitz, E.R., Maggiolini, M. Mechanisms of estrogen signaling and gene expression *via* gpr30. Mol. Cell. Endocrinol. **2009**, *308*: 32-38.

[109] Levin, E.R. G protein-coupled receptor 30: Estrogen receptor or collaborator? Endocrinology **2009**, *150*: 1563-1565.

[110] Razandi, M., Pedram, A., Levin, E.R. Plasma membrane estrogen receptors signal to antiapoptosis in breast cancer. Mol. Endocrinol. **2000**, *14*: 1434-1447.

[111] Hisamoto, K., Ohmichi, M., Kurachi, H., Hayakawa, J., Kanda, Y., Nishio, Y., *et al.* Estrogen induces the akt-dependent activation of endothelial nitric-oxide synthase in vascular endothelial cells. J. Biol. Chem. **2001**, *276*: 3459-3467.

[112] Monje, P., Boland, R. Expression and cellular localization of naturally occurring beta estrogen receptors in uterine and mammary cell lines. J. Cell. Biochem. **2002**, *86*: 136-144.

[113] Chen, D.B., Bird, I.M., Zheng, J., Magness, R.R. Membrane estrogen receptor-dependent extracellular signal-regulated kinase pathway mediates acute activation of endothelial nitric

oxide synthase by estrogen in uterine artery endothelial cells. Endocrinology **2004**, *145*: 113-125.

[114] Belcher, S.M., Le, H.H., Spurling, L., Wong, J.K. Rapid estrogenic regulation of extracellular signal- regulated kinase 1/2 signaling in cerebellar granule cells involves a g protein- and protein kinase a-dependent mechanism and intracellular activation of protein phosphatase 2a. Endocrinology **2005**, *146*: 5397-5406.

[115] Mhyre, A.J., Shapiro, R.A., Dorsa, D.M. Estradiol reduces nonclassical transcription at cyclic adenosine 3',5'-monophosphate response elements in glioma cells expressing estrogen receptor alpha. Endocrinology **2006**, *147*: 1796-1804.

[116] Simoncini, T., Scorticati, C., Mannella, P., Fadiel, A., Giretti, M.S., Fu, X.-D., *et al.* Estrogen receptor {alpha} interacts with g{alpha}13 to drive actin remodeling and endothelial cell migration *via* the rhoa/rho kinase/moesin pathway. Mol. Endocrinol. **2006**, *20*: 1756-1771.

[117] Filardo, E., Quinn, J., Pang, Y., Graeber, C., Shaw, S., Dong, J., *et al.* Activation of the novel estrogen receptor g protein-coupled receptor 30 (gpr30) at the plasma membrane. Endocrinology **2007**, *148*: 3236-3245.

[118] Jaubert, A.-M., Mehebik-Mojaat, N., Lacasa, D., Sabourault, D., Giudicelli, Y., Ribiere, C. Nongenomic estrogen effects on nitric oxide synthase activity in rat adipocytes. Endocrinology **2007**, *148*: 2444-2452.

[119] Wang, C., Prossnitz, E.R., Roy, S.K. G protein-coupled receptor 30 expression is required for estrogen stimulation of primordial follicle formation in the hamster ovary. Endocrinology **2008**, *149*: 4452-4461.

[120] Graber, R., Sumida, C., Vallette, G., Nunez, E.A. Rapid and long-term effects of 17 beta-estradiol on pip2-phospholipase c-specific activity of mcf-7 cells. Cell. Signal. **1993**, *5*: 181-186.

[121] Improta-Brears, T., Whorton, A.R., Codazzi, F., York, J.D., Meyer, T., McDonnell, D.P. Estrogen-induced activation of mitogen-activated protein kinase requires mobilization of intracellular calcium. Proc. Natl. Acad. Sci. USA **1999**, *96*: 4686-4691.

[122] Levin, E.R. Cellular functions of plasma membrane estrogen receptors. Steroids **2002**, *67*: 471-475.

[123] Razandi, M., Oh, P., Pedram, A., Schnitzer, J., Levin, E.R. Ers associate with and regulate the production of caveolin: Implications for signaling and cellular actions. Mol. Endocrinol. **2002**, *16*: 100-115.

[124] Song, R.X.-D., McPherson, R.A., Adam, L., Bao, Y., Shupnik, M., Kumar, R., *et al.* Linkage of rapid estrogen action to mapk activation by er{alpha}-shc association and shc pathway activation. Mol. Endocrinol. **2002**, *16*: 116-127.

[125] Razandi, M., Alton, G., Pedram, A., Ghonshani, S., Webb, P., Levin, E.R. Identification of a structural determinant necessary for the localization and function of estrogen receptor alpha at the plasma membrane. Mol. Cell. Biol. **2003**, *23*: 1633-1646.

[126] Razandi, M., Pedram, A., Park, S.T., Levin, E.R. Proximal events in signaling by plasma membrane estrogen receptors. J. Biol. Chem. **2003**, *278*: 2701-2712.

[127] Bulayeva, N.N., Gametchu, B., Watson, C.S. Quantitative measurement of estrogen-induced erk 1 and 2 activation *via* multiple membrane-initiated signaling pathways. Steroids **2004**, *69*: 181-192.

[128] Fernando, R.I., Wimalasena, J. Estradiol abrogates apoptosis in breast cancer cells through inactivation of bad: Ras-dependent nongenomic pathways requiring signaling through erk and akt. Mol. Biol. Cell **2004**, *15*: 3266-3284.

[129] Razandi, M., Pedram, A., Merchenthaler, I., Greene, G.L., Levin, E.R. Plasma membrane estrogen receptors exist and functions as dimers. Mol. Endocrinol. **2004**, *18*: 2854-2865.

[130] Razandi, M., Pedram, A., Rosen, E.M., Levin, E.R. Brca1 inhibits membrane estrogen and growth factor receptor signaling to cell proliferation in breast cancer. Mol. Cell. Biol. **2004**, *24*: 5900-5913.

[131] Santen, R.J., Song, R.X., Zhang, Z., Yue, W., Kumar, R. Adaptive hypersensitivity to estrogen: Mechanism for sequential responses to hormonal therapy in breast cancer. Clin. Cancer Res. **2004**, *10*: 337S-345S.

[132] Simoncini, T., Mannella, P., Fornari, L., Caruso, A., Willis, M.Y., Garibaldi, S., *et al.* Differential signal transduction of progesterone and medroxyprogesterone acetate in human endothelial cells. Endocrinology **2004**, *145*: 5745-5756.

[133] Song, R.X., Barnes, C.J., Zhang, Z., Bao, Y., Kumar, R., Santen, R.J. The role of shc and insulin-like growth factor 1 receptor in mediating the translocation of estrogen receptor {alpha} to the plasma membrane. PNAS **2004**, *101*: 2076-2081.

[134] Toran-Allerand, C.D. Minireview: A plethora of estrogen receptors in the brain: Where will it end? Endocrinology **2004**, *145*: 1069-1074.

[135] Pietras, R.J., Marquez, D.C., Chen, H.W., Tsai, E., Weinberg, O., Fishbein, M. Estrogen and growth factor receptor interactions in human breast and non-small cell lung cancer cells. Steroids **2005**, *70*: 372-381.

[136] Watson, C.S., Bulayeva, N.N., Wozniak, A.L., Finnerty, C.C. Signaling from the membrane *via* membrane estrogen receptor-[alpha]: Estrogens, xenoestrogens, and phytoestrogens. Steroids **2005**, *70*: 364-371.

[137] Cui, Y., Parra, I., Zhang, M., Hilsenbeck, S.G., Tsimelzon, A., Furukawa, T., *et al.* Elevated expression of mitogen-activated protein kinase phosphatase 3 in breast tumors: A mechanism of tamoxifen resistance. Cancer Res. **2006**, *66*: 5950-5959.

[138] Riggins, R.B., Thomas, K.S., Ta, H.Q., Wen, J., Davis, R.J., Schuh, N.R., *et al.* Physical and functional interactions between cas and c-src induce tamoxifen resistance of breast cancer cells through pathways involving epidermal growth factor receptor and signal transducer and activator of transcription 5b. Cancer Res. **2006**, *66*: 7007-7015.

[139] Greger, J.G., Fursov, N., Cooch, N., McLarney, S., Freedman, L.P., Edwards, D.P., *et al.* Phosphorylation of mnar promotes estrogen activation of phosphatidylinositol 3-kinase. Mol. Cell. Biol. **2007**, *27*: 1904-1913.

[140] Song, R.X.D., Zhang, Z., Chen, Y., Bao, Y., Santen, R.J. Estrogen signaling *via* a linear pathway involving insulin-like growth factor i receptor, matrix metalloproteinases, and epidermal growth factor receptor to activate mitogen-activated protein kinase in mcf-7 breast cancer cells. Endocrinology **2007**, *148*: 4091-4101.

[141] Stoica, G.E., Franke, T.F., Wellstein, A., Czubayko, F., List, H.J., Reiter, R., *et al.* Estradiol rapidly activates akt *via* the erbb2 signaling pathway. Mol. Endocrinol. **2003**, *17*: 818-830.

[142] Rajhans, R., Nair, S., Holden, A.H., Kumar, R., Tekmal, R.R., Vadlamudi, R.K. Oncogenic potential of the nuclear receptor co-regulator proline-, glutamic acid-, leucine-rich protein 1/modulator of the nongenomic actions of the estrogen receptor. Cancer Res. **2007**, *67*: 5505-5512.

[143] Ghisletti, S., Meda, C., Maggi, A., Vegeto, E. 17{beta}-estradiol inhibits inflammatory gene expression by controlling nf-{kappa}b intracellular localization. Mol. Cell. Biol. **2005**, *25*: 2957-2968.

[144] Lu, Q., Pallas, D.C., Surks, H.K., Baur, W.E., Mendelsohn, M.E., Karas, R.H. Striatin assembles a membrane signaling complex necessary for rapid, nongenomic activation of endothelial no synthase by estrogen receptor {alpha}. PNAS **2004**: 0407492101.

[145] Sun, M., Yang, L., Feldman, R.I., Sun, X.M., Bhalla, K.N., Jove, R., *et al.* Activation of phosphatidylinositol 3-kinase/akt pathway by androgen through interaction of p85alpha, androgen receptor, and src. J. Biol. Chem. **2003**, *278*: 42992-43000.

[146] Kinoshita, Y., Chen, S. Induction of aromatase (cyp19) expression in breast cancer cells through a nongenomic action of estrogen receptor alpha. Cancer Res. **2003**, *63*: 3546-3555.

[147] Pietras, R.J., Szego, C.M. Specific binding sites for oestrogen at the outer surfaces of isolated endometrial cells. Nature **1977**, *265*: 69-72.

[148] Watson, C.S., Campbell, C.H., Gametchu, B. The dynamic and elusive membrane estrogen receptor-alpha. Steroids **2002**, *67*: 429-437.

[149] Sheldahl, L.C., Shapiro, R.A., Bryant, D.N., Koerner, I.P., Dorsa, D.M. Estrogen induces rapid translocation of estrogen receptor [beta], but not estrogen receptor [alpha], to the neuronal plasma membrane. Neuroscience **2008**, *153*: 751-761.

[150] Li, L., Haynes, M.P., Bender, J.R. Plasma membrane localization and function of the estrogen receptor alpha variant (er46) in human endothelial cells. Proc. Natl. Acad. Sci. U. S. A. **2003**, *100*: 4807-4812.

[151] Acconcia, F., Ascenzi, P., Fabozzi, G., Visca, P., Marino, M. S-palmitoylation modulates human estrogen receptor-alpha functions. Biochem. Biophys. Res. Commun. **2004**, *316*: 878-883.

[152] Pedram, A., Razandi, M., Sainson, R.C.A., Kim, J.K., Hughes, C.C., Levin, E.R. A conserved mechanism for steroid receptor translocation to the plasma membrane. J. Biol. Chem. **2007**, *282*: 22278-22288.

[153] Moriarty, K., Kim, K.H., Bender, J.R. Estrogen receptor-mediated rapid signaling. Endocrinology **2006**, *147*: 5557-5563.

[154] Wong, C.W., McNally, C., Nickbarg, E., Komm, B.S., Cheskis, B.J. Estrogen receptor-interacting protein that modulates its nongenomic activity-crosstalk with src/erk phosphorylation cascade. Proc. Natl. Acad. Sci. U. S. A. **2002**, *99*: 14783-14788.

[155] Edwards, D.P., Boonyaratanakornkit, V. Rapid extranuclear signaling by the estrogen receptor (er): Mnar couples er and src to the map kinase signaling pathway. Mol Interv **2003**, *3*: 12-15.

[156] Barletta, F., Wong, C.W., McNally, C., Komm, B.S., Katzenellenbogen, B., Cheskis, B.J. Characterization of the interactions of estrogen receptor and mnar in the activation of csrc. Mol. Endocrinol. **2004**, *18*: 1096-1108.

[157] Cheskis, B.J. Regulation of cell signalling cascades by steroid hormones. J. Cell. Biochem. **2004**, *93*: 20-27.

[158] Mishra, S.K., Balasenthil, S., Nguyen, D., Vadlamudi, R.K. Cloning and functional characterization of pelp1/mnar promoter. Gene **2004**, *330*: 115-122.

[159] Mishra, S.K., Talukder, A.H., Gururaj, A.E., Yang, Z., Singh, R.R., Mahoney, M.G., *et al.* Upstream determinants of estrogen receptor-{alpha} regulation of metastatic tumor antigen 3 pathway. J. Biol. Chem. **2004**, *279*: 32709-32715.

[160] Nair, S.S., Mishra, S.K., Yang, Z., Balasenthil, S., Kumar, R., Vadlamudi, R.K. Potential role of a novel transcriptional co-activator pelp1 in histone h1 displacement in cancer cells. Cancer Res. **2004**, *64*: 6416-6423.

[161] Kim, K.H., Bender, J.R. Rapid, estrogen receptor-mediated signaling: Why is the endothelium so special? Sci STKE **2005**, *2005*: pe28.

[162] Greger, J.G., Guo, Y., Henderson, R., Ross, J.F., Cheskis, B.J. Characterization of mnar expression. Steroids **2006**, *71*: 317-322.

[163] Li, L., Hisamoto, K., Kim, K.H., Haynes, M.P., Bauer, P.M., Sanjay, A., *et al.* Variant estrogen receptor c-src molecular interdependence and c-src structural requirements for endothelial no synthase activation. Proceedings of the National Academy of Sciences **2007**, *104*: 16468-16473.

[164] Marquez-Garban, D.C., Chen, H.-W., Fishbein, M.C., Goodglick, L., Pietras, R.J. Estrogen receptor signaling pathways in human non-small cell lung cancer. Steroids **2007**, *72*: 135-143.

[165] Cheskis, B.J., Greger, J., Cooch, N., McNally, C., McLarney, S., Lam, H.-S., *et al.* Mnar plays an important role in era activation of src/mapk and pi3k/akt signaling pathways. Steroids **2008**, *In Press, Corrected Proof.*

[166] Rajhans, R., Nair, H.B., Nair, S.S., Cortez, V., Ikuko, K., Kirma, N.B., *et al.* Modulation of *in situ* estrogen synthesis by proline-, glutamic acid-, and leucine-rich protein-1: Potential estrogen receptor autocrine signaling loop in breast cancer cells. Mol. Endocrinol. **2008**, *22*: 649-664.

[167] Hisamoto, K., Bender, J.R. Vascular cell signaling by membrane estrogen receptors. Steroids **2005**, *70*: 382-387.

[168] Flototto, T., Niederacher, D., Hohmann, D., Heimerzheim, T., Dall, P., Djahansouzi, S., *et al.* Molecular mechanism of estrogen receptor (er)[alpha]-specific, estradiol-dependent expression of the progesterone receptor (pr) b-isoform. The Journal of Steroid Biochemistry and Molecular Biology **2004**, *88*: 131-142.

[169] Filardo, E.J. Epidermal growth factor receptor (egfr) transactivation by estrogen *via* the g-protein-coupled receptor, gpr30: A novel signaling pathway with potential significance for breast cancer. J. Steroid Biochem. Mol. Biol. **2002**, *80*: 231-238.

[170] Revankar, C.M., Cimino, D.F., Sklar, L.A., Arterburn, J.B., Prossnitz, E.R. A transmembrane intracellular estrogen receptor mediates rapid cell signaling. Science **2005**, *307*: 1625-1630.

[171] Thomas, P., Pang, Y., Filardo, E.J., Dong, J. Identity of an estrogen membrane receptor coupled to a g protein in human breast cancer cells. Endocrinology **2005**, *146*: 624-632.

[172] Funakoshi, T., Yanai, A., Shinoda, K., Kawano, M.M., Mizukami, Y. G protein-coupled receptor 30 is an estrogen receptor in the plasma membrane. Biochem. Biophys. Res. Commun. **2006**, *346*: 904-910.

[173] Wang, C., Dehghani, B., Magrisso, I.J., Rick, E.A., Bonhomme, E., Cody, D.B., *et al.* Gpr30 contributes to estrogen-induced thymic atrophy. Mol. Endocrinol. **2008**, *22*: 636-648.

[174] Prossnitz, E.R., Oprea, T.I., Sklar, L.A., Arterburn, J.B. The ins and outs of gpr30: A transmembrane estrogen receptor. The Journal of Steroid Biochemistry and Molecular Biology **2008**, *109*: 350-353.

[175] Otto, C., Rohde-Schulz, B., Schwarz, G., Fuchs, I., Klewer, M., Brittain, D., *et al.* Gpr30 localizes to the endoplasmic reticulum and is not activated by estradiol. Endocrinology **2008**: en.2008-0269.

[176] Filardo, E.J., Quinn, J.A., Sabo, E. Association of the membrane estrogen receptor, gpr30, with breast tumor metastasis and transactivation of the epidermal growth factor receptor. Steroids **2008**, *73*: 870-873.

[177] Sakamoto, H., Matsuda, K.-i., Hosokawa, K., Nishi, M., Morris, J.F., Prossnitz, E.R., *et al.* Expression of g protein-coupled receptor-30, a g protein-coupled membrane estrogen receptor, in oxytocin neurons of the rat paraventricular and supraoptic nuclei. Endocrinology **2007**, *148*: 5842-5850.

[178] Thomas, P., Dong, J. Binding and activation of the seven-transmembrane estrogen receptor gpr30 by environmental estrogens: A potential novel mechanism of endocrine disruption. The Journal of Steroid Biochemistry and Molecular Biology **2006**, *102*: 175-179.

[179] Filardo, E.J., Graeber, C.T., Quinn, J.A., Resnick, M.B., Giri, D., DeLellis, R.A., *et al.* Distribution of gpr30, a seven membrane-spanning estrogen receptor, in primary breast cancer and its association with clinicopathologic determinants of tumor progression. Clin. Cancer Res. **2006**, *12*: 6359-6366.

[180] Kang, L., Zhang, X., Xie, Y., Tu, Y., Wang, D., Liu, Z., *et al.* Involvement of estrogen receptor variant er-{alpha}36, not gpr30, in nongenomic estrogen signaling. Mol. Endocrinol. **2010**, *24*: 709-721.

[181] Wang, Z., Zhang, X., Shen, P., Loggie, B.W., Chang, Y., Deuel, T.F. Identification, cloning, and expression of human estrogen receptor-[alpha]36, a novel variant of human estrogen receptor-[alpha]66. Biochem. Biophys. Res. Commun. **2005**, *336*: 1023-1027.

[182] Lin, S.-L., Yan, L.-Y., Liang, X.-W., Wang, Z.-B., Wang, Z.-Y., Qiao, J., *et al.* A novel variant of er-alpha, er-alpha36 mediates testosterone-stimulated erk and akt activation in endometrial cancer hec1a cells. Reproductive Biology and Endocrinology **2009**, *7*: 102.

[183] Shi, L., Dong, B., Li, Z., Lu, Y., Ouyang, T., Li, J., *et al.* Expression of er-{alpha}36, a novel variant of estrogen receptor {alpha}, and resistance to tamoxifen treatment in breast cancer. J. Clin. Oncol. **2009**, *27*: 3423-3429.

[184] ENCODE, project, consortium. Identification and analysis of functional elements in 1% of the human genome by the encode pilot project. Nature **2007**, *447*: 799-816.

[185] Kapranov, P., Cheng, J., Dike, S., Nix, D.A., Duttagupta, R., Willingham, A.T., *et al.* Rna maps reveal new rna classes and a possible function for pervasive transcription. Science **2007**, *316*: 1484-1488.

[186] Pennisi, E. Genomics: DNA study forces rethink of what it means to be a gene. Science **2007**, *316*: 1556-1557.

[187] Kapranov, P., St Laurent, G., Raz, T., Ozsolak, F., Reynolds, C.P., Sorensen, P., *et al.* The majority of total nuclear-encoded non-ribosomal rna in a human cell is 'dark matter' un-annotated rna. BMC Biol. **2010**, *8*: 149.

[188] Wery, M., Kwapisz, M., Morillon, A. Noncoding rnas in gene regulation. Wiley Interdiscip Rev Syst Biol Med **2011**, *3*: 728-738.

[189] Taft, R.J., Pang, K.C., Mercer, T.R., Dinger, M., Mattick, J.S. Non-coding rnas: Regulators of disease. The Journal of Pathology **2010**, *220*: 126-139.

[190] Ambros, V. The evolution of our thinking about micrornas. Nat. Med. **2008**, *14*: 1036-1040.

[191] Ruvkun, G. The perfect storm of tiny rnas. Nat. Med. **2008**, *14*: 1041-1045.

[192] Baulcombe, D.C. Short silencing rna: The dark matter of genetics? Cold Spring Harb. Symp. Quant. Biol. **2006**, *71*: 13-20.

[193] Saini, H.K., Griffiths-Jones, S., Enright, A.J. Genomic analysis of human microrna transcripts. Proceedings of the National Academy of Sciences **2007**, *104*: 17719-17724.

[194] Amaral, P.P., Dinger, M.E., Mercer, T.R., Mattick, J.S. The eukaryotic genome as an rna machine. Science **2008**, *319*: 1787-1789.

[195] Verghese, E.T., Hanby, A.M., Speirs, V., Hughes, T.A. Small is beautiful: Micrornas and breast cancer-where are we now? J. Pathol. **2008**, *215*: 214-221.

[196] Thomson, J.M., Newman, M., Parker, J.S., Morin-Kensicki, E.M., Wright, T., Hammond, S.M. Extensive post-transcriptional regulation of micrornas and its implications for cancer. Genes Dev. **2006**, *20*: 2202-2207.

[197] Yamagata, K., Fujiyama, S., Ito, S., Ueda, T., Murata, T., Naitou, M., *et al.* Maturation of microrna is hormonally regulated by a nuclear receptor. Mol. Cell **2009**, *36*: 340-347.

[198] Hock, J., Meister, G. The argonaute protein family. Genome Biol **2008**, *9*: 210.

[199] Pasquinelli, A.E., Hunter, S., Bracht, J. Micrornas: A developing story. Curr. Opin. Genet. Dev. **2005**, *15*: 200-205.

[200] Sen, G.L., Blau, H.M. A brief history of rnai: The silence of the genes. FASEB J. **2006**, *20*: 1293-1299.

[201] Berkhout, B., Jeang, K.-T. Riscy business: Micrornas, pathogenesis, and viruses. J. Biol. Chem. **2007**, *282*: 26641-26645.

[202] Cuellar, T.L., McManus, M.T. Micrornas and endocrine biology. J. Endocrinol. **2005**, *187*: 327-332.

[203] Chendrimada, T.P., Finn, K.J., Ji, X., Baillat, D., Gregory, R.I., Liebhaber, S.A., *et al.* Microrna silencing through risc recruitment of eif6. Nature **2007**, *447*: 823-828.

[204] Perron, M.P., Provost, P. Protein interactions and complexes in human microrna biogenesis and function. Front. Biosci. **2008**, *13*: 2537-2547.

[205] Friedman, R.C., Farh, K.K.-H., Burge, C.B., Bartel, D. Most mammalian mrnas are conserved targets of micrornas. Genome Res. **2008**: gr.082701.082108.

[206] Vasudevan, S., Tong, Y., Steitz, J.A. Switching from repression to activation: Micrornas can up-regulate translation. Science **2007**, *318*: 1931-1934.

[207] Tsuchiya, Y., Nakajima, M., Takagi, S., Taniya, T., Yokoi, T. Microrna regulates the expression of human cytochrome p450 1b1. Cancer Res. **2006**, *66*: 9090-9098.

[208] Adams, B.D., Furneaux, H., White, B.A. The micro-ribonucleic acid (mirna) mir-206 targets the human estrogen receptor-{alpha} (er{alpha}) and represses er{alpha} messenger rna and protein expression in breast cancer cell lines. Mol. Endocrinol. **2007**, *21*: 1132-1147.

[209] Corney, D.C., Flesken-Nikitin, A., Godwin, A.K., Wang, W., Nikitin, A.Y. Microrna-34b and microrna-34c are targets of p53 and cooperate in control of cell proliferation and adhesion-independent growth. Cancer Res. **2007**, *67*: 8433-8438.

[210] Gaur, A., Jewell, D.A., Liang, Y., Ridzon, D., Moore, J.H., Chen, C., *et al.* Characterization of microrna expression levels and their biological correlates in human cancer cell lines. Cancer Res. **2007**, *67*: 2456-2468.

[211] Gramantieri, L., Ferracin, M., Fornari, F., Veronese, A., Sabbioni, S., Liu, C.-G., *et al.* Cyclin g1 is a target of mir-122a, a microrna frequently down-regulated in human hepatocellular carcinoma. Cancer Res. **2007**, *67*: 6092-6099.

[212] Hurteau, G.J., Carlson, J.A., Spivack, S.D., Brock, G.J. Overexpression of the microrna hsa-mir-200c leads to reduced expression of transcription factor 8 and increased expression of e-cadherin. Cancer Res. **2007**, *67*: 7972-7976.

[213] Johnson, C.D., Esquela-Kerscher, A., Stefani, G., Byrom, M., Kelnar, K., Ovcharenko, D., *et al.* The let-7 microrna represses cell proliferation pathways in human cells. Cancer Res. **2007**, *67*: 7713-7722.

[214] Lujambio, A., Ropero, S., Ballestar, E., Fraga, M.F., Cerrato, C., Setien, F., *et al.* Genetic unmasking of an epigenetically silenced microrna in human cancer cells. Cancer Res. **2007**, *67*: 1424-1429.

[215] Meng, F., Henson, R., Wehbe-Janek, H., Ghoshal, K., Jacob, S.T., Patel, T. Microrna-21 regulates expression of the pten tumor suppressor gene in human hepatocellular cancer. Gastroenterology **2007**, *133*: 647-658.

[216] Osada, H., Takahashi, T. Micrornas in biological processes and carcinogenesis. Carcinogenesis **2007**, *28*: 2-12.

[217] Sampson, V.B., Rong, N.H., Han, J., Yang, Q., Aris, V., Soteropoulos, P., *et al.* Microrna let-7a down-regulates myc and reverts myc-induced growth in burkitt lymphoma cells. Cancer Res. **2007**, *67*: 9762-9770.

[218] Sylvestre, Y., De Guire, V., Querido, E., Mukhopadhyay, U.K., Bourdeau, V., Major, F., *et al.* An e2f/mir-20a autoregulatory feedback loop. J. Biol. Chem. **2007**, *282*: 2135-2143.

[219] Thai, T.-H., Calado, D.P., Casola, S., Ansel, K.M., Xiao, C., Xue, Y., *et al.* Regulation of the germinal center response by microrna-155. Science **2007**, *316*: 604-608.

[220] Woods, K., Thomson, J.M., Hammond, S.M. Direct regulation of an oncogenic micro-rna cluster by e2f transcription factors. J. Biol. Chem. **2007**, *282*: 2130-2134.

[221] Zhang, W., Dahlberg, J.E., Tam, W. Micrornas in tumorigenesis: A primer. Am. J. Pathol. **2007**, *171*: 728-738.

[222] Zhou, Y., Yau, C., Gray, J.W., Chew, K., Dairkee, S.H., Moore, D.H., *et al.* Enhanced nf kappa b and ap-1 transcriptional activity associated with antiestrogen resistant breast cancer. BMC Cancer **2007**, *7*: 59.

[223] Kozomara, A., Griffiths-Jones, S. Mirbase: Integrating microrna annotation and deep-sequencing data. Nucleic Acids Res. **2011**, *39*: D152-D157.

[224] Volinia, S., Calin, G.A., Liu, C.-G., Ambs, S., Cimmino, A., Petrocca, F., *et al.* A microrna expression signature of human solid tumors defines cancer gene targets. Proceedings of the National Academy of Sciences **2006**, *103*: 2257-2261.

[225] Zhang, L., Huang, J., Yang, N., Greshock, J., Megraw, M.S., Giannakakis, A., *et al.* Micrornas exhibit high frequency genomic alterations in human cancer. PNAS **2006**, *103*: 9136-9141.

[226] Griffiths-Jones, S., Grocock, R.J., van Dongen, S., Bateman, A., Enright, A.J. Mirbase: Microrna sequences, targets and gene nomenclature. Nucleic Acids Res **2006**, *34*: D140-144.

[227] Okamura, K., Phillips, M.D., Tyler, D.M., Duan, H., Chou, Y.-t., Lai, E.C. The regulatory activity of microrna[ast] species has substantial influence on microrna and 3[prime] utr evolution. Nat Struct Mol Biol **2008**, *15*: 354-363.

[228] Tanzer, A., Stadler, P.F. Molecular evolution of a microrna cluster. J Mol Biol **2004**, *339*: 327-335.

[229] Gregory, R.I., Yan, K.-p., Amuthan, G., Chendrimada, T., Doratotaj, B., Cooch, N., *et al.* The microprocessor complex mediates the genesis of micrornas. Nature **2004**, *432*: 235-240.

[230] Han, J., Pedersen, J.S., Kwon, S.C., Belair, C.D., Kim, Y.-K., Yeom, K.-H., *et al.* Posttranscriptional crossregulation between drosha and dgcr8. Cell **2009**, *136*: 75-84.

[231] Davis, B.N., Hilyard, A.C., Lagna, G., Hata, A. Smad proteins control drosha-mediated microrna maturation. Nature **2008**, *454*: 56-61.

[232] Sakamoto, S., Aoki, K., Higuchi, T., Todaka, H., Morisawa, K., Tamaki, N., *et al.* The nf90-nf45 complex functions as a negative regulator in the microrna processing pathway. Mol. Cell. Biol. **2009**, *29*: 3754-3769.

[233] Watanabe, M., Yanagisawa, J., Kitagawa, H., Takeyama, K.-i., Ogawa, S., Arao, Y., *et al.* A subfamily of rna-binding dead-box proteins acts as an estrogen receptor [alpha] co-activator through the n-terminal activation domain (af-1) with an rna co-activator, sra. EMBO J **2001**, *20*: 1341-1352.

[234] Melo, S.A., Moutinho, C., Ropero, S., Calin, G.A., Rossi, S., Spizzo, R., *et al.* A genetic defect in exportin-5 traps precursor micrornas in the nucleus of cancer cells. Cancer Cell **2010**, *18*: 303-315.

[235] Lee, Y., Hur, I., Park, S.Y., Kim, Y.K., Suh, M.R., Kim, V.N. The role of pact in the rna silencing pathway. EMBO J **2006**, *25*: 522-532.

[236] Chendrimada, T.P., Gregory, R.I., Kumaraswamy, E., Norman, J., Cooch, N., Nishikura, K., *et al.* Trbp recruits the dicer complex to ago2 for microrna processing and gene silencing. Nature **2005**, *436*: 740-744.

[237] Paroo, Z., Ye, X., Chen, S., Liu, Q. Phosphorylation of the human microrna-generating complex mediates mapk/erk signaling. Cell **2009**, *139*: 112-122.

[238] Cochrane, D., Cittelly, D., Howe, E., Spoelstra, N., McKinsey, E., LaPara, K., *et al.* Micrornas link estrogen receptor alpha status and dicer levels in breast cancer. Hormones and Cancer **2010**, *1*: 306-319.

[239] Cifuentes, D., Xue, H., Taylor, D.W., Patnode, H., Mishima, Y., Cheloufi, S., *et al.* A novel mirna processing pathway independent of dicer requires argonaute2 catalytic activity. Science **2010**, *328*: 1694-1698.

[240] Adams, B.D., Claffey, K.P., White, B.A. Argonaute-2 expression is regulated by epidermal growth factor receptor and mitogen-activated protein kinase signaling and correlates with a transformed phenotype in breast cancer cells. Endocrinology **2009**, *150*: 14-23.

[241] Chan, J.A., Krichevsky, A.M., Kosik, K.S. Microrna-21 is an antiapoptotic factor in human glioblastoma cells. Cancer Res. **2005**, *65*: 6029-6033.

[242] Hammond, S.M. Micrornas as oncogenes. Curr. Opin. Genet. Dev. **2005**, *16*: 4-9.

[243] Hayashita, Y., Osada, H., Tatematsu, Y., Yamada, H., Yanagisawa, K., Tomida, S., *et al.* A polycistronic microrna cluster, mir-17-92, is overexpressed in human lung cancers and enhances cell proliferation. Cancer Res. **2005**, *65*: 9628-9632.

[244] Johnson, S.M., Grosshans, H., Shingara, J., Byrom, M., Jarvis, R., Cheng, A., *et al.* Ras is regulated by the let-7 microrna family. Cell **2005**, *120*: 635-647.

[245] Diederichs, S., Haber, D.A. Sequence variations of micrornas in human cancer: Alterations in predicted secondary structure do not affect processing. Cancer Res. **2006**, *66*: 6097-6104.

[246] Scott, G.K., Mattie, M.D., Berger, C.E., Benz, S.C., Benz, C.C. Rapid alteration of microrna levels by histone deacetylase inhibition. Cancer Res. **2006**, *66*: 1277-1281.

[247] Giannakakis, A., Coukos, G., Hatzigeorgiou, A., Sandaltzopoulos, R., Zhang, L. Mirna genetic alterations in human cancers. Expert Opinion on Biological Therapy **2007**, *7*: 1375-1386.

[248] Sempere, L.F., Christensen, M., Silahtaroglu, A., Bak, M., Heath, C.V., Schwartz, G., *et al.* Altered microrna expression confined to specific epithelial cell subpopulations in breast cancer. Cancer Res. **2007**, *67*: 11612-11620.

[249] Gusev, Y., Schmittgen, T.D., Lerner, M., Postier, R., Brackett, D. Computational analysis of biological functions and pathways collectively targeted by co-expressed micrornas in cancer. BMC Bioinformatics **2007**, *8 Suppl 7*: S16.

[250] Lu, Z., Liu, M., Stribinskis, V., Klinge, C.M., Ramos, K.S., Colburn, N.H., *et al.* Microrna-21 promotes cell transformation by targeting the programmed cell death 4 gene. Oncogene **2008**, *27*: 4373-4379.

[251] Zhu, S., Si, M.-L., Wu, H., Mo, Y.-Y. Microrna-21 targets the tumor suppressor gene tropomyosin 1 (tpm1). J. Biol. Chem. **2007**, *282*: 14328-14336.

[252] Lu, Z., Liu, M., Stribinskis, V., Klinge, C.M., Ramos, K.S., Colburn, N.H., *et al.* Microrna-21 promotes cell transformation by targeting the programmed cell death 4 gene. Oncogene **2008**, *27*: 4373-4379.

[253] Wickramasinghe, N.S., Manavalan, T.T., Dougherty, S.M., Riggs, K.A., Li, Y., Klinge, C.M. Estradiol downregulates mir-21 expression and increases mir-21 target gene expression in mcf-7 breast cancer cells. Nucl. Acids Res. **2009**: gkp117.

[254] Wang, Z.X., Lu, B.B., Wang, H., Cheng, Z.X., Yin, Y.M. Microrna-21 modulates chemosensitivity of breast cancer cells to doxorubicin by targeting pten. Arch Med Res **2011**, *42*: 281-290.

[255] Ferracin, M., Querzoli, P., Calin, G.A., Negrini, M. Micrornas: Toward the clinic for breast cancer patients. Seminars in Oncology **2011**, *38*: 764-775.

[256] Scott, G.K., Goga, A., Bhaumik, D., Berger, C.E., Sullivan, C.S., Benz, C.C. Coordinate suppression of erbb2 and erbb3 by enforced expression of micro-rna mir-125a or mir-125b. J. Biol. Chem. **2007**, *282*: 1479-1486.

[257] Park, S.M., Shell, S., Radjabi, A.R., Schickel, R., Feig, C., Boyerinas, B., *et al.* Let-7 prevents early cancer progression by suppressing expression of the embryonic gene hmga2. Cell Cycle **2007**, *6*: 2585-2590.

[258] Paterson, E.L., Kolesnikoff, N., Gregory, P.A., Bert, A.G., Khew-Goodall, Y., Goodall, G.J. The microrna-200 family regulates epithelial to mesenchymal transition. ScientificWorldJournal **2008**, *8*: 901-904.

[259] Gregory, P.A., Bert, A.G., Paterson, E.L., Barry, S.C., Tsykin, A., Farshid, G., *et al.* The mir-200 family and mir-205 regulate epithelial to mesenchymal transition by targeting zeb1 and sip1. Nat Cell Biol **2008**, *10*: 593-601.

[260] Peter, M.E. Let-7 and mir-200 micrornas: Guardians against pluripotency and cancer progression. Cell Cycle **2009**, *8*: 843-852.

[261] Guttilla, I.K., Adams, B.D., White, B.A. Erα, micrornas, and the epithelial–mesenchymal transition in breast cancer. Trends in Endocrinology & Metabolism **2012**, *23*: 73-82.

[262] Burk, U., Schubert, J., Wellner, U., Schmalhofer, O., Vincan, E., Spaderna, S., *et al.* A reciprocal repression between zeb1 and members of the mir-200 family promotes emt and invasion in cancer cells. EMBO Rep **2008**, *9*: 582-589.

[263] Cochrane, D.R., Howe, E.N., Spoelstra, N.S., Richer, J.K. Loss of mir-200c: A marker of aggressiveness and chemoresistance in female reproductive cancers. J Oncol **2010**, *2010*: 821717.

[264] Uhlmann, S., Zhang, J.D., Schwager, A., Mannsperger, H., Riazalhosseini, Y., Burmester, S., *et al.* Mir-200bc/429 cluster targets plc[gamma]1 and differentially regulates proliferation and egf-driven invasion than mir-200a/141 in breast cancer. Oncogene **2010**, *29*: 4297-4306.

[265] Ma, L., Reinhardt, F., Pan, E., Soutschek, J., Bhat, B., Marcusson, E.G., *et al.* Therapeutic silencing of mir-10b inhibits metastasis in a mouse mammary tumor model. Nat Biotechnol **2010**, *28*: 341-347.

[266] Huang, Q., Gumireddy, K., Schrier, M., le Sage, C., Nagel, R., Nair, S., *et al.* The micrornas mir-373 and mir-520c promote tumour invasion and metastasis. Nat Cell Biol **2008**, *10*: 202-210.
[267] Sun, L., Yan, W., Wang, Y., Sun, G., Luo, H., Zhang, J., *et al.* Microrna-10b induces glioma cell invasion by modulating mmp-14 and upar expression *via* hoxd10. Brain Res **2011**, *1389*: 9-18.
[268] Hossain, A., Kuo, M.T., Saunders, G.F. Mir-17-5p regulates breast cancer cell proliferation by inhibiting translation of aib1 mrna. Mol. Cell. Biol. **2006**, *26*: 8191-8201.
[269] Volinia, S., Galasso, M., Costinean, S., Tagliavini, L., Gamberoni, G., Drusco, A., *et al.* Reprogramming of mirna networks in cancer and leukemia. Genome Res. **2010**, *20*: 589-599.
[270] Xiang, J., Wu, J. Feud or friend? The role of the mir-17-92 cluster in tumorigenesis. Curr Genomics **2010**, *11*: 129-135.
[271] Li, H., Bian, C., Liao, L., Li, J., Zhao, R.C. Mir-17-5p promotes human breast cancer cell migration and invasion through suppression of hbp1. Breast Cancer Res Treat **2010**.
[272] Kim, K., Chadalapaka, G., Lee, S.O., Yamada, D., Sastre-Garau, X., Defossez, P.A., *et al.* Identification of oncogenic microrna-17-92/zbtb4/specificity protein axis in breast cancer. Oncogene **2012**, *31*: 1034-1044.
[273] Iorio, M.V., Ferracin, M., Liu, C.G., Veronese, A., Spizzo, R., Sabbioni, S., *et al.* Microrna gene expression deregulation in human breast cancer. Cancer Res **2005**, *65*: 7065-7070.
[274] Lowery, A.J., Miller, N., Devaney, A., McNeill, R.E., Davoren, P.A., Lemetre, C., *et al.* Microrna signatures predict oestrogen receptor, progesterone receptor and her2/neu receptor status in breast cancer. Breast Cancer Res **2009**, *11*: R27.
[275] Roth, C., Rack, B., Muller, V., Janni, W., Pantel, K., Schwarzenbach, H. Circulating micrornas as blood-based markers for patients with primary and metastatic breast cancer. Breast Cancer Res **2010**, *12*: R90.
[276] Hu, Z., Dong, J., Wang, L.E., Ma, H., Liu, J., Zhao, Y., *et al.* Serum microrna profiling and breast cancer risk: The use of mir-484/191 as endogenous controls. Carcinogenesis **2012**.
[277] Heneghan, H.M., Miller, N., Kelly, R., Newell, J., Kerin, M.J. Systemic mirna-195 differentiates breast cancer from other malignancies and is a potential biomarker for detecting noninvasive and early stage disease. Oncologist **2010**, *15*: 673-682.
[278] Klinge, C.M. Estrogen regulation of microrna expression. Curr. Genomics **2009**, *10*: 169-183.
[279] Klinge, C.M. Mirnas and estrogen action. Trends Endocrinol. Metab. **2012**, *miRNAs and estrogen action*.
[280] Creighton, C.J., Nagaraja, A.K., Hanash, S.M., Matzuk, M.M., Gunaratne, P.H. A bioinformatics tool for linking gene expression profiling results with public databases of microrna target predictions. RNA **2008**, *14*: 2290-2296.
[281] Cohen, A., Shmoish, M., Levi, L., Cheruti, U., Levavi-Sivan, B., Lubzens, E. Alterations in micro-ribonucleic acid expression profiles reveal a novel pathway for estrogen regulation. Endocrinology **2008**, *149*: 1687-1696.
[282] Kovalchuk, O., Tryndyak, V.P., Montgomery, B., Boyko, A., Kutanzi, K., Zemp, F., *et al.* Estrogen-induced rat breast carcinogenesis is characterized by alterations in DNA methylation, histone modifications and aberrant microrna expression. Cell Cycle **2007**, *6*: 2010-2018.

[283] Dai, R., Phillips, R.A., Zhang, Y., Khan, D., Crasta, O., Ahmed, S.A. Suppression of lps-induced ifn{gamma} and nitric oxide in splenic lymphocytes by select estrogen-regulated mirna: A novel mechanism of immune modulation. Blood **2008**, *112*: 4591-4597.

[284] Hong, X., Luense, L.J., McGinnis, L.K., Nothnick, W.B., Christenson, L.K. Dicerl is essential for female fertility and normal development of the female reproductive system. Endocrinology **2008**: en.2008-0294.

[285] Carletti, M.Z., Christenson, L.K. Microrna in the ovary and female reproductive tract. J. Anim Sci. **2008**: jas.2008-1331.

[286] Hu, S.-J., Ren, G., Liu, J.-L., Zhao, Z.-A., Yu, Y.-S., Su, R.-W., *et al.* Microrna expression and regulation in mouse uterus during embryo implantation. J. Biol. Chem. **2008**, *283*: 23473-23484.

[287] Nothnick, W.B., Healy, C. Estrogen induces distinct patterns of microrna expression within the mouse uterus. Reprod Sci **2010**, *17*: 987-994.

[288] Henderson, B.E., Feigelson, H.S. Hormonal carcinogenesis. Carcinogenesis **2000**, *21*: 427-433.

[289] Russo, J., Tahin, Q., Lareef, M.H., Hu, Y.F., Russo, I.H. Neoplastic transformation of human breast epithelial cells by estrogens and chemical carcinogens. Environ. Mol. Mutagen. **2002**, *39*: 254-263.

[290] Russo, J., Lareef, M.H., Tahin, Q., Hu, Y.F., Slater, C., Ao, X., *et al.* 17beta-estradiol is carcinogenic in human breast epithelial cells. J. Steroid Biochem. Mol. Biol. **2002**, *80*: 149-162.

[291] Russo, J., Fernandez, S.V., Russo, P.A., Fernbaugh, R., Sheriff, F.S., Lareef, H.M., *et al.* 17-beta-estradiol induces transformation and tumorigenesis in human breast epithelial cells. FASEB J. **2006**, *20*: 1622-1634.

[292] Wiseman, R.A. Breast cancer: Critical data analysis concludes that estrogens are not the cause, however lifestyle changes can alter risk rapidly. J. Clin. Epidemiol. **2004**, *57*: 766-772.

[293] Hah, N., Danko, Charles G., Core, L., Waterfall, Joshua J., Siepel, A., Lis, John T., *et al.* A rapid, extensive, and transient transcriptional response to estrogen signaling in breast cancer cells. Cell **2011**, *145*: 622-634.

[294] Cicatiello, L., Mutarelli, M., Grober, O.M., Paris, O., Ferraro, L., Ravo, M., *et al.* Estrogen receptor alpha controls a gene network in luminal-like breast cancer cells comprising multiple transcription factors and micrornas. Am J Pathol **2010**, *176*: 2113-2130.

[295] Harris, H.A., Katzenellenbogen, J.A., Katzenellenbogen, B.S. Characterization of the biological roles of the estrogen receptors, eralpha and erbeta, in estrogen target tissues *in vivo* through the use of an eralpha-selective ligand. Endocrinology **2002**, *143*: 4172-4177.

[296] Meyers, M.J., Sun, J., Carlson, K.E., Marriner, G.A., Katzenellenbogen, B.S., Katzenellenbogen, J.A. Estrogen receptor-beta potency-selective ligands: Structure-activity relationship studies of diarylpropionitriles and their acetylene and polar analogues. J. Med. Chem. **2001**, *44*: 4230-4251.

[297] Ferguson, A.T., Lapidus, R.G., Baylin, S.B., Davidson, N.E. Demethylation of the estrogen receptor gene in estrogen receptor-negative breast cancer cells can reactivate estrogen receptor gene expression. Cancer Res. **1995**, *55*: 2279-2283.

[298] Yang, X., Ferguson, A.T., Nass, S.J., Phillips, D.L., Butash, K.A., Wang, S.M., *et al.* Transcriptional activation of estrogen receptor alpha in human breast cancer cells by histone deacetylase inhibition. Cancer Res. **2000**, *60*: 6890-6894.

[299] Yang, X., Phillips, D.L., Ferguson, A.T., Nelson, W.G., Herman, J.G., Davidson, N.E. Synergistic activation of functional estrogen receptor (er)-alpha by DNA methyltransferase and histone deacetylase inhibition in human er-alpha-negative breast cancer cells. Cancer Res. **2001**, *61*: 7025-7029.

[300] Sharma, D., Blum, J., Yang, X., Beaulieu, N., Macleod, A.R., Davidson, N.E. Release of methyl cpg binding proteins and histone deacetylase 1 from the estrogen receptor {alpha} (er) promoter upon reactivation in er-negative human breast cancer cells. Mol. Endocrinol. **2005**, *19*: 1740-1751.

[301] Sharma, D., Saxena, N.K., Davidson, N.E., Vertino, P.M. Restoration of tamoxifen sensitivity in estrogen receptor-negative breast cancer cells: Tamoxifen-bound reactivated er recruits distinctive corepressor complexes. Cancer Res. **2006**, *66*: 6370-6378.

[302] Lam, E.W.-F., Shah, K., Brosens, J.J. The diversity of sex steroid action: The role of micro-rnas and foxo transcription factors in cycling endometrium and cancer. J. Endocrinol. **2012**, *212*: 13-25.

[303] Pan, Q., Luo, X., Chegini, N. Differential expression of micrornas in myometrium and leiomyomas and regulation by ovarian steroids. J. Cell. Mol. Med. **2008**, *12*: 227-240.

[304] Castellano, L., Giamas, G., Jacob, J., Coombes, R.C., Lucchesi, W., Thiruchelvam, P., *et al.* The estrogen receptor-alpha-induced microrna signature regulates itself and its transcriptional response. Proc Natl Acad Sci U S A **2009**, *106*: 15732-15737.

[305] Maillot, G., Lacroix-Triki, M., Pierredon, S., Gratadou, L., Schmidt, S., Benes, V., *et al.* Widespread estrogen-dependent repression of micrornas involved in breast tumor cell growth. Cancer Res **2009**, *69*: 8332-8340.

[306] Manavalan, T.T., Teng, Y., Appana, S.N., Datta, S., Kalbfleisch, T.S., Li, Y., *et al.* Differential expression of microrna expression in tamoxifen-sensitive mcf-7 *versus* tamoxifen-resistant ly2 human breast cancer cells. Cancer Lett. **2011**, *313*: 26-43.

[307] Wickramasinghe, N., Manavalan, T., Dougherty, S., Riggs, K., Li, Y., Klinge, C. Estradiol downregulates mir-21 expression and increases mir-21 target gene expression in mcf-7 breast cancer cells. Nucleic Acids Res **2009**, *37*: 2584-2595

[308] Bhat-Nakshatri, P., Wang, G., Collins, N.R., Thomson, M.J., Geistlinger, T.R., Carroll, J.S., *et al.* Estradiol-regulated micrornas control estradiol response in breast cancer cells. Nucleic Acids Res **2009**, *37*: 4850-4861.

[309] Tilghman, S.L., Bratton, M.R., Segar, H.C., Martin, E.C., Rhodes, L.V., Li, M., *et al.* Endocrine disruptor regulation of microrna expression in breast carcinoma cells. PLoS ONE **2012**, *7*: e32754.

[310] Kondo, N., Toyama, T., Sugiura, H., Fujii, Y., Yamashita, H. Mir-206 expression is down-regulated in estrogen receptor {alpha}-positive human breast cancer. Cancer Res. **2008**, *68*: 5004-5008.

[311] Zhao, J.-J., Lin, J., Yang, H., Kong, W., He, L., Ma, X., *et al.* Microrna-221/222 negatively regulates eralpha and associates with tamoxifen resistance in breast cancer. J. Biol. Chem. **2008**, *283*: 31079-31086.

[312] Cochrane, D.R., Cittelly, D.M., Howe, E.N., Spoelstra, N.S., McKinsey, E.L., LaPara, K., *et al.* Micrornas link estrogen receptor alpha status and dicer levels in breast cancer. Horm Cancer **2010**, *1*: 306-319.

[313] Yoshimoto, N., Toyama, T., Takahashi, S., Sugiura, H., Endo, Y., Iwasa, M., *et al.* Distinct expressions of micrornas that directly target estrogen receptor alpha in human breast cancer. Breast Cancer Res Treat **2011**, *130*: 331-339.

[314] Lu, Y., Roy, S., Nuovo, G., Ramaswamy, B., Miller, T., Shapiro, C., *et al.* Anti-mir-222 and -181b suppresses growth of tamoxifen resistant xenografts in mouse by targeting timp3 and modulating mitogenic signal. Journal of Biological Chemistry **2011**.

[315] Xiong, J., Yu, D., Wei, N., Fu, H., Cai, T., Huang, Y., *et al.* An estrogen receptor alpha suppressor, microrna-22, is downregulated in estrogen receptor alpha-positive human breast cancer cell lines and clinical samples. FEBS J **2010**.

[316] Pandey, D.P., Picard, D. Mir-22 inhibits estrogen signaling by directly targeting the estrogen receptor {alpha} mrna. Mol. Cell. Biol. **2009**, *29*: 3783-3790.

[317] Sun, M., Estrov, Z., Ji, Y., Coombes, K.R., Harris, D.H., Kurzrock, R. Curcumin (diferuloylmethane) alters the expression profiles of micrornas in human pancreatic cancer cells. Mol Cancer Ther **2008**, *7*: 464-473.

[318] Leivonen, S.K., Makela, R., Ostling, P., Kohonen, P., Haapa-Paananen, S., Kleivi, K., *et al.* Protein lysate microarray analysis to identify micrornas regulating estrogen receptor signaling in breast cancer cell lines. Oncogene **2009**, *28*: 3926-3936.

[319] Foley, N.H., Bray, I., Watters, K.M., Das, S., Bryan, K., Bernas, T., *et al.* Micrornas 10a and 10b are potent inducers of neuroblastoma cell differentiation through targeting of nuclear receptor corepressor 2. Cell Death Differ. **2011**, *18*: 1089-1098.

[320] Wu, J., Bao, J., Wang, L., Hu, Y., Xu, C. Microrna-184 downregulates nuclear receptor corepressor 2 in mouse spermatogenesis. BMC Dev. Biol. **2011**, *11*: 64.

[321] Bui-Nguyen, T.M., Pakala, S.B., Sirigiri, D.R., Martin, E., Murad, F., Kumar, R. Stimulation of inducible nitric oxide by hepatitis b virus transactivator protein hbx requires mta1 co-regulator. J. Biol. Chem. **2010**, *285*: 6980-6986.

[322] Jiang, A., Zhang, S., Li, Z., Liang, R., Ren, S., Li, J., *et al.* Mir-615-3p promotes the phagocytic capacity of splenic macrophages by targeting ligand-dependent nuclear receptor corepressor in cirrhosis-related portal hypertension. Exp. Biol. Med. **2011**, *236*: 672-680.

[323] Buchan, J.R., Parker, R. Molecular biology: The two faces of mirna. Science **2007**, *318*: 1877-1878.

[324] Cheng, C., Fu, X., Alves, P., Gerstein, M. Mrna expression profiles show differential regulatory effects of micrornas between estrogen receptor-positive and estrogen receptor-negative breast cancer. Genome Biol **2009**, *10*: R90.

[325] Zhao, J.J., Lin, J., Yang, H., Kong, W., He, L., Ma, X., *et al.* Microrna-221/222 negatively regulates estrogen receptor alpha and is associated with tamoxifen resistance in breast cancer. J Biol Chem **2008**, *283*: 31079-31086.

[326] Rao, X., Di Leva, G., Li, M., Fang, F., Devlin, C., Hartman-Frey, C., *et al.* Microrna-221/222 confers breast cancer fulvestrant resistance by regulating multiple signaling pathways. Oncogene **2011**, *30*: 1082-1097.

[327] Shimono, Y., Zabala, M., Cho, R.W., Lobo, N., Dalerba, P., Qian, D., *et al.* Downregulation of mirna-200c links breast cancer stem cells with normal stem cells. Cell **2009**, *138*: 592-603.

[328] Cittelly, D.M., Das, P.M., Salvo, V.A., Fonseca, J.P., Burow, M.E., Jones, F.E. Oncogenic her2{delta}16 suppresses mir-15a/16 and deregulates bcl-2 to promote endocrine resistance of breast tumors. Carcinogenesis **2010**.

[329] Cittelly, D.M., Das, P.M., Spoelstra, N.S., Edgerton, S.M., Richer, J.K., Thor, A.D., *et al.* Downregulation of mir-342 is associated with tamoxifen resistant breast tumors. Mol Cancer **2010**, *9*: 317.

[330] Bergamaschi, A., Katzenellenbogen, B.S. Tamoxifen downregulation of mir-451 increases 14-3-3zeta and promotes breast cancer cell survival and endocrine resistance. Oncogene **2011**.

[331] Fowler, A.M., Santen, R.J., Allred, D.C. "Dwarf" estrogen receptor in breast cancer and resistance to tamoxifen. J. Clin. Oncol. **2009**, *27*: 3413-3415.

[332] Zhao, Y., Deng, C., Lu, W., Xiao, J., Ma, D., Guo, M., *et al.* Let-7 micrornas induce tamoxifen sensitivity by downregulation of estrogen receptor alpha signaling in breast cancer. Mol. Med. **2011**, *17*: 1233-1241.

[333] Manavalan, T.T., Teng, Y., Litchfield, L.M., Muluhngwi, P., Al-Rayyan, N., Klinge, C.M. Reduced expression of mir-200 family members contributes to antiestrogen resistance in ly2 human breast cancer cells. PLoS ONE **2013**, *8*: e62334.

[334] Klinge, C.M., Wickramasinghe, N.S., Ivanova, M.M., Dougherty, S.M. Resveratrol stimulates nitric oxide production by increasing estrogen receptor {alpha}-src-caveolin-1 interaction and phosphorylation in human umbilical vein endothelial cells. FASEB J. **2008**, *22*: 2185-2197.

[335] Chambliss, K.L., Simon, L., Yuhanna, I.S., Mineo, C., Shaul, P.W. Dissecting the basis of nongenomic activation of endothelial nitric oxide synthase by estradiol: Role of eralpha domains with known nuclear functions. Mol. Endocrinol. **2005**, *19*: 277-289.

[336] Klinge, C.M. Estrogenic control of mitochondrial function and biogenesis. J. Cell. Biochem. **2008**, *105*: 1342-1351.

[337] Chen, J.-Q., Russo, P.A., Cooke, C., Russo, I.H., Russo, J. Er[beta] shifts from mitochondria to nucleus during estrogen-induced neoplastic transformation of human breast epithelial cells and is involved in estrogen-induced synthesis of mitochondrial respiratory chain proteins. Biochimica et Biophysica Acta (BBA) - Molecular Cell Research **2007**, *1773*: 1732-1746.

[338] Mertens-Talcott, S.U., Chintharlapalli, S., Li, X., Safe, S. The oncogenic microrna-27a targets genes that regulate specificity protein transcription factors and the g2-m checkpoint in mda-mb-231 breast cancer cells. Cancer Res. **2007**, *67*: 11001-11011.

[339] Yu, F., Yao, H., Zhu, P., Zhang, X., Pan, Q., Gong, C., *et al.* Let-7 regulates self renewal and tumorigenicity of breast cancer cells. Cell **2007**, *131*: 1109-1123.

[340] Tsang, W.P., Kwok, T.T. Let-7a microrna suppresses therapeutics-induced cancer cell death by targeting caspase-3. Apoptosis **2008**, *13*: 1215-1222.

[341] Zhao, Y., Deng, C., Wang, J., Xiao, J., Gatalica, Z., Recker, R.R., *et al.* Let-7 family mirnas regulate estrogen receptor alpha signaling in estrogen receptor positive breast cancer. Breast Cancer Res. Treat. **2011**, *127*: 69-80.

[342] Ovcharenko, D., Kelnar, K., Johnson, C., Leng, N., Brown, D. Genome-scale microrna and small interfering rna screens identify small rna modulators of trail-induced apoptosis pathway. Cancer Res. **2007**, *67*: 10782-10788.

[343] Volinia, S., Galasso, M., Sana, M.E., Wise, T.F., Palatini, J., Huebner, K., *et al.* Breast cancer signatures for invasiveness and prognosis defined by deep sequencing of microrna. Proceedings of the National Academy of Sciences **2012**.

[344] Garzon, R., Pichiorri, F., Palumbo, T., Visentini, M., Aqeilan, R., Cimmino, A., *et al.* Microrna gene expression during retinoic acid-induced differentiation of human acute promyelocytic leukemia. Oncogene **2007**, *26*: 4148-4157.

[345] Ivanovska, I., Ball, A.S., Diaz, R.L., Magnus, J.F., Kibukawa, M., Schelter, J.M., *et al.* Micrornas in the mir-106b family regulate p21/cdkn1a and promote cell cycle progression. Mol. Cell. Biol. **2008**, *28*: 2167-2174.

[346] Calin, G.A., Croce, C.M. Microrna-cancer connection: The beginning of a new tale. Cancer Res. **2006**, *66*: 7390-7394.

[347] Cimmino, A., Calin, G.A., Fabbri, M., Iorio, M.V., Ferracin, M., Shimizu, M., *et al.* Mir-15 and mir-16 induce apoptosis by targeting bcl2. Proceedings of the National Academy of Sciences **2005**, *102*: 13944-13949.

[348] Finnerty, J.R., Wang, W.-X., Hébert, S.S., Wilfred, B.R., Mao, G., Nelson, P.T. The mir-15/107 group of microrna genes: Evolutionary biology, cellular functions, and roles in human diseases. J. Mol. Biol. **2010**, *402*: 491-509.

[349] Sun, L., Yao, Y., Liu, B., Lin, Z., Lin, L., Yang, M., *et al.* Mir-200b and mir-15b regulate chemotherapy-induced epithelial-mesenchymal transition in human tongue cancer cells by targeting bmi1. Oncogene **2012**, *31*: 432-445.

[350] Linsley, P.S., Schelter, J., Burchard, J., Kibukawa, M., Martin, M.M., Bartz, S.R., *et al.* Transcripts targeted by the microrna-16 family cooperatively regulate cell cycle progression. Mol. Cell. Biol. **2007**, *27*: 2240-2252.

[351] Liu, Q., Fu, H., Sun, F., Zhang, H., Tie, Y., Zhu, J., *et al.* Mir-16 family induces cell cycle arrest by regulating multiple cell cycle genes. Nucl. Acids Res. **2008**, *36*: 5391-5404.

[352] Mihelich, B.L., Khramtsova, E.A., Arva, N., Vaishnav, A., Johnson, D.N., Giangreco, A.A., *et al.* Mir-183-96-182 cluster is overexpressed in prostate tissue and regulates zinc homeostasis in prostate cells. J. Biol. Chem. **2011**, *286*: 44503-44511.

[353] Kulshreshtha, R., Ferracin, M., Wojcik, S.E., Garzon, R., Alder, H., Agosto-Perez, F.J., *et al.* A microrna signature of hypoxia. Mol. Cell. Biol. **2007**, *27*: 1859-1867.

[354] Vrba, L., Garbe, J.C., Stampfer, M.R., Futscher, B.W. Epigenetic regulation of normal human mammary cell type–specific mirnas. Genome Res. **2011**, *21*: 2026-2037.

[355] Howe, E., Cochrane, D., Richer, J. Targets of mir-200c mediate suppression of cell motility and anoikis resistance. Breast Cancer Research **2011**, *13*: R45.

[356] Pan, Q., Luo, X., Toloubeydokhti, T., Chegini, N. The expression profile of micro-rna in endometrium and endometriosis and the influence of ovarian steroids on their expression. Mol. Hum. Reprod. **2007**, *13*: 797-806.

[357] Cloonan, N., Brown, M.K., Steptoe, A.L., Wani, S., Chan, W.L., Forrest, A.R., *et al.* The mir-17-5p microrna is a key regulator of the g1/s phase cell cycle transition. Genome Biol **2008**, *9*: R127.

[358] Farazi, T.A., Horlings, H.M., ten Hoeve, J.J., Mihailovic, A., Halfwerk, H., Morozov, P., *et al.* Microrna sequence and expression analysis in breast tumors by deep sequencing. Cancer Res. **2011**, *71*: 4443-4453.

[359] Yan, L.-X., Huang, X.-F., Shao, Q., Huang, M.-Y., Deng, L., Wu, Q.-L., *et al.* Microrna mir-21 overexpression in human breast cancer is associated with advanced clinical stage, lymph node metastasis and patient poor prognosis. RNA **2008**, *14*: 2348-2360.

[360] Bronisz, A., Godlewski, J., Wallace, J.A., Merchant, A.S., Nowicki, M.O., Mathsyaraja, H., *et al.* Reprogramming of the tumour microenvironment by stromal pten-regulated mir-320. Nat Cell Biol **2012**, *14*: 159-167.

[361] Gao, W., Shen, H., Liu, L., Xu, J., Xu, J., Shu, Y. Mir-21 overexpression in human primary squamous cell lung carcinoma is associated with poor patient prognosis. J. Cancer Res. Clin. Oncol. **2011**, *137*: 557-566.

[362] Wang, C.H., Lee, D.Y., Deng, Z., Jeyapalan, Z., Lee, S.C., Kahai, S., *et al.* Microrna mir-328 regulates zonation morphogenesis by targeting cd44 expression. PLoS ONE **2008**, *3*: e2420.

[363] Li, X., Pan, Y.-Z., Seigel, G.M., Hu, Z.-H., Huang, M., Yu, A.-M. Breast cancer resistance protein bcrp/abcg2 regulatory micrornas (hsa-mir-328, -519c and -520h) and their

differential expression in stem-like abcg2+ cancer cells. Biochem. Pharmacol. **2011**, *81*: 783-792.

[364] Lin, J., Huang, S., Wu, S., Ding, J., Zhao, Y., Liang, L., *et al.* Microrna-423 promotes cell growth and regulates g1/s transition by targeting p21cip1/waf1 in hepatocellular carcinoma. Carcinogenesis **2011**, *32*: 1641-1647.

[365] Kong, D., Piao, Y.S., Yamashita, S., Oshima, H., Oguma, K., Fushida, S., *et al.* Inflammation-induced repression of tumor suppressor mir-7 in gastric tumor cells. Oncogene **2011**.

Send Orders for Reprints to reprints@benthamscince.net
Advances in Genome Science, Vol. 3, **2014, 83-126**

CHAPTER 3

Update on Basic and Applied Aspects of Genome Analysis of Lactic Acid-Producing Bacteria, Including Bifidobacteria

Baltasar Mayo[1,*], Douwe van Sinderen[2] and Marco Ventura[3]

[1]Department of Microbiology and Biochemistry, Dairy Research Institute of Asturias Principality (IPLA-CSIC), Villaviciosa, Asturias, Spain; [2]Alimentary Pharmabiotic Centre and Department of Microbiology, University College Cork, Cork, Ireland and [3]Laboratory of Probiogenomics, Department of Genetics, Biology of Microorganisms, Anthropology and Evolution, University of Parma, Italy

Abstract: The rapidly increasing number of genome sequences provides an important resource to study the genetics, physiology and biochemical capabilities of lactic acid bacteria (LAB) and bifidobacteria. Despite the fact that they are phylogenitally unrelated, bifidobacteria are usually included under the term of LAB, as they are found within the same natural environments and are frequently used for the same purpose, *i.e.* as probiotics to exert health-promoting effects on the gastrointestinal and/or genitourinary tracts of animals and humans. The most updated sequence information (GOLD genome online database; http://www.genomesonline.org) summarizes 280 completed LAB genomes (14 *Lactococcus*, 165 *Lactobacillus*, 18 *Leuconostoc*, 6 *Streptococcus thermophilus*, 13 *Oenococcus*, 6 *Pediococcus*, 4 *Weissella*, and 54 *Bifidobacterium*) and more than 500 in progress. The typical LAB genome is relatively small (from 1.7 Mb in the case of *S. thermophilus* to 3.9 Mb for *Lactobacillus pentosus*), thus harboring a limited assortment of genes (from around 1,600 to over 3,000). These small genomes code for a broad array of transporters for efficient carbon and nitrogen assimilation from the nutritionally-rich niches they usually inhabit, and specify a rather limited range of biosynthetic and catabolic capabilities. The variation in the number of genes even within strains of the same species suggests that the genome of LAB is rather plastic and dynamic. The genome diversification of LAB from their ancestors is thought to be driven by niche adaptation through extensive gene loss, gene duplication, and acquisition of key functions *via* lateral gene transfer. The availability of genome sequences is expected to revolutionize the exploitation of the metabolic potential of LAB, improving their use in bioprocessing and their utilization in biotechnological and health-related applications.

Keywords: Adjunct cultures, bifidobacteria, *Bifidobacterium, Bifidobacterium.*

***Address correspondence to Baltasar Mayo:** Department of Microbiology and Biochemistry, Dairy Research Institute of Asturias Principality (IPLA-CSIC), Villaviciosa, Asturias, Spain; Tel: +34 985 89 21 31; Fax: +34 985 89 22 33; E-mail: baltasar.mayo@ipla.csic.es

Christian Neri (Ed)

animalis, Bifidobacterium lactis, *Bifidobacterium longum,* cheese, dairy cultures, dairy microbiology, dairy products, fermented dairy products, gastrointestinal tract, genome analysis, genome sequencing, genomics, intestinal microbiology, LAB, lactic acid bacteria, lactobacilli, *Lactobacillus, Lactobacillus delbrueckii, Lactobacillus helveticus, Lactobacillus plantarum,* lactococci, *Lactococcus, Lactococcus lactis, Leuconostoc,* probiotics, starters, *Streptococcus thermophilus,* yogurt.

INTRODUCTION

Lactic acid bacteria (LAB) encompass a heterogeneous group of microorganisms, which have as a common metabolic property the production of lactic acid from the fermentation of carbohydrates [1]. LAB are Gram positive, non-sporulating and acid tolerant bacteria, belonging to the phylum *Firmicutes*. The most typical LAB members are included within the genera *Lactobacillus*, *Lactococcus*, *Leuconostoc, Pediococcus, Weissella,* and *Oenococcus*, and the species *Streptococcus thermophilus*. From a biochemical perspective LAB include both homofermenters, which mainly produce lactic acid, and heterofermenters, which, apart from lactic acid, yield a variety of fermentation products, such as acetic acid, ethanol, carbon dioxide and formic acid [2]. The ecological distribution of LAB is vast; they are found in a large variety of environments, including milk and dairy products, vegetable and plants, cereals and meat. Many LAB species are used for the manufacture and preservation of fermented feed and foods from raw agricultural materials in which they are either present as contaminants or deliberately added as starters in order to specifically perform the fermentation process. Development of these bacteria contributes to the final organoleptic, rheological and nutritional properties of fermented products [3]. Members of the genus *Bifidobacterium* are also Gram-positive, non-sporulating bacteria which produce lactic and acetic acid as their major fermentation end-products [4]. However, taxonomically speaking they are members of the phylum *Actinobacteria*, but will be discussed as part of this review as they are frequently found within the same natural ecosystems and used in association with LAB species in many fermented dairy products. For the purpose of this review, phylogenetically 'true' LAB and bifidobacteria will be included under the general term of LAB. Various LAB species are also commonly found among the resident microbiota of the gastrointestinal tract (GIT) and genitourinary tract (GUT) of human and animals [5, 6]. In these environments they are considered to be beneficial components of the microbiota, playing a large variety of health-promoting functions, such as immunomodulation, intestinal integrity and pathogen resistance [4, 7]. For this reason, strains of certain species (the majority belonging to the *Bifidobacterium* and *Lactobacillus* genera) have traditionally been used as probiotics, and added as functional components to various food products [3, 4]. Thus, the commercial exploitation of LAB species as starter and as probiotic cultures is economically very significant. Consequently, over the last 30 years research on genetics, physiology and applications of LAB has enjoyed explosive growth [8, 9].

One particular area that has undergone very rapid progress over the last two decades is the characterization of LAB genomes. The availability of these genome sequences has improved our understanding of the fermentation pathways and biochemical routes involved in industrial and probiotic applications. This knowledge will ultimately allow a full exploitation of their fermentative capabilities, facilitating, at the same time, the genetic manipulation of these bacteria, which would lead to the use of LAB species as a cell factory for new biotechnological applications, such as expression of heterologous proteins, synthesis of food-grade additives and nutraceuticals, or as vaccine and therapeutic delivery systems [10-13].

GENERAL FEATURES OF LAB GENOMES

At the time of writing (October, 2013), nearly 300 complete genome sequences of species of LAB strains are publicly available, while more than 500 are at various stages of completion (http://www.genomesonline.org). These bacteria have relatively small genomes (average genome size of around 2 Mb, with a coding capacity of approximately 2,000 genes), with the number of genes found in a given genome ranging from 1,600 to 3,200 [14, 15]. With punctual exceptions, all genomes display architectural features of a typical bacterial chromosome; *i.e.*, co-orientation between gene transcription and DNA replication and an asymmetric bias in nucleotide composition of leading and lagging DNA strands.

Only a few genetic traits appear to be universally conserved among the different LAB genomes, including enzymes involved in glycolysis in 'true' LAB or those of the bifido's shunt in bifidobacteria [15, 16]. Different genetic events (*i.e.*, mutation, gene duplication, horizontal gene transfer (HGT), gene decay, gene loss and genome rearrangements) have all been considered to contribute to the present genome shape and structure of LAB species. In fact, adaptation to nutritionally rich environments (*e.g.*, milk, plant material, human and animal GIT) has promoted progressive gene decay but also acquisition of genes involved in key traits through HGT [17, 18]. Evidence for genome decay has been observed for most sequenced LAB species, particularly in genes involved in protein and carbohydrate metabolism, uptake and utilization. Notably, genome simplification and decay of anabolic and catabolic routes was particularly obvious from the genome analysis of the yoghurt-associated, cooperative bacteria *Streptococcus thermophilus* [19] and *Lactobacillus delbrueckii* subsp. *bulgaricus* (*Lactobacillus bulgaricus*) [20], as well as in the cheese starter culture *Lactobacillus helveticus* [17, 21]. As compared to their closest relatives, around 10-12% of the coding genes in these aforementioned bacteria appeared to be present as pseudogenes; *i.e.* non-functional genes, due to frame shifts, non-sense mutations, deletions or truncations.

In the following sections, a short description of the principal features of the genome sequences of a number of selected LAB species among those frequently used as starters and adjunct cultures in food fermentations and/or as probiotics in functional ingredients is presented. Some basic and applied aspects gathered as a result of the analysis of these genomes, in particular those of industrial, probiotic and/or biotechnological interest will also be discussed.

Bifidobacteria

Bifidobacteria are Gram-positive prokaryotes that naturally colonize the human and animal GIT. Bifidobacteria are among the first colonizers of the intestine of neonates and are believed to represent the largest fraction of the infant microbiota, contributing to the maturation of the developing immune system [22]. Although not numerically dominant in adults, they are considered key commensals in promoting health and well-being [6]. This is the reason for their extensive use as probiotics. Among the approximately 30 recognized species of bifidobacteria (Fig. **1**) [23], complete genome sequences are now (March, 2012) available for 18 species

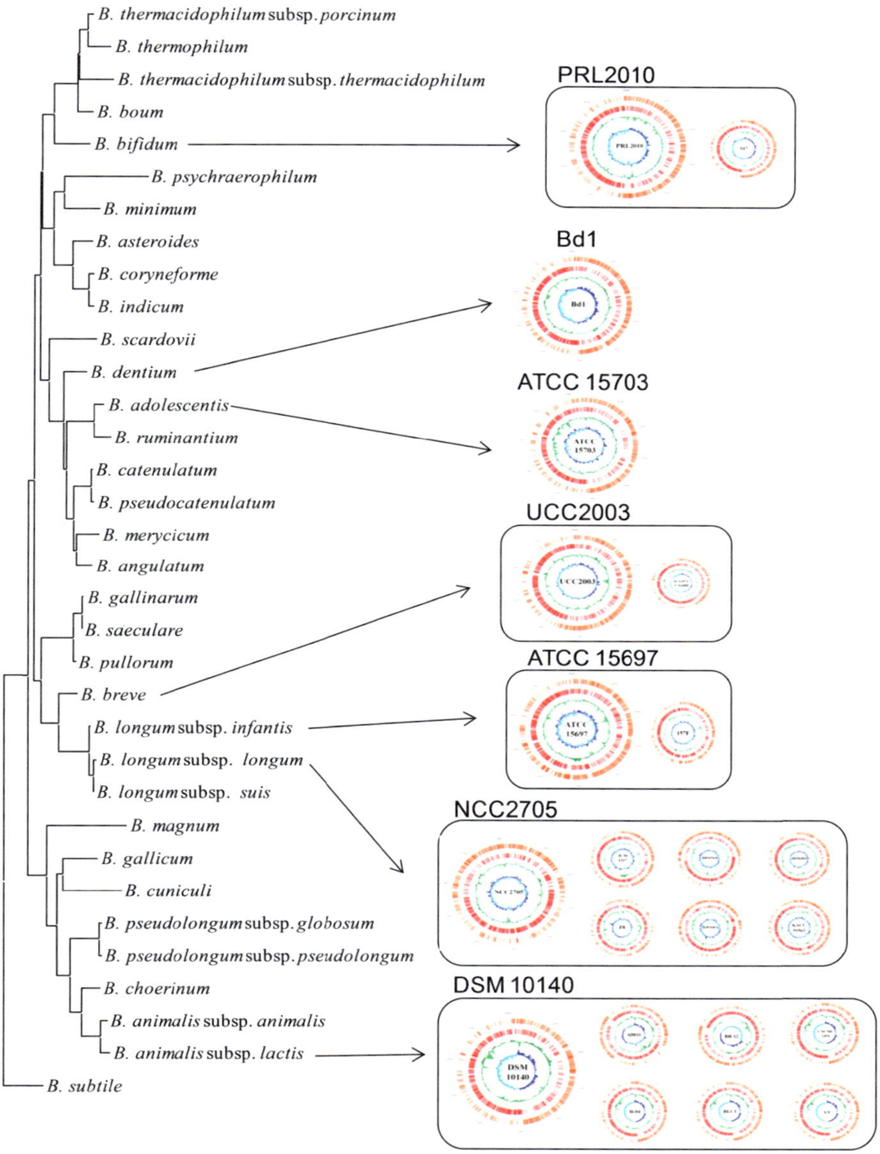

Figure 1: Bifidobacterial species and strains whose genome sequences have been fully decoded. On the left, image of the current bifidobacteria phylogenetic lineages delineating the six different groups of bifidobacteria species based on 16S rRNA gene sequences (*B. longum, B. pullorum, B. adolescentis, B. asteroides, B. boum,* and *B. pseudolongum*). On the right, schematic genome atlas of the various bifidobacterial genome sequences. Codes of the strains of reference genomes are highlighted.

(http://www.genomesonly.org). Completely sequenced strains belong to *Bifidobacterium adolescentis* (1), *Bifidobacterium animalis* subsp. *lactis* (6), *Bifidobacterium bifidum* (2), *Bifidobacterium breve* (2), *Bifidobacterium dentium* (1), *Bifidobacterium longum* subsp. *longum* (6), and *B. longum* subsp. *infantis* (4) (Fig. **1**). Colonization of the human GIT at birth has been related to their ability to grow in human milk oligosaccharides (HMOs), glycans and mucin through an arsenal of fucosidases, sialidases, β-galactosidases, and N-acetyl-β-hexosaminidases. This property has recently been investigated by *in silico* genomic analysis and by proteomics [24, 25].

Bifidobacterium longum subsp. *longum* NCC2705

The 2.26-Mb genome of NCC2705 strain is organized as a 60%-GC circular chromosome, containing four *rrn* operons, 57 tRNAs, 16 intact insertion sequence (IS) elements, integrated plasmid sequences [18], and a prophage-like element [26]. The genome is predicted to encode 1,727 proteins. Several physiological traits that may explain the successful adaptation of this bacterium to the environment of the colon have been found. In particular, an unexpectedly large amount of coding capacity (>8% of the genome) is dedicated to the transport (mainly ABC-type transporters) and catabolism (glycosyl hydrolases) of (mono-, oligo- and poly-) saccharides, features shared with other colonic inhabitants, such as *Enterococcus faecium* and *Bacteroides fragilis* [27]. Many of these genes are present in seemingly self-regulated modules that appear to have arisen from gene duplication or horizontal acquisition, suggesting that *B. longum* is under strong selective pressure to acquire catabolic diversity in order to successfully compete for nutritients in the GIT ecosystem [18, 24]. Complete pathways for most amino acids, all nucleotides, and some key vitamins were identified; however, routes for the biosynthesis of aspartic acid and cysteine were atypical. More importantly, genome analysis provided insights into the reciprocal interactions of bifidobacteria with their hosts. Polypeptides showing homology to proteins needed for the production of glycoprotein-binding fimbriae, which are structures that may be important for adhesion and persistence in the GIT, have been identified. Furthermore, a eukaryotic-type serine protease inhibitor (serpin) that may be involved in the reported immunomodulatory activity of bifidobacteria has been identified and characterized [28].

Bifidobacterium breve UCC2003

The *B. breve* UCC2003 genome is one of the larger bifidobacterial genomes so far sequenced (2,422,668 bps) [29]. The circular chromosome of *B. breve* UCC2003 contains 1,985 predicted genes and two *rrn* operons, 54 tRNAs, 26 intact insertion sequence (IS) elements, and a single prophage-like element and one CRISPR (clustered regulatory interspaced short palindromic repeats) [26, 29]. Similar to *B. longum* subsp. *longum* NCC2705, a significant proportion of the genome of *B. breve* UCC2003 encodes enzymes involved in carbohydrate metabolism including 40 glycosyl hydrolases whose assumed substrates represent a wide range of oligo- and poly-saccharides. Several of these glycosyl hydrolases are assumed to exert their activity outside the cytoplasm, such as an amylopullulanase, which allows growth of UCC2003 on starch and related sugar polymers [30]. The *B. breve* UCC2003 genome also contains a *fos* operon, which encodes a putative permease, a conserved hypothetical protein, and a β-fructofuranosidase. Transcriptional analysis of this operon in *B. breve* grown in the presence of different carbohydrate sources revealed its involvement in the breakdown of short-chain fructo-oligosaccharides (FOS) [30]. Furthermore, analysis of the *B. breve* UCC2003 genome showed that this organism encodes a relatively small number of phosphoenolpyruvate phosphotransferase systems (PEP-PTS), which are typically involved in the internalization and metabolism of monosaccharides [31].

The genome of UCC2003, but also all other examined bifidobacterial genomes, was shown to encode a novel type of pilus, *i.e.*, the type IVb tight adherence (*tad*) pilus-encoding gene cluster, which has been shown to be essential for efficient *in vivo* murine gut colonization [29] (see below).

Bifidobacterium longum subsp. *infantis* ATCC15697

This microorganism possesses the largest genome (2,832,748 bp) so far described within the genus *Bifidobacterium* [32]. It contains four *rrn* operons, 79 tRNAs and five prophage sequences [26, 32]. The chromosome of *B. longum* subsp. *infantis* ATCC 15697 displays properties sustaining a specific adaptation to the infant gut ecological niche, which consists in the ability of this strain to consume specific human milk carbohydrates known as Human Milk Oligosaccharides (HMO). In particular, the *B. longum* subsp. *infantis* ATCC 15697 genome contains a gene cluster of 43 Kb encoding various glycosyl hydrolases and carbohydrate transporters necessary for importing and metabolizing HMOs [32]. This 43 Kb large gene cluster specifies a variety of catabolic enzymes, such as fucosidase, sialidase, β-hexosaminidase and β-galactosidase activities, as well as extracellular solute binding proteins and permeases predicted to be active on HMOs [32]. In addition, the genome of ATCC15697 contains additional genetic loci specifying fucosidases and sialidases, as well as a complete urease operon, predicted to be involved in the utilization of urea, which represents an important nitrogen source of milk.

Bifidobacterium bifidum PRL2010

Recently the genome of another bifidobacterial strain originally isolated from infant stool, *Bifidobacterium bifidum* PRL2010, has been fully decoded [33]. The analysis of the PRL2010 2,214,650 bp chromosome revealed a competitive nutrient-utilization strategy that specifically targets host-derived glycans [33, 34]. In fact, classification according to the Carbohydrate Active Enzymes (CAZy) system demonstrated that the PRL2010 genome specifies 77 carbohydrate-active genes including glycosyl hydrolases (GH), glycosyl transferases (GT) and glycosyl esterases (CE), which are distributed among 20 GH families, five GT, three CE families and 5 carbohydrate-binding module (CBM) families [33]. Almost 60% of the identified GH enzymes from PRL2010 are associated with the degradation of mucin-derived oligosaccharides, most of which are uniquely present in the *B. bifidum* chromosome and absent in the other bifidobacterial genomes so far sequenced. Furthermore, the genome sequences of this microorganism revealed the presence of genetic elements such as sortase-dependent pili that might play a role in the colonization and/or microbe-host communication of this microorganism within the infant gut [35].

Bifidobacterium dentium Bd1

Bifidobacteria are generally considered as “safe microorganisms”. However, in the *Bifidobacterium* genus, *Bifidobacterium dentium* has been detected in high numbers in infected dentine from carious lesions in children and thus has been considered as an opportunistic pathogen [36]. The genome of *B. dentium* Bd1, a strain isolated from dental caries, was sequenced to completion to reveal a single circular 2,636,368 bp chromosome with 2143 predicted open reading frames. The chromosome encompasses four *rrn* operons, 55 tRNAs and two prophage sequences [37, 38]. *In silico* analyses of the genome sequence revealed multiple ways in which Bd1 has adapted to the oral environment through specialized nutrient acquisition, defences against antimicrobials and gene products that increase fitness and competitiveness within the oral niche [37]. *B. dentium* Bd1 was shown to degrade a wide variety of carbohydrates, consistent with genome-based predictions. Furthermore, colonization and persistence factors implicated in tissue adhesion, acid tolerance, and the metabolism of human saliva-derived compounds were also identified through the dissection of the genome sequences of *B. dentium* Bd1 [37].

Bifidobacterium animalis subsp. *lactis* DSM10140

This microorganism is widely exploited by food industry as a probiotic bacterium. Notably, genome sequencing of *B. animalis* subsp. *lactis* DSM10140 revealed a relative small chromosome size of this strain, *i.e.*, 1,938,483 bp [39], including 2143 genes and two *rrn* operons, 52 tRNAs, and a single prophage remnant [38, 39]. Such findings may indicate that this genome has undergone evolution through genome decay, which might have occurred as a consequence of its intensive application in industrial processes (*e.g.*, production of large microbial biomasses in fermentation processes). Notably, the genes involved in the catabolism of human milk oligosaccharides in *B. longum* subsp. infantis [31] and in the degradation and utilization of mucin in *B. bifidum* PRL2010 [32] were not identified in the chromosome of *B. animalis* subsp. *lactis* DSM10140 [38]. In addition, the simplification process to which the chromosome of *B. animalis* subsp. *lactis* DSM10140 has been subjected may be responsible for the significant reduction of sugar transporters compared to other bifidobacterial genomes, including the complete absence of PTS systems in the chromosome of DSM10140.

The genome sequences of various other strains belonging to the *B. animalis* subsp. *lactis* species have been fully decoded [40, 41]. The high degree of genomic conservation in terms of size, organization, and sequence noticed between these *B. animalis* subsp. *lactis* strains is indicative of a genomically monomorphic subspecies.

Lactococci

Lactococci are nonpathogenic, AT-rich, Gram-positive LAB commonly dominant in natural niches such as spontaneous milk fermentations, on cattle, and on plant material. *Lactococcus lactis* strains, particularly representatives of the *lactis* and *cremoris* subspecies, are the main components of starters used in the economically important fermentation of milk into cheese, contributing to acidification, prevention of undesirable bacterial growth, and flavor formation through their proteolytic and amino acid conversion pathways [3]. *L. lactis* is probably the best characterized LAB species and, for this reason, has become a paradigm for fundamental physiological and genetic research. Lactococci have recently been targeted for novel biotechnological applications, such as expression of heterologous proteins, synthesis of food-grade additives and nutraceuticals [42], and in vaccine delivery [43]. At present, the complete sequences of nine *L. lactis* strains are publicly available. Among others, those of the plasmid-cured strains *L. lactis* subsp. *lactis* IL1403 [44] and *L. lactis* subsp. *cremoris* MG1363 [45], and the plasmid-containing starter strain *L. lactis* subsp. *cremoris* SK11 [15]. Recently, the complete genome sequence of a non-dairy strain, *L. lactis* subsp. *lactis* KF147, isolated from mung bean sprouts has also been reported [46].

Lactococcus lactis IL1403, MG1363 and SK11 Genomes

The genomes of the *L. lactis* strains range from 2.3 to 2.6-Mb in size, encoding 2,300 to 2,500 proteins, 6 *rrn* operons and 62 tRNA genes. A variable number of prophages (4 to 6) were encountered in these strains, as well as an extremely rich panoply of IS elements (ranging in number from 43 to 130). These elements do not appear to be distributed at random. In fact, the non-random distribution of IS in *L. lactis* IL1403 suggests that its chromosome is the product of a recent recombination event between two closely related genomes [44]. In contrast, the concentration of one-fifth of the 71 IS elements in a specific 56-kb region in the *L. lactis* MG1363 strain was interpreted as an integration hotspot region [45]. This integration hotspot carries genes that are typically associated with lactococcal plasmids and a repeat sequence specifically found on *L. lactis* plasmids and in the so-called "lateral gene transfer hotspot" of the *S. thermophilus* genome [19]. The chromosomes of MG1363, IL1403 and SK11 show extensive gene synteny, and even more so if a large chromosomal inversion previously described in strain

MG1363 [47] is ignored. The parent strain of MG1363 was subjected to a prophage curing strategy [48], but it still carries four remnant/satellite phages and two apparently complete prophages [45]. A complete set of late competence genes was found in all sequenced strains, including a gene equivalent to *comX* of *Streptococcus pneumoniae*, which encodes the ECF-type σ-factor necessary for transcription of competence genes [45]. This suggests that this species possesses the ability to develop natural competence for transformation, provided that appropriate physiological conditions are met. Surprisingly, the functions necessary for aerobic respiration were also found to be encoded on the genome. Thus, *L. lactis* may carry out oxidative phosphorylation if exogenous protoporphyrinogen is provided. This capability, which has been analyzed in detail [49, 50], is the basis for a completely new industrial process of producing lactic starters [51].

Interest in non-dairy *L. lactis* is increasing due to the search for unique flavor-forming capabilities and production of novel broad-range antimicrobials [52]. Complete and low-coverage genome sequencing of two *L. lactis* strains isolated from plants has allowed a comparison of their genomes to those of dairy strains [4, 53]. Adaptation to grow on plant substrates was evident from the presence of gene sets for the uptake and degradation of complex plant polymers such as xylan, arabinan, glucans, fructans, but also for the uptake and utilization of typical plant cell wall components such as α-galactosides, β-glucosides, arabinose, xylose, galacturonate, glucuronate and gluconate [53].

Lactobacilli

Lactobacilli are a broad, morphologically-defined group of LAB characterized by the formation of lactic acid as a sole or main end product of carbohydrate metabolism. The lactobacilli are Gram-positive, non-spore-forming rods or coccobacilli with a GC content usually below 50 mol %. The nutritional requirements of lactobacilli are a reflection of their various habitats, which are typically rich in carbohydrate- and protein-containing substrates. They are found on plants, dairy and meat products, and as components of the microbiota of the animal and human GIT and GUT.

Lactobacilli are important starters and adjunct cultures in the production of foods that require lactic acid fermentation, notably dairy products (yogurt and cheese), fermented vegetables (olives, pickles, and sauerkraut), fermented meats (salami, sausages), and sourdough bread and other cereal-based food commodities [3]. Although less numerous than bifidobacteria, lactobacilli inhabiting the GIT of animals are thought to exert pivotal roles in the establishment and maintenance of a properly functioning GIT [7], and a number of strains have been used as probiotics for more than 75 years [54]. Beneficial effects attributed to probiotic lactobacilli include colonization of intestinal and genital mucosa [55, 56], inhibition of pathogens [57, 58], immunomodulation [59] and cholesterol assimilation [60]. It is therefore not surprising that most of the 39 genomes from lactobacilli that have so far been sequenced are of human intestinal origin.

Lactobacillus acidophilus NCFM

L. acidophilus NCFM is a probiotic human isolate surviving GIT passage, which has been produced commercially since 1972 [61]. The complete genome is 1,993,564 bp and devoid of plasmids. The average GC content is 34.71% with 1,864 predicted ORFs, of which 72.5% have been functionally classified [62]. Four *rrn* operons were found on the NCFM genome sequence and 61 tRNA genes. Nine phage-related integrases were also predicted, but no complete prophages were found. However, three unique regions designated as potential autonomous units (PAUs) were identified. These units resemble a singular structure, bearing characteristics of both plasmids and phages. Analysis of the three PAUs revealed the presence of two R/M systems and a killer protein of a possible prophage-maintenance system. A Clustered, Regularly Interspaced Short Palindromic Repeat (CRISPR) locus containing 32 nearly perfect 29-bp repeats was discovered. Similar DNA spacers have been found in more than different 40 microorganisms including *E. coli* [63]. These have been shown to be involved in resistance to bacteriophages. Furthermore, the chromosomal locus for lactacin B, a class II bacteriocin previously isolated and characterized from this strain [64], was identified. *In silico* analysis of the genome of the NCFM strain indicated that this strain has the potential to synthesize just three amino acids (cysteine, serine and aspartate). Congruently, the genome is predicted to encode an array of ABC-transporters (nine) for amino acids and oligopeptides, separate di- and oligopeptide transport systems, and many (*i.e.* 22) amino acid permeases [62]. For protein degradation and peptide utilization, the organism is predicted to encode 20 putative peptidases/proteases, including homologs of PrtP and PrtM, a proteinase system for extracellular casein degradation in *L. lactis*. Various gene clusters coding for the metabolism of a variety of carbohydrates, including FOS and raffinose, are present on the NCFM chromosome, often accompanied by transcriptional regulator-encoding genes belonging to the *lacI* family. Furthermore, several genes that specify mucus- and fibronectin-binding proteins, presumably implicated in adhesion to human intestinal cells, were also identified. These features are likely to contribute to the organisms' gastric survival, promoting interactions with the intestinal mucosa.

Lactobacillus delbrueckii subsp. *bulgaricus* ATTC 11842

L. bulgaricus is a representative of the LAB species used worldwide for yogurt production in association with *S. thermophilus*. The size of the *L. bulgaricus* genome is around 1.8 Mb with an overall GC content of 49.7% [20]. An intriguing feature of the replication terminus region of the ATTC 11842 genome is the presence of a 47.5-kbp inverted repeat, representing an extremely rare structure in bacterial genomes. In contrast to many other LAB species, the *L. bulgaricus* genome does not contain any prophage. A relatively high number of *rrn* operons (9) and tRNA genes (95) were identified, suggesting that the genome has undergone a recent size reduction. This is further supported by the presence of a substantial number of pseudogenes (270 in total), various incomplete metabolic pathways, and relatively few regulatory components [20]. A much higher GC content at codon position three than expected on the basis of the overall GC content suggested that the composition of the genome is evolving towards a higher GC content. This may be the result of its adaptation to milk from a plant-associated habitat, as suggested by the presence of complete and incomplete PTS and other sugar transport systems and hydrolytic enzymes. In protocooperation with *S. thermophilus,* the loss of superfluous functions might be advantageous in the stable protein and lactose-rich milk environment.

Lactobacillus helveticus DPC 4571

L. helveticus belongs to the *L. acidophilus-L. delbrueckii* group and is frequently used in dairy technology as a starter or adjunct culture to reduce bitterness and to increase flavor notes in cheese [65]. The complete genome sequence of the cheese culture *L. helveticus* DPC 4571 consists of 2.08 Mb with an average GC content of 37.73% [21]. It contains four *rrn* operons, 73 tRNA genes, about 1,600 protein-encoding genes and a large number of pseudogenes. Seventy five percent of the predicted ORFs in DPC 4571 have orthologues in the *L. acidophilus* NCFM genome. Not surprisingly, these two species share 98.4% of their 16S rRNA gene sequences. The same relatedness was observed by constructing a phylogenetic supertree with 47 ribosomal proteins [21]. A remarkable and rather astonishing feature of the *L. helveticus* DPC 4571 genome is the presence of 213 IS elements, belonging to 21 different classes [21]. In spite of this, IS-associated gene deletion and decay was not clearly evident for the majority of genes lost. Furthermore, the extensive whole-genome conservation between *L. helveticus* and other sequenced intestinal lactobacilli suggests that the IS elements did not promote (frequent) genomic rearrangements. A genomic island of 100-kbp characterized by a GC content of 42% and flanked by IS elements and unique 12-bp direct repeats is a firm candidate of adaptative HGT in this bacterium. Although non probiotic, certain *L. helveticus* strains have been shown to exert beneficial effects by production during milk fermentation of bioactive peptides with antihypertensive and immunomodulatory properties [66]. The DPC 4571 proteolytic system was found to be represented by more than 24 genes. The products of these genes showed significant homology to known peptidases and, while clear homologs of several of these had been reported previously, novel enzymes of interest to the dairy industry (*i.e.* those with similarity to PepE, PepQ, PepT or PepD) were also identified [21].

Lactobacillus johnsonii NCC 533

L. johnsonii NCC 533 is a member of the "acidophilus" group of intestinal lactobacilli that has been extensively studied for their probiotic activities, including pathogen inhibition, epithelial cell attachment, and immunomodulation. The *L. johnsonii* NCC 533 genome is 1.99-Mb in size with a GC content of 34.6% [67]. It contains six *rrn* operons at four loci, 79 tRNAs, 14 complete IS elements from three known families, and two complete prophages. Strikingly, the organism completely lacks genes encoding biosynthetic pathways for amino acids, purine nucleotides, and most cofactors. In apparent compensation, a remarkable number of uncommon and often duplicated amino acid permeases, peptidases, and PEP-PTSs were discovered, suggesting a strong dependency of strain NCC 533 on the host or other intestinal microbes to provide simple monomeric nutrients. Thus, competition of *L. johnsonii* with bifidobacteria and bacteroides seems unlikely; instead this species appears to be better adapted to the upper part (*i.e.* the small intestine) of the GIT, where amino acids, peptides, and mono- and oligosaccharides are abundant. Genome analysis also predicted an abundance (>12) of large and unusual cell-surface proteins, including fimbrial subunits similar to those of pathogens, which may be involved in adhesion to glycoproteins or other components of mucin [67], a characteristic expected to support persistence in the GIT. In this stable environment, *L. jonhsonii* was found to direct transcription through a single (primary) sigma factor. Three bile salt hydrolases and two bile acid transporters, which might also be critical for GIT survival, were also detected. *In silico* genome comparison with the genome sequence of the closely related *L. gasseri* revealed extensive synteny. Moreover, the DNA sequence of many housekeeping genes of these two bacteria showed a high degree of similarity.

Lactobacillus plantarum WCFS1

L. plantarum WCFS1 is a single colony isolate from *L. plantarum* NCIMB 8826, originally isolated from human saliva. The genome of the WCFS1 strain consists of a circular chromosome and three plasmids (of 1.9, 2.3 and 36.0 kbp) [68]. It is 3.3-Mb in size with an overall GC content of 44.5%, and contains 3,052 predicted protein-encoding genes, five *rrn* operons evenly distributed around the chromosome and 62 tRNA genes [68]. The *L. plantarum* genome encodes complete pathways for biosynthesis of most amino acids, except for branched-chain amino acids (valine, leucine and isoleucine). The genome also lacks an extracellular protease equivalent to PrtP of *L. lactis*. Consistent with the classification of *L. plantarum* as a facultative heterofermentative lactic acid bacterium, its genome encodes all enzymes required for the glycolysis and phosphoketolase pathways; components of these two routes appear to belong to the class of highly expressed genes in this organism (evident from their codon-adaptation index). Moreover, *L. plantarum* encodes a large pyruvate-dissipating potential, leading to various fermentation end-products. *L. plantarum* is encountered in many different environmental niches, and this flexible and adaptive behaviour is reflected by the relatively large number of regulatory, transport (including 25 complete PEP-PTS sugar transport systems), and stress-related proteins. More than 200 extracellular proteins are predicted to be encoded by the *L. plantarum* genome, of which many are assumed to be bound to the cell envelope. A large proportion of the genes encoding sugar transport and utilization, as well as genes encoding extracellular functions, are clustered in a 600-kb region near the origin of replication. Many of these genes display deviation of nucleotide composition, consistent with a foreign origin. Kleerebezem *et al.* [68] called this chromosomal segment "a lifestyle adaptation region", suggesting that many of the proteins encoded by this region provide adaptive properties to *L. plantarum*. For the first time in an LAB, a gene cluster predicted to be involved in the synthesis of a non-ribosomal peptide was found, although the assumed peptide-like product(s) remain uncharacterized.

Lactobacillus rhamnosus GG

L. rhamnosus GG isolated from the gut microbiota of a healthy human is one of the best studied probiotic lactobacilli. Various health effects have been attributed to this strain, such as prevention and treatment of acute diarrhea in children, prevention of antibiotic-associated diarrhea, and prevention and treatment of allergy [69]. The *L. rhamnosus* GG genome has been recently published and compared to that of the dairy adjunct starter strain *L. rhamnosus* Lc705 [70]. Both genomes display high sequence identity and synteny, except in a few interspersed strain-specific genomic islands (five in *L. rhamnosus* GG and four in *L. rhamnosus* Lc705). Genes in these regions encode proteins for exopolysaccharide (EPS) biosynthesis, specific sugar utilization and bacteriophage production. The genome of GG contains a single circular chromosome of 3.01 Mb in size with a GC content of 47%, which is predicted to encode 2,944 proteins [70]. It furthermore contains five rRNA operons, 57 tRNA genes, three prophage clusters and 69 transposases belonging to different IS elements. Strain GG is equipped with a vast array of genes involved in carbohydrate utilization, a metabolic activity considered advantageous for bacteria residing in the proximal region of the small intestine. Unlike *L. rhamnosus* Lc705, strain GG does not use lactose, a feature exploited in industrial applications. This negative property is caused by the presence of frameshifts in *lacT* (antiterminator) and *lacG* (6-phospho-β-galactosidase) genes. Both strains encode a cell envelope serine protease (PrtP), its maturation protein (PrtM), a proteinase (PrtR), and a similar set of 25 peptidases. However, strain GG cannot degrade casin, a property that may be associated to an additional secreted subtilisin-like serine protease in *L. rhamnosus* Lc705 (LC705_02680). Among the components involved in host-microbe interaction in strain GG, a long galactose-rich EPS has been reported to modulate biofilm formation [71], a type IIb bacteriocin operon, and multiple proteins with domains related to adhesion and colonization factors [70]. In addition, two predicted pilus-encoding gene clusters each encompassing genes for three putative pilin subunits and a single sortase were identified in the *L. rhamnosus* GG genome [70] (see below).

Lactobacillus sakei 23K

L. sakei is a psychrotrophic lactic acid bacterium which is naturally found on fresh meat and fish, but especially on fermented meat products where it is widely used as a starter [3]. The *L. sakei* 23K genome is 1.8-Mb in size with a GC content of 41.25 % [72]. It contains seven *rrn* operons and 63 tRNA genes. Sequences of 12 IS elements from four families and a prophage remnant were also observed. Consistent with life in a protein-rich environment, the genome sequence of 23K revealed auxotrophy for all amino acids except aspartate and glutamic acid. Its specialized metabolic repertoire includes scavenging of purine nucleosides that may improve its competiveness on raw meat products. Many genes appear to be responsible for coping with the harsh conditions of food processing (antimicrobial substances, high salt concentration, changing redox conditions and oxygen levels) [72]. In comparison to intestinal lactobacilli, *L. sakei* and *L. plantarum* are far better equipped to cope with changing redox conditions and oxygen content, although *L. sakei* appears to deal more effectively with toxic oxygen reactive compounds. Apparently, iron and heme acquisition correlates with this resistance, an ability characteristic of pathogenic bacteria. Genes potentially responsible for biofilm formation and cellular aggregation have been found in the *L. sakei* genome (including four proteins with a LPXTG motif and 15 having a WXL-like domain) [72], which may assist the organism in colonizing meat surfaces, were also identified.

Lactobacillus salivarius UCC118

L. salivarius is a mucosa-associated bacterium present in the faeces, intestinal mucosa, tongue and rectum of around 10% of the human infant and adults [73]. *L. salivarius* is part of a distinct clade at the periphery of the genus *Lactobacillus* [74]. The UCC118 strain belonging to the *salivarius* subspecies was isolated from the terminal ileum is a bacteriocin-producing strain with probiotic characteristics. The genome of *L. salivarius* subsp. *salivarius* UCC118 is 2.13-Mb in size, and is comprised of two large replicons [a 1.83 Mb chromosome and a 242-kb megaplasmid (pMP118)] and two smaller plasmids [of 20.4 kbp (pSF118-20), and 44 kbp (pSF118-44)] [75]. Seven *rrn* operons and 78 tRNAs, representing all 20 amino acids, were also detected on the genome. Two intact prophages and two remnant phage-related sequences and 43 copies representing 16 different classes of IS elements are also present. The genome sequence indicates an intermediate level of auxotrophy compared to other sequenced lactobacilli. The circular chromosome of *L. salivarius* UCC118 is the smallest *Lactobacillus* chromosome so far sequenced (57.6 kbp smaller than that of *L. sakei* and 165.5 kbp smaller than that of *L. johnsonii*), however this bacterium contains a number of plasmids, one of which, pMP118, has a size of 242 kbp and was for this reason referred to as a megaplasmid. Megaplasmids of this size had not previously been characterized in LAB, but were shown to be widely distributed among different *L. salivarius* strains belonging to both *salivarius* and *salicinius* subspecies (sizes ranging from 100- to 380-kbp) [75, 76]. More recently, even bigger plasmids have been identified in other intestinal lactobacilli [76]. No single-copy essential genes were present on the megaplasmid of UCC118 strain. However, contingency amino acid metabolic genes and carbohydrate utilization genes, including two genes for completion of the pentose phosphate pathway (a finding that groups the species among the facultative heterofermentative lactobacilli), were encoded by the megaplasmid [75]. Furthermore, this plasmid harbors genes for production of the ABP118 bacteriocin, a bile salt hydrolase, a presumptive conjugation locus, and other genes potentially relevant for probiotic properties. Notably, bacteriocin production was recently demonstrated to be the primary mediator of *in vivo L. salivarius* UCC118 protection in mice against the pathogenic organism *Listeria monocytogenes* [77].

Leuconostoc

The genus *Leuconostoc* comprises Gram-positive cocci that are claimed to be taxonomically and ecologically related to *Lactococcus* species. However, *Leuconostoc* are obligate heterofermenters and produce EPSs in sugar-rich media. Two species are of industrial significance, namely *Leuconostoc mesenteroides* subsp. *cremoris* and *Leuconostoc lactis*, which are both frequently included as starters and adjunct cultures in the production of cheese and butter. Leuconostocs grow slowly, especially in milk, so they are not important for lactose fermentation, but they are critical for the production of flavor compounds through utilization of citrate to produce diacetyl [78].

Leuconostoc citreum KM20

L. citreum is a dextran-producing species that can be found, as many other *Leuconostoc* species, in fermented foods and feeds of plant and dairy origin, such as cheese, pickles, sauerkraut, and cabbage. The K20 strain was originally isolated as a dominant microorganism from Kimchy, a popular fermented Korean commodity made from a variety of spicy vegetables, for which it is used as a starter. *L. citreum* KM20 can suppress growth of many pathogenic microorganisms and has been found to be cytotoxic to HT-29 cells. Consequently, this strain has also been considered as a probiotic. The *L. citreum* KM20 genome consists of 1,796,248 bp with a collective 39% GC content [79]. Besides the single chromosome KM20 harbors four circular plasmids of 38.7, 31.5, 18.0, and 12.2 kbp. The entire genome showed 1,820 protein-coding genes (1,702 on the chromosome and 118 on plasmids), four rRNA operons and 67 tRNA genes. Complete phages were not found, but the genome contains five copies each of IS elements belonging to the IS*3* and IS*30* families [79]. Genome analysis revealed a complete gene set for heterolactic fermentation *via* the phosphoketolase pathway with an incomplete tricarboxylic acid cycle [79]. A vast array of hydrolases and carbohydrate transport genes were observed, which agrees well with its association with plant-associated material. In addition, multiple genes for dextransucrases and alternansucrases have been shown, as well as a plasmid-encoded, cell wall-anchored protein with five putative mucus-binding domains. This protein may be related to the probiotic properties of KM20 strain.

Oenococcus oeni

O. oeni is the only species of the genus *Oenococcus*. It is an acidophilic member of the *Leuconostoc* branch, indigenous to grapes and other plant-related environments [80]. Together with lactobacilli, pediococci, and *Leuconostoc* species, *O. oeni* is responsible for the malolactic fermentation in wine and cider [81]. This bacterium has rarely been associated with off-flavors and undesirable metabolites in fermented beverages. Consequently, it is frequently used as a malolactic starter.

O. Oeni PSU-1

The complete genome of the natural plasmid-free *O. oeni* PSU-1 strain is approximately 1.7-Mb in size with a GC content of 38%. Around 1,700 ORFs have been predicted from the sequence, of which 75% have been functionally classified [82]. The genome size is very similar to that of strain IOE 8413 (ATCC BAA-1163), whose genome sequencing is almost complete [14]. Only two *rrn* operons in opposite orientation have been found and 43 tRNA genes scattered through the chromosome (although 15 of them are clustered at one location), representing all 20 amino acids. As an obligate heterofermentative lactic acid bacterium, the genome of *O. oeni* encodes all enzymes for the phosphoketolase pathway. Moreover, genes related to flavor modification in wine, in particular those involved in the malolactic fermentation capacity (*mleAP* and *mleR*) and citrate utilization (citrate lyase gene cluster, the butanediol pathway and other genes) have been identified [82]. Indeed, five genes homologous to the malate permease and one to the malate decarboxylase can be found on the PSU-1 genome. Thus, *O. oeni* possesses more genes in this cluster of orthologous groups of proteins (COG) than any other LAB, probably reflecting its adaptation to malate rich environments. Various stress-related systems have also been identified which are thought to assist *O. oeni* to survive the harsh environment of wine and cider, including malate, citrate and amino acid conversion enzymes, class I heat shock genes (*groESL* and *dnaK* operons), a putative F_0-F_1 ATPase system (*atpBDFHAGDC*), and other genes involved in stress responses (*clpX, clpLP, trxA, ftsH*, and *omrA*).

Streptococcus thermophilus

The genus *Streptococcus* comprises several harmful pathogenic species, such as *Streptococcus pyogenes* and *S. pneumoniae*, but also a single Generally Regarded As Safe (GRAS) species, *S. thermophilus*. *S. thermophilus* is closely related to the *S. salivarius* found in the human oral cavity, of which it was considered until recently a subspecies. *S. thermophilus* is a relatively coherent and homogeneous species with a low level of nucleotide polymorphism among strains, suggesting it has recently emerged [83]. At present, it is used extensively in the food industry for the manufacture of many dairy products (yogurt, hard cooked cheeses of the Italian and Swiss types, soft cheeses, *etc.*), and it is considered the second most important industrial dairy starter after *L. lactis* [3].

S. thermophilus CNRZ 1066 and LMG 13811

The genome sequences of two *S. thermophilus* strains, that of CNRZ 1066 and LMG 13811, have been determined and analyzed [19]. These have been compared to genome sequence of *S. thermophilus* LMD9 [15]. The genome size of all three strains is nearly identical (around 1.8-Mb) with a GC content of 39%. Sixty-seven tRNA genes were found and six *rrn* operons. Access to these three genome sequences of *S. thermophilus* has allowed a better understanding of the evolutionary path followed by this species [83]. *S. thermophilus* and its pathogenic relatives still share a substantial part of their overall physiology and metabolism. *S. thermophilus* seems to have evolved mainly through loss-of-function events mirroring the dairy niche, resulting in the absence of most streptococcal virulence-related genes such as those involved in cell adhesion, and host invasion or escape from the immune system [19]. However, the presence of numerous pseudogenes (around 10%) also suggests an ongoing regressive evolution. The detailed *in silico* investigation of its cellular metabolism illustrates that evolution has shaped the *S. thermophilus* genome by selection for optimal growth in milk. Notably, *S. thermophilus* has maintained a well-developed nitrogen metabolism, while its sugar catabolic abilities are strongly degenerated (two of the most highly decaying functional groups relate to carbohydrate hydrolysis, uptake, and fermentation). Additionally, *S. thermophilus* shares its ecological niche with other LAB such as *L. bulgaricus*, resulting in specific metabolic cooperation, which is either revealed by the maintenance of dedicated pathways (*e.g.*, folate and formate production) or by the loss of key metabolic functions provided by the symbiotic partner (*e.g.*, extracellular proteolytic activity for casein hydrolysis). Although gene decay is obvious in the *S. thermophilus* genome, numerous small genomic islands seem to have been acquired by HGT process. These regions encode a number of important adaptive traits, which are of industrial relevance such as polysaccharide biosynthesis (*eps, rgp*), bacteriocin production (*blp, lab*), restriction-modification systems or oxygen tolerance. *S. thermophilus* is considered a non-competent organism, although the *S. thermophilus* genome contains all late competence genes, a situation that is similar to that observed for *L. lactis*. In addition, it contains a *comX*-like gene coding for a typical peptide pheromone-dependent two-component system that is similar to the competence control loci of *Streptococcus mutans* and *S. pyogenes*. The regulatory pathway controlling expression of key components of competence in *S. thermophilus* has recently been reported [84] (see also below for a general discussion on natural transformation in LAB).

EVOLUTIONARY ASPECTS

Selective pressure and competitiveness have been driving the evolution of LAB species in the nutrient-rich environments they inhabit [17]. Not surprisingly, the analysis of different LAB genomes has revealed well equipped organisms for a wide range of metabolic activities, defense and stress responses, specifically needed to live, reproduce and survive in plant-derived materials, diary and meat products, and animal and human mucosa. Given the close phylogenetic relationships of these organisms, comparison of gene content across the species and reconstruction of ancestral gene sets indicate that adaptation to nutrient-rich environments has for most species promoted a combination of extensive gene loss (or degradation of dispensable genes) from ancestral bacterial types and acquisitions of key genes *via* HGT.

Ancestral gene loss and metabolic simplification contrasts with lineage-specific duplication and/or acquisitions of unique (key) genes. Plasmid-encoded genes, which, for instance, account for up to 5% of the total gene content in *L. lactis* [85], are essential for growth of LAB species in specific environments. All LAB genomes contain IS elements and transposons, varying from approximately 0.2% of the genome in *L. gasseri* [15] up to nearly 10% in *L. helveticus* [21]. Sometimes, perhaps simplistically, viewed as genomic parasites, IS and transposons are thought to contribute substantially to the generation of genetic diversity, thus promoting adaptation of bacteria [86].

From an evolutionary perspective, milk in particular is a very recently developed niche as compared to plants, and adaptation of what probably were plant-associated bacteria to the dairy environment seems to be a plausible evolutionary event. The analysis of the genomes of the two *L. lactis* plant isolates and their comparison to those of dairy strains has provided a first view of the molecular basis of adaptation of this bacterium to plant and milk environments [44-46]. A high synteny was found between the genomes of dairy and plant isolates, but numerous genes were identified in the sequences of the plant isolates that do not have homologs in the gene complement of dairy *L. lactis* strains. By calculating the GC content of unique genes and gene clusters of the plant isolates and comparing them to their best hits, Siezen *et al.* [53] showed that most of the genes have a GC content close to or slightly lower than the 35% average for *L. lactis* genomes, while their best BLAST hits generally have a higher GC content (except for the clostridial best hits, which have a lower GC content). This strongly suggests that most of these genes and gene clusters that appear to be specifically associated with the genomes of lactococcal plant isolates were in fact not acquired by lateral gene transfer. Rather, they seemed to be more ancient and appear to have been lost from the genomes of the diary lactococci.

Certain practices in the dairy industry appear to have caused specific adaptation of some bacteria to particular milk products. This is thought to be the case for *L. delbrueckii, L. helveticus* and *S. thermophilus*, whose closest relatives are represented by commensal or pathogenic bacteria from the human and animal GIT. Strains of these species are adapted to grow in milk under the stringent conditions that are used to make dairy products such as yoghurt and cooked cheeses. In fact, they have never been isolated from traditional dairy products that do not undergo heating [87]. Adaptation to milk of *S. thermophilus* has resulted in genome degradation of genes that are dispensable for growth in this medium (including many involved in pathogenesis) and acquisition by horizontal gene transfer of genetic traits dedicated to efficient exploitation of milk's nutrients [83]. As an example, a specific synporter for lactose found in *S. thermophilus* is absent from its pathogenic relatives. In common with *S. thermophilus*, adaptation of *L. helveticus* to the dairy niche involved gene decay of transport proteins, energy metabolism genes, regulators and amino acid metabolic genes, while lateral gene transfer events provided this bacterium with specific genes for fatty acid biosynthesis, restriction endonucleases and amino acid metabolism genes [21]. Indeed, a nine niche-specific gene set, enconding components of the proteolytic system and restriction endonucleases, has been identified in *L. helveticus* but is absent in its gut inhabiting relative *L. acidophilus* [17]. Gene gain has also been proposed for the acquisition by *L. bulgaricus* and *L. lactis* of a 17 kbp region of DNA, containing multiple copies of IS*1191* and a mosaic of genes devoted to generate essential nutrients from milk components [88]. A unique copy of *metC* was identified among the genes identified on this 17 kbp fragment. This gene endows LAB species with the ability to synthesize methionine, a rare amino acid in milk. Many LAB strains of different species harbor plasmids of sizes ranging from about 2-kbp to more than 300-kbp [76, 85], which are considered to represent adaptative tools. In addition, being different from plasmids, most species seem to harbor chromosomal regions dedicated to the incorporation of adaptative (exogenous) DNA (the so called "life style island" or "hotspot islands") [19, 21, 68].

IDENTIFICATION OF EXISTING AND NOVEL BIOCHEMICAL ROUTES

Manipulation of fermentation pathways of lactic acid-producing bacteria to obtain a better understanding of their biochemical routes, or to improve the bacterium's efficiency (robustness) in industrial applications, was undertaken soon after the development of the first genetic engineering tools [8, 9]. As microbial genomes encode the various options determining metabolism, physiology, biosynthetic capabilities, and adaptability to varying conditions and environments of the organisms, the availability of genome sequences has significantly expanded our capabilities to exploit these microorganisms metabolic and bioprocessing potential and their contribution to health and well-being [14]. The metabolic abilities of *B. longum* subsp. *longum* NCC2705 have been reconstructed *in silico* from its genome sequence. This analysis revealed that this strain possesses genes for the synthesis of at least 19 amino acids from ammonium, pyrimidine and purine nucleotides from glutamine, and most enzymes for the synthesis of folic acid, thiamin and nicotinate. The genome sequence of *B. breve* UCC2003 has been used to formulate a chemically-defined medium for this bacterium to undertake fundamental studies [89]. In that sense, genomic information can also be useful for the formulation of species- or strain-specific prebiotics to enhance growth of beneficial populations (see below).

Glucose Uptake Systems in *L. lactis*

Galactose and lactose are not always desirable in food products, because of lactose intolerance, and the involvement of galactose in cataract occurrence in certain individuals in combination with alcohol intake. These considerations led to an attempt to engineer *L. lactis* into a glucose-producing cell factory. A strategy was designed to remove all glucose-PTS import activities [88]. The glucokinase gene was also deleted to prevent glucose (entering the cell through non-PTS mechanisms) from being phosphorylated and then processed *via* glucolysis. However, a glucokinase (Δ*glk*), $EII^{man/glc}$ (Δ*ptnABCD*) double mutant was still capable of growth in glucose, which thus suggested that a third transport system is responsible for glucose uptake. DNA microarrays were used to investigate the transcriptomes of *L. lactis* wild type and mutant strains [88]. This technique unambiguously demonstrated that EII^{cel}

(*ptcBAC*), previously thought to be involved in the transport of cellobiose, was the only up-regulated operon in the double knock-out strain. A triple mutant was constructed and selected on galactose. This strain was indeed shown to be incapable of growth on glucose. Subsequently, the lactose plasmid pMG820, carrying the genes for the lactose-PTS and the tagatose-6-phosphate pathway, was introduced by conjugation into the mutant. The new strain was shown to grow slowly on lactose, but it efficiently fermented the galactose moiety, while excreting glucose at equimolecular levels to the lactose internalized [88]. No galactose was detectable in the growth medium, as opposed to the wild type strain which produced low levels of this sugar. Indeed, the strain was able of growing in skim milk where it produced up to 22 mM of glucose.

Carbohydrate Degradation and Utilization

As obligate fermenters, LAB need to use carbohydrates as a source of energy for growth and consequent acidification. Not surprisingly, a vast array of diverse carbohydrate-degrading (glycosyl hydrolases) enzymes have been encountered scattered throughout their genomes. However, as a result of specialization and adaptation, significant differences exist in the number and types of glycosyl hydrolases among the different LAB species. β-galactosidases, β-phospho-galactosidases and β-glucosidases are involved in the utilization of lactose in the milk environment, while α-galactosidases are responsible in plant-derived environments for the breakdown of α-galactosides, compounds that can not be degraded by monogastric animals, thus causing flatulence and other intestinal malfunctions [90]. Furthermore, oligosaccharide-type galactosides are recognized as bifidogenic factors and prebiotics, meaning that LAB and bifidobacteria must contain specific-degrading machineries. Moreover, carbohydrate hydrolases are also involved in the intestinal tournover of the glycocalix material, which may represent colonizing and competence factors for bacteria [27]. Glycosydases from intestinal microorganisms may also participate in the deglycosylation of phytoestrogens from plant materials of the diet (isoflavones, lignans, and fenylflavonoids), which in several cases is an essential step in the formation of biologically-active compounds, such as equol, enterodiol, enterolactone and 8-phenylnaringenine [91, 92]. Alternatively, glycosidases may also contribute to the activation of pro-carcinogenic molecules that may ultimately be harmful for the host [91]. In this respect, β-glycosidases release aglycones with mutagenic and/or carcinogenic activities and β-glucuronidases may re-activate in the intestine conjugated carcinogens that are produced by the liver.

As a striking example of carbohydrate-degrading activities, it is predicted that bifidobacterial genomes encode a large number (more than 40) of glycosyl hydrolases, some of which are assumed to exert their activity outside the cytoplasm; *e.g.* amylopullulanase that allows growth of UCC2003 on starch and related sugar polymers [84]. β-fructofuranosidases are among the enzymes reported to allow the digestion of a particular group of bifidogenic carbohydrates, the fructo-oligosaccharides (FOS) [30]. These enzymes specifically catalyze the hydrolysis of the β-(2-1)-glycosidic bond between glucose and its neighboring fructose moiety in sucrose, and the glycosidic connection between two fructose moieties in short FOS. As mentioned, the *B. breve* UCC2003 genome contains a FOS operon, which encodes a putative permease, a conserved hypothetical protein and a β-fructofuranosidase [30]. Transcriptional analysis of this operon in *B. breve* UCC2003 grown in the presence of different carbohydrate sources showed its involvement in the hydrolysis of short-chain FOS [93].

Another class of enzymes involved in the utilization of prebiotics by bifidobacteria includes β-galactosidases. These enzymes are essential for bifidobacteria to be able to grow in milk or milk-components like lactose and lactose-derived transgalacto-oligosaccharides (TOS). Arabinofuranosyl-containing oligosaccharides derived from plant cell-wall constituents, such as arabinan and arabinoxylan, can be fermented by bifidobacteria through arabinoxylan- and arabinofurano-hydrolases [94]. However, the rate of degradation of these compounds is rather low and it has been assumed that other bacteria (*e.g. Bacteroides*) are needed for the complete hydrolysis of these polymers [25]. Bifidobacteria are also known to grow rapidly on soymilk substrates containing large amounts of α-galactosyl-oligosaccharides (such raffinoase and stachyose) [95]. Enzymes such as α-galactosidases are responsible for the hydrolysis of these types of substrates as well as for the utilization of melibiose and galactomannan. Furthermore, genome analyses of *B. longum* subsp. *infantis* ATCC15697 and *B. bifidum* PRL2010 have revealed a rich arsenal of genes encoding enzymes involved in the complete hydrolysis of host-glycans such as HMOs and mucin [32, 33].

Respiration in *L. lactis*

Sugar fermentation was long considered to represent the sole means of energy metabolism available to lactic acid bacteria, producing organic acids (mainly lactic acid) as a final product. While this description is still true, some LAB species exhibit a respiratory capacity in the presence of oxygen and exogenous heme, resulting in the production of greatly reduced amounts of lactic acid.

Early evidence for the respiratory capability of LAB was largely overlooked [96], and the initial reports on the respiration in *L. lactis* [49] were essentially based on the analysis of the complete genome sequence of *L. lactis* IL1403 [44]. Subsequent research confirmed that this species has indeed the ability to respire in the presence of oxygen [49, 50], provided the growth medium contains heme because this bacterium does not have a functional biosynthetic pathway for this compound. Heme is an iron containing porphyrin, which is an essential co-factor of the cytochrome oxidase system. An important consequence of respiration is a more efficient conversion of the carbon source into biomass, resulting in higher cell yields and increased survival following growth. Under respiratory conditions acetate, acetoin and diacetyl are produced from pyruvate at the expense of lactic acid [49]. Transcriptomic analysis of *L. lactis* MG1363 showed that the pyruvate dehydrogenase complex (encoded by the *pdhABCD* operon) was up-regulated 4-fold under these circumstances [97]. Acetolactate syntase (*als*) and α-acetolactate decarboxylase (*aldC*) genes were also up-regulated, facilitating the synthesis of both diacetyl and acetoin. The most highly up-regulated gene under respiratory conditions was *ygfC*, encoding a putative regulatory protein, which increased almost 100-fold. In contrast, the expression of the pyruvate formate lyase (*pfl*) and the alcohol dehydrogenase (*adhE*) was reduced 2.5- and 50-fold, respectively [97].

These results are of industrial significance and allowed the development of a patented process for the production of LAB starter cultures [51]. The patent was licensed to Chr. Hansen in 1999, and the initial results obtained with the well-characterized laboratory strain *L. lactis* IL1403 were optimized for different industrial *L. lactis* strains and also for *Leuconostoc* species [98]. Industrial *L.*

lactis strains strains were also assayed in aeration in the absence of added heme to distinguish simply aeration from true respiration [98]. Numerous genes were differentially expressed under these two different conditions. Approximately half of these genes have an unknown function, indicating that more research is needed to fully understand the physiology of respiration in this species. Starter cultures obtained by the respiration technology have been assayed in pilot scale tests for Cheddar cheese production. Manufacture parameters were all within the normal range, and differences between respiration-grown cells and the fermentation-grown cells were not observed [98]. Parameters such as moisture content, total soluble nitrogen or pH did not significantly change after two or six months of ripening. Indeed, sensory differences were not perceived by two trained sensory panels after these periods. Industrial scale trials of Cheddar, Feta and cottage cheese have already been performed, and again no significant differences were seen in the manufacturing parameters, cheese microbiology, chemistry, texture or flavor development [as reported in reference 98].

Respiration in *L. plantarum* needs the exogenous addition of both heme and vitamin K2, which act as a source of menaquinone [99]. As this bacterium lacks superoxide dismutase, high levels of manganese are also needed for its aerobic growth. Under respiratory contidions, growth of *L. plantarum* results in higher biomass yields. It also impacts on the robustness of the cells, as cells from aerobic cultures have been shown to be more resistant to stress conditions such as those encountered during industrial fermentations [100].

Increased biomass after aerated incubation in the presence of heme was not obtained for *S. thermophilus, L. bulgaricus* and *L. helveticus*. Analysis of the complete genome sequences of these species [19-21] neither revealed the presence of genes for cytochrome oxidase nor for the biosynthesis of quinones, features which are believed to be essential for respiration.

Mode of Action of Probiotics

The molecular mechanisms by which probiotic LAB colonize and persist in the GIT, and the ways by which they positively impact on host health are far from completely understood. The genome analysis of well-know probiotic strains may give some clues that should then be experimentally tested under laboratory conditions and in clinical trials. This approach has recently received the name of Probiogenomics [101]. This discipline uses the genomic data to investigate the mechanisms by which probiotic bacteria adapt to its specific environment, while revealing genetic functions that mediate the specific host-microbe interactions.

In this way, transcriptome analysis of the *B. breve* UCC2003 genome in a murine colonization model has recently revealed differential expression of a type IVb tight adherence (*tad*) pilus-encoding gene cluster, which proved to be essential for efficient *in vivo* murine gut colonization [29]. The Tad pilus-encoding locus was shown to be conserved among different *Bifidobacteirum* species, thus supporting the general involvement of pili in gut colonization and persistence for bifidobacteria [35].

In a similar approach, pili-like structures have been reported for the first time in lactobacilli [70]. Pili are cell surface-localized protrusions that have been well characterized in Gram-positive pathogens. Analysis of the *L. rhamnosus* GG genome identified two separate pilus clusters in its genome (*spaCBA* and *spaFED*). The latter operon was also shown to be present in the genome of *L. rhamnosus* Lc705, while the former is encoded in a GG-specific genomic island. The purified *spaC* and *spaB* components of the first cluster and *spaF* of the second have been demonstrated to bind mucus [70, 102]. Functional analysis using knockout mutants further proved that the SpaCBA pili, the only ones produced under *in vivo* conditions [103], was involved in efficient adherence to Caco-2 cells and biofilm formation [104]. In addition, reduced expression of interleukin-8 (IL-8) mRNA was induced when Caco-2 cells were incubated with the wild type strain as compared with the mutant. This suggests that while providing mucus-binding ability, which may explain colonization and persistence of *L. rhamnosus* GG in the intestine, pili migh also modulate IL-8 expression through interaction with surface molecules of the host cells [104]. This system represents a previously undescribed mechanism for the interaction of probiotics with host tissue.

GENOME DIVERSITY AND STRAIN COMPARISON

Whole genome sequencing, genome data mining, and comparative genomics provide insights into genetic content, differences and similarities among species and strains, and offer important clues into possible gene functions [105]. A powerful method to quickly determine the genome content of a bacterial strain whose genome sequence is not known is comparative genome hybridization (CGH). By this technique, chromosomal DNA fragments of test and sequenced strains, carrying differential fluorescent labels, are hybridized against a DNA microarray that represents all identified genes of the latter strain. CGH is commonly used to determine the genomic composition and genome plasticity of bacteria [106]. The technique is also considered as an alternative to the complete sequencing of genomes, especially in closely related species or in strains of the same species [16, 106, 107], because, as the microarray analysis is based on DNA-DNA hybridization, distantly related species lacking enough DNA sequence identity can not be evaluated by this technique. Two different types of genomic diversity are expected, the selfish mobile DNA (referred to as the mobilome), which enters and leaves the genome, and remnants of ancestral DNA, that has not been lost during genome reduction.

At present, only a few studies describe the use of CGH to determine the genome composition in LAB species. In one of these studies, DNA microarrays of the *L. lactis* IL1403 were hybridized with a mixture of randomly Cy3- or Cy5-labeled DNA fragments from *L. lactis* IL1403 and *L. lactis* MG1363 [88]. As the genomes of these two strains are available, the competitive hybridization behaviour of the whole gene complements could give some clues about the differences between similarity levels, copy number and hybridization efficiency. A clear positive correlation between gene similarity and the hybridization signal was obtained for genes with a similarity of 75% or higher. However, below this critical level of similarity a relative large scatter in ratios between different genes was observed [88]. Microarrays have also been used to explore the genome diversity of *L. plantarum* strains [106]. DNA microarrays based on *L. plantarum* WCFS1 were hybridized against total DNA from 20 strains. Genes that were present in the WCFS1 strains but not in others were analyzed with respect to their chromosome location, base composition, and putative functions. A high degree of gene content variability was observed among the *L. plantarum* strains examined. The majority of genes found in the variable regions were seen to be involved in sugar metabolism, and based on their unusual base composition and overrepresentation these genes were thought to constitute lifestyle adaptation regions in this bacterium [106].

Similarly, the technique has been used for a comparison of several members of the *L. acidophilus* group against microarrays based on the *L. johnsonii* NCC 533 genome, addressing both intra- and interspecies diversity [107]. A clear stepwise decrease in similarity between members of the *L. acidophilus* complex was found, suggesting the component species belong to a natural phylogenetic group. Thus, this technology may also be exploited to clarify the taxonomical relationships in problematic bacterial groups. The intraspecies differences were analyzed by comparing four different strains with the *L. johnsonii* NCC 533 genome. Overall, DNA from the test strains failed to hybridize with 8% to 17% of the ORFs from the NCC 533 reference strain [107]. In contrast to *L. plantarum*, the region around the origin of replication proved to be the largest genome segment of gene conservation. In *L. johnsonii* the region around the replication terminus was shown to be a major area of genetic diversity. Moreover, DNA microarray based on the NCC533 genome revealed that a large proportion of the NCC 533 strain-specific DNA sequences are represented by phage DNA [108].

Genome-wide comparison of nine different *B. longum* strains has recently been reported by this technique [16]. Seven major regions of non-conserved sequences were identified, of which four have a low GC content and had previously been identified by comparison of the genome sequence of NCC 2705 and DJO10A.

Availability of genome sequences of many *Lactobacillus* species has recently allowed the proposal of a new discipline, phylogenomics, which consist on phylogenetic analysis based on whole-genome comparison [109]. More than one hundred species of lactobacilli are recognized at present (http://www.ncbi.nlm.nih.gov/Taxonomy). They are strictly fermentative (either homo- or heterofermenters), aerotolerant or anaerobic, aciduric or acidophilic, and have complex nutritional requirements (carbohydrates, amino acids, peptides, fatty acid esters, salts, nucleic acid derivatives, vitamins). Phylogenomics can serve as a tool for the revision of the currently unsatisfactory taxonomy of the subgeneric divisions of lactobacilli and other LAB species [109].

Availability of whole genome sequences further allows the construction of genome-scale metabolic models, such as those developed for *L. plantarum* [110], *L. lactis* [111], and *S. thermophilus* [112]. These models would permit to simulate growth of strains under different conditions (pH, lactose content) and to compare predicted and experimental data for the study of industrially relevant features.

GLOBAL ANALYSIS OF LAB PROTEINS, PROTEOMICS

The proteome is the total protein complement of a genome under specific conditions. In most cases, it is investigated by combining 2D gel electrophoresis and protein identification by peptide mass spectrometry and peptide sequencing using an MS-MS approach. Only in combination with the genome sequence can the proteomics potential be fully exploited [113], although the peptides obtained can also be directly compared to the ever-increasing protein databases. Like transcriptomic analyses, proteomic studies can generate quantitative measures, but of protein levels in the latter case. Combining transcriptomic and proteomic data it is possible to determine whether regulation takes place at the transcriptional or post-transcriptional level.

Reference maps can also be produced for a mixed bacterial population. For example, Gagnaire *et al.* [114] identified the proteins released into Emmental cheese by strains of *L. helveticus, L. delbrueckii* subsp. *lactis* and *S. thermophilus*. The analysis showed that some peptidases from *L. helveticus* and *S. thermophilus* are released into the cheese matrix, indicating that these may contribute to peptide degradation during ripening.

Proteomics may also be helpful in uncovering the relationships among the main energy generating pathways (amino acid decarboxylation, malate decarboxylation and arginine deimination), which may be regulated by means of sophisticated networks of proteins [115]. It has also been applied to study the complex interactions and mutualistic relationships between LAB species and their environment [116], including the interactive effects between LAB and animal or human intestinal cells [117], which may ultimately lead to a more rational design of probiotics.

NATURAL TRANSFORMATION OF CERTAIN LAB

Progress in the molecular characterization of important LAB properties to be used as starters or probiotics, and their further improvement by genetic engineering has been traditionally hampered by a lack of genetic tools [118]. Indeed, the genetics of these microbes is poorly developed relative to other microorganisms of industrial significance. Some plasmids were soon recognized as conjugative [119], a capability that was utilized to construct robust industrial starter strains equipped with many phage resistance systems, bacteriocin production, *etc.* [9]. Transduction was discovered in *L. lactis* even before conjugation [120]. In contrast, natural transformation has never been observed for any of the typical LAB species. (However, as judging for the present shape of the genomes, these two processes might have had a significant role in determination of the gene content of these genomes). Thus, the introduction of DNA into LAB species initially relied on the formation and regeneration of protoplasts, and subsequently on electrotransformation methods [9]. These were developed for other bacterial types, and then adapted to LAB species [9, 119]. Homologous and heterologous plasmids were developed into plasmid vectors, opening the way to an array of important findings in plasmid- and chromosomally-encoded properties, such as the metabolic abilities that allow these bacteria to utilize lactose and casein, phage resistance, bacteriocin production and immunity, and EPS production [9].

Bacteria that are competent for natural genetic transformation are able to take up naked DNA from the environment, and incorporating it into their genomes by (homologous) recombination. Competence is a transient, tightly regulated process, involving two sets of genes: the early and late competence genes. One of the best-studied naturally competent bacteria is *S. pneumoniae* [121]. Early competence genes in this bacterium comprise a quorum-sensing system consisting of ComABCDE, in which *comC* encodes the precursor of a secreted peptide pheromone, the competence-stimulating peptide (CSP), which triggers development of the competent state when its external concentration in the culture reaches a critical threshold. The CSP is secreted by the ABC-type transporter ComAB, acting through a two-component signal transduction pathway consisting of the histidine kinase ComD and the cognate response regulator ComE. Fourteen proteins are known in pneumococci to be necessary for uptake of extracellular

DNA and its subsequent incorporation into the recipient's genome. These 14 proteins are all encoded by so-called late genes. Late genes share an 8-bp sequence in their promoter regions, which are recognized by an alternative sigma factor (ComX). Circumstantial evidence indicates that ComX is encoded by one of the early genes and therefore depends on ComE for expression [122].

Interestingly, recent genome sequencing has shown that the ComX regulon appears to be present in different LAB species, including *L. lactis* [44], *L. plantarum* [68], and all *S. thermophilus* species [19]. This finding suggests that many LAB might actually be naturally transformable provided that growth conditions promoting the development of competence can be identified. Alternatively, these genes may have other functions or represent nonfunctional relics inherited from a shared competent ancestor.

In an elegant study, Blomqvist *et al.* [84] analyzed the expression of the different components of the late competence genes in *S. thermophilus* LMG 18311. Under normal laboratory conditions this strain is not transformable. Blomqvist *et al.* [84] demonstrated that the level of expression of ComX was too low to turn on the competence state. As it had been reported before in *S. pneumoniae* [122], transient overexpression of ComX by an inducible promoter was sufficient to attain transformation efficiencies, using total genomic DNA from a streptomycin-resistant strain, higher than 10^{-3} cfu per donor strain [84]. The mechanism for controlling the development of competence could have degenerated during adaptation of *S. thermophilus* to the dairy niche, but it could also be likely that spontaneous development of competence in this species may require special and as-yet-undiscovered growth conditions. Indeed, evidence of HGT from other dairy bacteria to *S. thermophilus* LMG 18311 has been speculated [19]. Thus, it seems plausible that at least some of these gene transfer events have taken place by natural genetic transformation.

In a similar way, a (dormant) natural transformation capacity may be embedded in those LAB species that possess the complete set of competence genes. In light of the promising results of Blomqvist *et al.* [84], natural transformation holds great promise as a novel tool allowing easy, rapid, and efficient construction of food-grade mutants of industrial LAB strains, which may serve for designing starter and probiotic strains with improved properties by attractive, non-controversial techniques for both consumers and dairy industry.

CONCLUSIONS

The first lactic-acid producing bacterial species for which the complete genome sequence was publicly available was *L. lactis* subsp. *lactis* IL1404 in 2001. The genomic information gathered during recent years may be useful to gain insights into LAB and bifidobacterial physiology, biochemical potential, as well as their evolution, unraveling how the different species evolved and adapted to their respective ecological niche. This information may be of direct industrial and/or biotechnological application, as it was the case for the respiration process in *L. lactis*. The availability of the genome sequences of many LAB and bifidobacteria will allow global research approaches, such as comparative genomics, transcriptomics, metagenomics, proteomics, and metabolomics, some of which were not covered in this review. It would also allow the appearance and development of new research disciplines, such as phylogenomics and probiogenomics. Existing and new technologies can be exploited to address a medley of fundamental and applied questions, such as the genetic basis and control of the nitrogen and carbon metabolism, factors involved in host colonization and competition, resistance to antimicrobials and stress responses, crucial traits for the selection of technologically and functionally robust commercial strains. The sequence data provided by the finalized and ongoing genome sequencing projects and the development of more robust high-throughput techniques of DNA and protein analysis will further allow deeper genomic and post-genomic studies, translating sequences into biologically-relevant information.

ACKNOWLEDGEMENTS

This work was partially supported by projects from the Spanish Ministry of Education and Science to B.M. (AGL2011-24300), and through a Science Foundation Ireland CSET grant that supports the Alimentary Pharmabiotic Centre of which D.v.S. is a member.

CONFLICT OF INTEREST

The authors confirm that this chapters contents have no conflict of interest.

DISCLOSURE

The chapter submitted for Advances in Genome Science, Vol. 3 entitled **"Genome Analysis of Food Grade Lactic Acid-Producing Bacteria: From Basics to Applications"** is an update of our article published in Current Genomics, 2008 May; 9(3): 169-183.

REFERENCES

[1] Carr, F.J., Chill, D., Maida, N. The lactic acid bacteria: a literature survey. *Crit. Rev. Microbiol.,* **2002,** *28:* 281-370.

[2] Kleerebezem, M., Hugenholtz, J. Metabolic pathway engineering in lactic acid bacteria. *Curr. Opin. Biotechnol.,* **2003,** *14:* 232-237.

[3] Leroy, F., de Vuyst, L. Lactic acid bacteria as functional starter cultures for the food fermentation industry. *Trends Food Sci. Technol.,* **2004,** *15:* 67-78.

[4] Leahy, S.C., Higgins, D.G., Fitzgerald, G.F., van Sinderen, D. Getting better with bifidobacteria. *J. Appl. Microbiol.,* **2005,** *98:* 1303-1315.

[5] Eckburg, P.B., Bik, E.M., Bernstein, C.N., Purdom, E., Dethlefsen, L., Sargent, M., *et al.* Diversity of the human intestinal microbial flora. *Science,* **2005,** *308:* 1635-1638.

[6] Marchesi, J., Shanahan, F. The normal intestinal microbiota. *Curr. Opin. Infect Dis.,* **2007,** *20:* 508-513.

[7] Vaughan, E.E., Heilig, H.G., Ben-Amor, K., de Vos, W.M. Diversity, vitality and activities of intestinal lactic acid bacteria and bifidobacteria assessed by molecular approaches. *FEMS Microbiol. Rev.,* **2005,** *29:* 477-490.

[8] Wood, B.J.B., Warner, P.J. Genetics of Lactic Acid Bacteria. Kluwer Academic/Plenum Publishers, New York, USA, **2003.**

[9] Gasson, M.J., de Vos, W.M. *Genetics and Biotechnology of Lactic Acid Bacteria.* Springer, Berlin, Germany, **2004.**

[10] Hugenholtz, J., Sybesma, W., Nierop Groot, M., Wisselink, W., Ladero, V., Burgess, K., *et al.* Metabolic Engineering of Lactic Acid Bacteria for the production of Nutraceuticals. *Antonie van Leeuwenhoek,* **2002,** *82:* 217-235.

[11] Wells, J.M., Mercenier, A. Lactic acid bacteria as mucosal delivery system. In, *Genetics of Lactic Acid Bacteria.* B. Wood and P.J. Warner (eds.), pp. 261-290. Kluwer Academic/Plenum Publishers, New York, USA, **2003.**

[12] Hanniffy, S., Wiedermann, U., Repa, A., Mercenier, A., Daniel, C., Fioramonti, J., *et al.* Potential and opportunities for use of recombinant lactic acid bacteria in human health. *Adv. Appl. Microbiol.,* **2004,** *56:* 1-64.

[13] Stanton, C., Ross, R.P., Fitzgerald, G.F and D. van Sinderen. Fermented functional foods. *Curr. Opin. Biotechnol.,* **2005,** *16:* 198-203.

[14] Klaenhammer, T.R., Barrangou, R., Buck, B.L., Azcaráte-Peril, M.A., Altermann, E. Genomic features of lactic acid bacteria effecting bioprocessing and health. *FEMS Microbiol. Rev.,* **2005,** *29:* 393-409.

[15] Makarova, K., Slesarev, A., Wolf, Y., Sorokin, A., Koonin, E., Pavlov, A., *et al.* Comparative genomics of the lactic acid bacteria. *Proc. Natl. Acad. Sci. USA,* **2006,** *103:* 15611-15616.

[16] Klijn, A., Mercenier, A., Arigoni, F. Lessons from the genomes of bifidobacteria. *FEMS Microbiol. Rev.,* **2005,** *29:* 491-509.

[17] O'Sullivan, O., O'Callaghan, J., Sangrador-Vegas, A., McAuliffe, O., Slattery, L., Kaleta, P., *et al.* Comparative genomics of lactic acid bacteria reveals a niche-specific gene set. *BMC Microbiol.,* **2009,** *9:* 50.

[18] Schell, M.A., Karmirantzou, M., Snel, B., Vilanova, D., Berger, B., Pessi, G., *et al.* The genome sequence of *Bifidobacterium longum* reflects its adaptation to the human gastrointestinal tract. *Proc. Natl. Acad. Sci. USA,* **2002,** *99:* 14422-14427.

[19] Bolotin, A., Quinquis, B., Renault, P., Sorokin, A., Ehrlich, S.D., Kulakauskas, S., *et al.* Complete sequence and comparative genome analysis of the dairy bacterium *Streptococcus thermophilus*. *Nat. Biotechnol.,* **2004,** *22:* 1554-1558.

[20] van de Guchte, M., Penaud, S., Grimaldi, C., Barbe, V., Bryson, K., Nicolas, P., *et al.* The complete genome sequence of *Lactobacillus bulgaricus* reveals extensive and ongoing reductive evolution. *Proc. Natl. Acad. Sci. USA,* **2006,** *103:* 9274-9279.

[21] Callanan, M., Kaleta, P., O'Callaghan, J., O'Sullivan, O., Jordan, K., McAuliffe, O., *et al.* Genome sequence of *Lactobacillus helveticus*, an organism distinguished by selective gene loss and insertion sequence element expansion. *J. Bacteriol.,* **2008,** *190:* 727-735.

[22] Mazmanian, S.K., Liu, C.H., Tzianabos, A.O., Kasper, D.L. An immunomodulatory molecule of symbiotic bacteria directs maturation of the host immune system. *Cell,* **2005,** *122:* 107-118.

[23] Ventura, M., van Sinderen, D., Fitzgerald, G.F., Zink, R. Insights into the taxonomy, genetics and physiology of bifidobacteria. Antonie van Leeuwenhoek, **2004,** *86:* 205-223.

[24] LoCascio, R.G., Desai, P., Sela, D.A., Weimer, B., Mills, D. A. Broad conservation of milk utilization genes in *Bifidobacterium longum* subsp. *infantis* as revealed by comparative genomic hybridization. *Appl. Environ. Microbiol.,* **2010,** *76:* 7373-7381.

[25] Turroni, F., Foroni, E., Serafini, F., Vappiani, A., Montanini, B., Bottacini, F, *et al.* Ability of *Bifidobacterium* breve to grow of different types of milk: exploring the metabolism of milk through genome analysis. *Appl. Environ. Microbiol.,* **2011,** *77:* 7408-7417.

[26] Ventura, M., F. Turroni, G.L. Mendez, E. Foroni, A. Zomer, S. Duranti, V., *et al.* Comparative analyses of prophage-like elements present in bifidobacterial genomes. *Appl. Environ. Microbiol.,* **2009,** *75:* 6929-6936.

[27] Flint, H.J., Bayer, E.A., Rincon, M.T., Lamed, R., White, B.A. Polysaccharide utilization by gut bacteria: potential for new insights from genomic analysis. *Nat. Rev. Microbiol.,* **2008,** *6:* 121-131.

[28] Ivanov, D., Emonet, C., Foata, F., Affolter, M., Delley, M., Fisseha, M., *et al.* A serpin from the gut bacterium *Bifidobacterium longum* inhibits eukaryotic elastase-like serine proteases. *J. Biol. Chem.,* **2006,** *281:* 17246-17252.

[29] O'Connell Motherway, M., Zomer, A., Leahy, S.C., Reunanen, J., Bottacini, F., Claesson, M.J., *et al.* Functional genome analysis of *Bifidobacterium breve* UCC2003 reveals type IVb tight adherence (Tad) pili as an essential and conserved host-colonization factor. *Proc Natl Acad Sci USA,* **2011,** *108:* 11217-11222.

[30] Ryan, S.M., Fitzgerald, G.F., Van Sinderen, D. Transcriptional regulation and characterization of a novel beta-fructofuranosidase-encoding gene from *Bifidobacterium breve* UCC2003. *Appl. Environ. Microbiol.,* **2005,** *71:* 3475-3482.

[31] Mazé, A., O'Connell Motherway, M., Fitzgerald, G.F., Deutscher, J., van Sinderen, D. Identification and characterization of a fructose phosphotransferase system in *Bifidobacterium breve* UCC2003. *Appl. Environ. Microbiol.,* **2007,** *73:* 545-553.

[32] Sela, D.A., Chapman, J., Adeuya, A., Kim, J.H., Chen, F., Whitehead, T.R., *et al.* The genome sequence of *Bifidobacterium longum* subsp. *infantis* reveals adaptations for milk utilization within the infant microbiome. *Proc Natl Acad Sci USA,* **2008,** *105:* 18964-18969.

[33] Turroni, F., Bottacini, F., Foroni, E., Mulder, I., Kim, J.H., Zomer, A., *et al.* Genome analysis of *Bifidobacterium bifidum* PRL2010 reveals metabolic pathways for host-derived glycan foraging. *Proc. Natl. Acad. Sci. USA,* **2010,** *107:* 19514-19519.

[34] Turroni, F., Milani, C., van Sinderen, D., Ventura, M. Genetic strategies for mucin metabolism in *Bifidobacterium bifidum* PRL2010: an example of possible human-microbe co-evolution. *Gut Microbes,* **2011,** *2:* 183-189.

[35] Foroni, E., Serafini, F., Amidani, D., Turroni, F., He, F., Bottacini, F., *et al.* Genetic analysis and morphological identification of pilus-like structures in members of the genus *Bifidobacterium. Microbial Cell Factories,* **2011,** *10:* S16.

[36] Mantzourani, M., Fenlon, M., Beighton, D. Association between *Bifidobacteriaceae* and the clinical severity of root caries lesions. *Oral Microbiol. Immunol.,* **2009,** *24:* 32-37.

[37] Ventura M., Turroni F., Zomer A., Foroni E., Giubellini V., Canchaya C., *et al.* The *Bifidobacterium dentium* Bd1 genome sequence reflects its genetic adaptation to the human oral cavity. *Plos Genetics,* **2009,** *12:* e1000785.

[38] Ventura, M., Turroni, F., Foroni, E., Durante, E., Guibellini, V., Bottacini, F., *et al.* Analysis of bifidobacterial prophage-like sequences. *Antonie van Leeuwenhoek,* **2010,** *98:* 39-50.

[39] Barrangou, R., Briczinski, E.P., Traeger, L.L., Loquasto, J.R., Richards, M., Horvath, P., *et al.* Comparison of the complete genome sequences of *Bifidobacterium animalis* subsp. *lactis* DSM 10140 and Bl-04. *J. Bacteriol.,* **2009,** *191:* 4144-4151.

[40] Garrigues C, Johansen, E., Pedersen, M.B. Complete genome sequence of *Bifidobacterium animalis* subsp. *lactis* BB-12, a widely consumed probiotic strain. *J. Bacteriol.,* **2010,** *192:* 2467-2468.

[41] Bottacini, F., Dal Bello, F., Turroni, F., Milani, C., Duranti, S., Foroni, E., *et al.* Complete genome sequence of *Bifidobacterium animalis* subsp. *lactis* BLC1. J. Bacteriol., **2011,** *193:* 6387-6388.

[42] Hols, P., Kleerebezem, M., Schanck, A.N., Ferain, T., Hugenholtz, J., Delcour, J., *et al.* Conversion of *Lactococcus lactis* from homolactic to homoalanine fermentation through metabolic engineering. *Nat. Biotechnol.,* **1999,** *17:* 588-592.

[43] Nouaille, S., Ribeiro, L.A., Miyoshi, A., Pontes, D., Le Loir, Y., Oliveira, S.C., *et al.* Heterologous protein production and delivery systems for *Lactococcus lactis*. *Genet. Mol. Res.,* **2003,** *2:* 102-111.

[44] Bolotin, A., Wincker, P., Mauger, S., Jaillon, O., Malarme, K., Weissenbach, J., *et al.* The complete genome sequence of the lactic acid bacterium *Lactococcus lactis* ssp. *lactis* IL1403. *Genome Res.,* **2001,** *11:* 731-753.

[45] Wegmann, U., O'Connell Motherway, M., Zomer, A., Buist, G., Shearman, C., Canchaya, C., *et al.* Complete genome sequence of the prototype lactic acid bacterium *Lactococcus lactis* subsp. *cremoris* MG1363. *J. Bacteriol.,* **2007,** *189:* 3256-3270.

[46] Siezen, R.J., Bayjanov, J., Renckens, B., Wels, M., van Hijum, S.A., Molenaar, D., *et al.* Complete genome sequence of *Lactococcus lactis* subsp. *lactis* KF147, a plant-associated lactic acid bacterium. *J. Bacteriol.,* **2010,** *192:* 2649-2650.

[47] Le Bourgeois, P., Lautier, M., van den Berghe, L., Gasson, M.J., Ritzenthaler, P. Physical and genetic map of the *Lactococcus lactis* subsp. *cremoris* MG1363 chromosome:

comparison with that of *Lactococcus lactis* subsp. *lactis* IL 1403 reveals a large genome inversion. *J. Bacteriol.,* **1995,** *177:* 2840-2850.

[48] Gasson, M.J. Plasmid complements of *Streptococcus lactis* NCDO 712 and other lactic streptococci after protoplast-induced curing. *J. Bacteriol.,* **1983,** *154*: 1-9.

[49] Duwat, P., Sourice, S., Cesselin, B., Lamberet, G., Vido, K., Gaudu, P., *et al.* Respiration capacity of the fermenting bacterium *Lactococcus lactis* and its positive effects on growth and survival. *J. Bacteriol.,* **2001,** *183:* 4509-4516.

[50] Gaudu, P., Vido, K., Cesselin, B., Kulakauskas, S., Tremblay, J., Rezaiki, L., *et al.* Respiration capacity and consequences in *Lactococcus lactis*. *Antonie van Leeuwenhoek,* **2002,** *82:* 263-269.

[51] Duwat, P., Sourice, S., Gruss, A. Process for preparing starter cultures of lactic acid bacteria. *French Patent Application,* **1998,** FR9809463.

[52] Ayad, E.H., Veheul, A., de Jong, D., Wouters, J.T., Smit, G. Flavor forming abilities and amino acid requirements of *Lactococcus lactis* strains isolated from artisanal and non-dairy origin. *Int. Dairy J.,* **1999,** *9:* 725-735.

[53] Siezen, R.J., Starrenburg, M.J., Boekhorst, J., Renckens, B., Molenaar, D., van Hylckama-Vlieg, J.E. Genome-scale genotype-phenotype matching of two *Lactococcus lactis* isolates from plants identifies mechanisms of adaptation to the plant niche. *Appl. Environ. Microbiol.,* **2008,** *74:* 424-236.

[54] Rettger, L.F., Levy, M.N., Weinstein, L., Weiss, J.E. *Lactobacillus acidophilus and its Therapeutic Application.* Yale University Press, New Haven, USA, **1935.**

[55] Johansson, M.L., Molin, G., Jeppsson, B., Nobaek, S., Ahrné, S., Bengmark, S. Administration of different *Lactobacillus* strains in fermented oatmeal soup: *in vivo* colonization of human intestinal mucosa and effect on the indigenous flora. *Appl. Environ. Microbiol.,* **1993,** *59:* 15-20.

[56] Kwok, L., Stapleton, A.E., Stamm, W.E., Hillier, S.L., Wobbe, C.L., Gupta, K. Adherence of *Lactobacillus crispatus* to vaginal epithelial cells from women with or without a history of recurrent urinary tract infection. *J. Urol.,* **2006,** *176:* 2050-2054.

[57] Hudault, S., Lievin, V., Bernet-Camard, M.F., Servin, A.L. Antagonistic activity exerted *in vitro* and *in vivo* by *Lactobacillus casei* (strain GG) against *Salmonella typhimurium* C5 infection. *Appl. Environ. Microbiol.,* **1993,** *63:* 513-518.

[58] Coconnier, M.H., Lievin, V., Hemery, E., Servin, A.L. Antagonistic activity against *Helicobacter* infection *in vitro* and *in vivo* by the human *Lactobacillus acidophilus* strain L. *Appl. Environ. Microbiol.,* **1998,** *64:* 4573-4580.

[59] Kimura, K., McCartney, A.L., McConnell, M.A., Tannock, G.W. Analysis of fecal populations of bifidobacteria and lactobacilli and investigation of the immunological responses of their human hosts to the predominant strains. *Appl. Environ. Microbiol.,* **1997,** *63:* 3394-3398.

[60] Pereira, D.I., Gibson, G.R. Cholesterol assimilation by lactic acid bacteria and bifidobacteria isolated from the human gut. *Appl. Environ. Microbiol.,* **2002,** *68:* 4689-4693.

[61] Sanders, M.E., Klaenhammer, T.R. Invited review: the scientific basis of *Lactobacillus acidophilus* NCFM functionality as a probiotic. *J. Dairy Sci.,* **2001,** *84:* 319-331.

[62] Altermann, E., Russell, W.M., Azcárate-Peril, M.A., Barrangou, R., Buck, B.L., McAuliffe, O., *et al.* Complete genome sequence of the probiotic lactic acid bacterium *Lactobacillus acidophilus* NCFM. *Proc. Natl. Acad. Sci. USA*, **2005,** *102:* 3906-3912.

[63] Barrangou, R., Fremaux, C., Deveau, H., Richards, M., Boyaval, P., Moineau, S., *et al.* CRISPR provides acquired resistance against viruses in prokaryotes. *Science,* **2007,** *315:* 1709-1712.

[64] Barefoot, S.F., Klaenhammer, T.R. Purification and characterization of the *Lactobacillus acidophilus* bacteriocin lactacin B. *Antimicrob. Agents Chemother.,* **1984,** *26:* 328-334.

[65] Slattery, L., O'Callaghan, J., Fitzgerald, G.F., Beresford, T., Ross, R.P. Invited review: *Lactobacillus helveticus*--a thermophilic dairy starter related to gut bacteria. *J. Dairy Sci.,* **2010,** *93:* 4435-4454.

[66] Takano, T. Anti-hypertensive activity of fermented dairy products containing biogenic peptides. *Antonie van Leeuwenhoek,* **2002,** *82:* 333-30.

[67] Pridmore, R.D., Berger, B., Desiere, F., Vilanova, D., Barretto, C., Pittet, A.C., *et al.* The genome sequence of the probiotic intestinal bacterium *Lactobacillus johnsonii* NCC 533. *Proc. Natl. Acad. Sci. USA,* **2004,** *101:* 2512-2517.

[68] Kleerebezem, M., Boekhorst, J., van Kranenburg, R., Molenaar, D., Kuipers, O.P., Leer, R., *et al.* Complete genome sequence of *Lactobacillus plantarum* WCFS1. *Proc. Natl. Acad. Sci. USA,* **2003,** *100:* 1990-1995.

[69] Doron, S., Snydman, D.R., Gorbach, S.L. *Lactobacillus* GG: bacteriology and clinical applications. *Gastroenterol. Clin. N. Am.,* **2005,** *34:* 483-498.

[70] Kankainen, M., Paulin, L., Tynkkynen, S., von Ossowski, I., Reunanen, J., Partanen, P., *et al.* Comparative genomic analysis of *Lactobacillus rhamnosus* GG reveals pili containing a human- mucus binding protein. *Proc. Natl. Acad. Sci. USA,* **2009,** *106:* 17193-17198.

[71] Lebeer, S., Verhoeven, T.L., Francius, G., Schoofs, G., Lambrichts, I., Dufrêne, Y., *et al.* Identification of a gene cluster for the biosynthesis of a long, galactose-rich exopolysaccharide in *Lactobacillus rhamnosus* GG and functional analysis of the priming glycosyltransferase. *Appl. Environ. Microbiol.,* **2009,** *75:* 3554-3563.

[72] Chaillou, S., Champomier-Verges, M.C., Cornet, M., Crutz-Le Coq, A.M., Dudez, A.M., Martin, V., *et al.* The complete genome sequence of the meat-borne lactic acid bacterium *Lactobacillus sakei* 23K. *Nat. Biotechnol.,* **2005,** *23:* 1527-1533.

[73] Molin, G., Jeppsson, B., Johansson, M.L., Ahrné S., Nobaek, S., Stahl, M., *et al.* Numerical taxonomy of *Lactobacillus* spp. associated with healthy and diseased mucosa of the human intestines. *J. Appl. Bacteriol.,* **1993,** *74:* 314-323.

[74] Makarova, K.S., Koonin, E.V. Evolutionary genomics of lactic acid bacteria. *J. Bacteriol.,* **2007,** *189:* 1199-1208.

[75] Claesson, M.J., Li, Y., Leahy, S., Canchaya, C., van Pijkeren, J.P., Cerdeño-Tárraga, A.M., *et al.* Multireplicon genome architecture of *Lactobacillus salivarius*. *Proc. Natl. Acad. Sci. USA,* **2006,** *103:* 6718-6723.

[76] Li, Y., Canchaya, C., Fang, F., Raftis, E., Ryan, K.A., van Pijkeren, J.P., *et al.* Distribution of megaplasmids in *Lactobacillus salivarius* and other lactobacilli. *J. Bacteriol.,* **2007,** *189:* 6128-6139.

[77] Corr, S.C., Li, Y., Riedel, C.U., O'Toole, P.W., Hill, C., Gahan, C.G. Bacteriocin production as a mechanism for the antiinfective activity of *Lactobacillus salivarius* UCC118. *Proc. Natl. Acad. Sci. USA,* **2007,** *104:* 7617-7621.

[78] Hugenholtz, J. Citrate metabolism in lactic acid bacteria. *FEMS Microbiol. Rev.,* **1993,** *12:* 165-178.

[79] Kim, J.F., Jeong, H., Lee, J.S., Choi, S.H., Ha, M., Hur, C.G., *et al.* The complete genome sequence of *Leuconostoc citreum* KM20. *J. Bacteriol.,* **2008,** *190:* 3093-3094.

[80] Dicks, L.M.T., Dellaglio, F., Collins, M.D. Proposal to reclassify *Leuconostoc oenus* as *Oenococcus oeni* (corrig.) gen nov, comb nov. *Int. J. Syst. Bacteriol.,* **1995,** *45:* 395-397.

[81] Versari, A., Parpinello, G.P., Cattaneo, M. *Leuconostoc oenos* and malolactic fermentation of wine: a review. *J. Ind. Microbiol. Biotechnol.,* **1999,** *23:* 447-455.

[82] Mills, D.A., Rawsthorne, H., Parker, C., Tamir, D., Makarova, K. Genomic analysis of *Oenococcus oeni* PSU-1 and its relevance to winemaking. *FEMS Microbiol. Rev.,* **2005,** *29:* 465-475.

[83] Hols, P., Hancy, F., Fontaine, L., Grossiord, B., Prozzi, D., Leblond-Bourget, N., *et al.* New insights in the molecular biology and physiology of *Streptococcus thermophilus* revealed by comparative genomics. *FEMS Microbiol. Rev.,* **2005,** *29:* 435-463.

[84] Blomqvist, T., Steinmoen, H., Håvarstein, L.S. Natural genetic transformation: a novel tool for efficient genetic engineering of the dairy bacterium *Streptococcus thermophilus*. *Appl. Environ. Microbiol.,* **2006,** *72:* 6751-6756.

[85] Siezen, R.J., Renckens, B., van Swam, I., Peters, S., van Kranenburg, R., Kleerebezem, M., *et al.* Complete sequences of four plasmids of *Lactococcus lactis* subsp. *cremoris* SK11 reveal extensive adaptation to the dairy environment. *Appl. Environ. Microbiol.,* **2005,** *71:* 8371-8382.

[86] Schneider, D., Lenski, R.E. Dynamics of insertion sequence elements during experimental evolution of bacteria. *Res. Microbiol.,* **2004,** *155:* 319-327.

[87] Cogan, T.M., Barbosa, M., Beuvier, E., Bianchi-Salvadori, S., Cocconcelli, P.S., Fernandes, I., *et al.* Characterization of the lactic acid bacteria in artisan dairy products. *J. Dairy Res.,* **1997,** *64:* 409-421.

[88] Kok, J., Buist, G., Zomer, A.L., van Hijum, S.A., Kuipers, O.P. Comparative and functional genomics of lactococci. *FEMS Microbiol. Rev.,* **2005,** *29:* 411-433.

[89] Ventura, M., O'Connell Motherway, M., Leahy, S., Moreno-Muñoz, J.A., Fitzgerald, G.F., van Sinderen, D. From bacterial genome to functionality; case bifidobacteria. *Int. J. Food Microbiol.,* **2007,** *120:* 2-12.

[90] Tomomatsu, H. Health effects of oligosaccharides. *Food Technol.,* **1994,** *48:* 61-65.

[91] Possemiers, S., Bolca, S., Eeckhaut, E., Depypere, H., Verstraete, W. Metabolism of isoflavones, lignans and prenylflavonoids by intestinal bacteria: producer phenotyping and relation with intestinal community. *FEMS Microbiol. Ecol.,* **2007,** *61:* 372-383.

[92] Blaut, M., Clavel, T. Metabolic diversity of the intestinal microbiota: implications for health and disease. *J. Nutr.,* **2007,** *137:* 751S-755S.

[93] Ryan, S.M., Fitzgerald, G.F., Van Sinderen, D. Screening and identification of starch, amylopectin and pullulan-degrading activities in bifidobacterial strains. *Appl. Environ. Microbiol.,* **2006,** *72:* 5289-5296.

[94] van den Broek, L.A.M., Hinz, S.W.A., Beldman, G., Doeswijk-Voragen, C.H.L., Vincken, J.-P., Voragen, A.G.J. Glycosyl hydrolases from *Bifidobacterium adolescentis* DSM20083. An overview. *Lait,* **2005,** *85:* 125-133.

[95] Garro, M.S., de Valdez, G.F., Oliver, G., de Giori, G.S. Hydrolysis of soya milk oligosaccharides by *Bifidobacterium longum* CRL 849, *Z. Lebensm. Unters. Forsch. A.,* **1999,** *208:* 57-59.

[96] Sijpesteijn, A.K. Induction of cytochrome formation and stimulation of oxidative dissimilation by hemin in *Streptococcus lactis* and *Leuconostoc mesenteroides*. *Antonie van Leeuwenhoek,* **1970,** *36:* 335-348.

[97] Vido, K., Le Bars, D., Mistou, M.Y., Anglade, P., Gruss, A., Gaudu, P. Proteome analyses of heme-dependent respiration in *Lactococcus lactis*: involvement of the proteolytic system. *J. Bacteriol.,* **2004,** *186:* 1648-1657.

[98] Pedersen, M.B., Iversen, S.L., Sorensen, K.I., Johansen, E. The long and winding road from the research laboratory to industrial applications of lactic acid bacteria. *FEMS Microbiol. Rev.,* **2005,** *29:* 611-624.

[99] Brooijmans, R.J., de Vos, W.M., Hugenholtz, J. *Lactobacillus plantarum* WCFS1 electron transport chains. *Appl. Environ. Microbiol.,* **2009,** *75:* 3580-3585.

[100] Watanabe, M., van der Veen, S., Nakajima, H., Abee, T. Effect of respiration and manganese on oxidative stress resistance of *Lactobacillus plantarum* WCFS1. *Microbiology,* **2012,** *158:* 293-300.

[101] Ventura, M., O'Flaherty, S., Claesson, M.J., Turroni, F., Klaenhammer, T.R., van Sinderen, *et al.* Genome-scale analyses of health-promoting bacteria: probiogenomics. *Nat. Rev. Microbiol.,* **2009,** *7:* 61-71.

[102] von Ossowski, I., Reunanen, J., Satokari, R., Vesterlund, S., Kankainen, M., Huhtinen, H., *et al.* Mucosal adhesion properties of the probiotic *Lactobacillus rhamnosus* GG SpaCBA and SpaFED pilin subunits. *Appl. Environ. Microbiol.,* **2010, 76:** 2049-2057.

[103] Reunanen, J., von Ossowski, I., Hendrickx, A.P., Palva, A., de Vos, W.M. Characterization of the SpaCBA pilus fibers in the probiotic *Lactobacillus rhamnosus* GG. *Appl. Environ. Microbiol.,* **2012,** *78:* 2337-2344.

[104] Lebeer, S., Claes, I., Tytgat, H.L., Verhoeven, T.L., Marien, E., von Ossowski, I., *et al.* Functional analysis of *Lactobacillus rhamnosus* GG pili in relation to adhesion and immunomodulatory interactions with intestinal epithelial cells. *Appl. Environ. Microbiol.,* **2012,** *78:* 185-193.

[105] Siezen, R.J., van Enckevort, F.H., Kleerebezem, M., Teusink, B. Genome data mining of lactic acid bacteria: the impact of bioinformatics. *Curr. Opin. Biotechnol.,* **2004,** *15:* 105-115.

[106] Molenaar, D., Bringel, F., Schuren, F.H., de Vos, W.M., Siezen, R.J., Kleerebezem, M. Exploring *Lactobacillus plantarum* genome diversity by using microarrays. *J. Bacteriol.,* **2005,** *187:* 6119-6127.

[107] Berger, B., Pridmore, R.D., Barretto, C., Delmas-Julien, F., Schreiber, K., Arigoni, F., *et al.* Similarity and differences in the *Lactobacillus acidophilus* group identified by polyphasic analysis and comparative genomics. *J. Bacteriol.,* **2007,** *189:* 1311-1321.

[108] Ventura, M., Canchaya, C., Bernini, V., Altermann, E., Barrangou, R., McGrath, S., *et al.* Comparative genomics and transcriptional analysis of prophages identified in the genomes of *Lactobacillus gasseri, Lactobacillus salivarius*, and *Lactobacillus casei*. *Appl. Environ. Microbiol.,* **2006,** *72:* 3130-3146.

[109] Claesson, M.J., van Sinderen, D., and O'Toole, P.W. *Lactobacillus* phylogenomics – towards a reclassification of the genus. *Int. J. Syst. Evolut. Microbiol.,* **2008,** *58:* 2945-2954.

[110] Teusink, B., van Enckvort, F.H., Francke, C., Wiersma, A., Wegkamp, A., Smid, E.J., *et al.* *In silico* reconstruction of the metabolic pathways of *Lactobacillus plantarum*: comparing predictions of nutrient requirements with those from growth experiments. *Appl. Environ. Microbiol.,* **2005,** *71:* 7253-7262.

[111] Notebaart, R.A., Teusink, B. Accelerating the reconstruction of genome-scale metabolic networks. *BMC Bioinformatis,* **2006,** *7:* 296.

[112] Pastink, M.I., Teusink, B., Hols, P., Visser, S., de Vos, W.M., Hugenholtz, J. Genome-scale model of *Streptococcus thermophilus* LMG18311 for metabolic comparison of lactic acid bacteria. *Appl. Environ. Microbiol.,* **2009,** *75:* 3627-3633.

[113] Agaton, C., Uhlen, M., Hober, S. Genome-based proteomics. *Electrophoresis*, **2004,** *25:* 1280-1288.

[114] Gagnaire, V., Piot, M., Camier, B., Vissers, J.P., Jan, G., Leonil, J. Survey of bacterial proteins released in cheese: a proteomic approach. *Int. J. Food Microbiol.,* **2004,** *94:* 185-201.

[115] Pessione, A., Lamberti, C., Pessione, E. Proteomics as a tool for studying energy metabolism in lactic acid bacteria. *Mol. Biosyst.,* **2010,** *6:* 1419-1430.

[116] Sánchez, B., Ruiz, L., de los Reyes-Gavilán, C.G., Margolles, A. Proteomics of stress response in *Bifidobacterium*. *Front. Biosci.,* **2008,** *13:* 6905-6919.

[117] Aires, J., Butel, M.J. Proteomics, human gut microbiota and probiotics. *Expert Rev. Proteomics.,* **2011,** *8:* 279-288.

[118] Kuipers, O.P., Buist, G., Kok, J. Current strategies for improving food bacteria. *Res. Microbiol.,* **2000,** *151:* 815-822.

[119] McKay, L.L. Functional properties of plasmids in lactic streptococci. *Antonie van Leeuwenhoek,* **1983,** *49:* 259-274.

[120] Fitzgerald, G.F., Gasson, M.J. *In vivo* gene transfer systems and transposons. *Biochimie,* **1988,** *70:* 489-502.

[121] Berge, M., Moscoso, M., Prudhomme, B., Martin, B., Claverys, J.P. Uptake of transforming DNA in Gram-positive bacteria: a view from *Streptococcus pneumoniae. Mol. Microbiol.,* **2002,** *45:* 411-421.

[122] Luo, P., Li, H., Morrison, D.A. ComX is a unique link between multiple quorum sensing outputs and competence in *Streptococcus pneumoniae. Mol. Microbiol.,* **2003,** *50:* 5004-5016.

Send Orders for Reprints to reprints@benthamscince.net
Advances in Genome Science, Vol. 3, **2014, 127-142** **127**

CHAPTER 4

Recent Genomic Advances for Weed Science

David Horvath[*]

USDA-ARS, Bioscience Research Laboratory, 1605 Albrecht Blvd. Fargo ND 58105, USA

Abstract: Numerous genomic-based studies have provided insight to the physiological and evolutionary processes involved in developmental and environmental processes of model plants such as arabidopsis and rice. However, far fewer efforts have been attempted to use genomic resources to study physiological and evolutionary processes of weedy plants. Genomics-based tools such as extensive EST databases and microarrays have been developed for a limited number of weedy species, although application of information and resources developed for model plants and crops are possible and have been exploited. These tools have just begun to provide insights into the response of these weeds to herbivore and pathogen attack, survival of extreme environmental conditions, and interaction with crops. The potential of these tools to illuminate mechanisms controlling the traits that allow weeds to invade novel habitats, survive extreme environments, and that make weeds difficult to eradicate have potential for both improving crops and developing novel methods to control weeds.

Keywords: AFLP, *Amaranthus hypochondriacus*, bacterial artificial chromosome libraries, candidate genes, *conyza canadensis*, copy number variation, *Euphorbia esula*, expressed sequence tags, genome wide association mapping, genomic libraries, genomics, *Helianthus ssp.,* herbicide resistance, invasiveness, ISSR, microarray, next generation sequencing, pacific biosiences sequencing, siRNA, targeted sequencing, two-hybrid libraries, weeds.

1. GENOMICS TOOLS AND PURPOSE

Genomics is the study of the organization, evolution, and function of the genes and non-coding regions of the genome [1]. Various sub-disciplines of genomics focus on specific attributes of genome analysis. For example, functional genomics is the study of genes function and regulation and often centers on expression

***Address correspondence to David Horvath:** USDA-ARS, Bioscience Research Laboratory, 1605 Albrecht Blvd. Fargo ND 58105, USA; Tel: 701-239-1255; Fax: 701-239-1252; E-mail: David.horvath@ars.usda.gov

Christian Neri (Ed)

analysis tools such as microarrays (see below), and development of large scale reverse genetic resources such as libraries of thousands of individuals with characterized mutations in numerous genes [2], or identification of natural mutations from laboratory or wild populations [3]. Likewise, comparative genomics focuses on comparing the organization and sequences variation between species to obtain a better understanding of the evolution of genes and organisms [4]. Obviously there is overlap between sub-disciplines of genomics. For example the aim of "structural genomics" is to identify conserved protein domains [5]. Obviously such information would be relevant to functional and comparative genomics. Population genomics is similar to comparative genomics, but is focused on evolution of gene structure and function within and between closely related species [6]. The basic tools needed for genomic analysis include whole genome sequences, large scale mapping of genes, physical maps, bacterial artificial chromosome (BAC) libraries, collections of sequences of expressed genes (development of EST databases), methods to assess gene expression (at transcriptomic, proteomic and metabolomics level), as well as the databases for storage and computational programs needed to analyze and compare these huge datasets.

De novo whole genome sequencing is the holy grail of genomics for any organism, especially for non-model plants such as weeds [7]. The data generated from such a project not only assists in the identification of all real and potential genes within an organism, but also identification of *cis*-acting regulatory sequences (non-protein coding sequences surrounding genes that often contain the information needed for turning the genes on and off). Whole genome sequencing also provides a detailed physical map of gene order and distance. However, the cost of such a project makes it outside the reach of most researchers studying weeds and other non-model plants. Fortunately, the advent of new technologies such as long read 454 Titanium platforms, Illumina paired end pyrosequencing, and SOLiD 3 can produce large amounts of sequence data for significantly less cost than traditional Sanger sequencing methods. However, these technologies are hindered somewhat due to the small read lengths they produce. For example, the Titanium 454 FLX technology produces read lengths greater than 500 bases and can also generate ~ up to 600 Mb of sequence data for under $8,000. The various paired end read technologies such as Illumina can produce up to 9 gigabases (Gb) of random 75 – 100 base sequences (for example, enough to provide nearly a 5X coverage of the ~2.2 Gb leafy spurge genome) for under $400 per Gb. Even greater cost reductions are achieved by the SOLiD 3 technology that boasts the generation of a Gb for under $300. The reduce the cost of such endeavors and bring the possibility of at least partial genome sequencing within the grasp of weed scientists [8]. However it should be noted that Ilumina and SOLiD3 generate very short sequence reads, and it is untenable to use them for *de novo* genome sequencing. Third generation sequencing technologies such as the MinIon sequencer produced by Oxford Nanopore Technologies and detection base additions from isolated DNA polymerases from technologies such as those developed by Pacific Biosciences offer both low cost high volume sequencing with read lengths between 3000 and 30,000 bases with 200 Mb to 1Gb of sequence for about $1,000 including sequencing and library preparation. However, both of these long-read technologies have problems with error rates (the Pacific Biosciences platform has error rates of about 15% and the MinIon platform has error rates of 4%). Neither is a significant issue provided there is sufficient redundancy in sequence data produced to allow building of consensus sequences.

Although full genome sequencing is ideal, obtaining genomic sequences of gene-rich regions could provide a more economical approach to provide important regulatory sequences without wasting resources on sequencing vast stretches of repetitive DNA. There are several approaches that could be used to isolate gene-rich regions of the genome. These include preparing genomic libraries using bacterial strains that destroy hypermethylated DNA sequences [9]. Most actively transcribed genomic sequences are hypomethylated and thus would be protected in the bacterial strains. Another possibility would be to enrich the genomic DNA for low copy number sequences. This has been done by CoT hybridization followed by separation of the DNA fragments on hydroxyapetite columns [10]. A final possibility lies in novel uses of magnetic bead purification technology. Comprehensive cDNA libraries (collection of DNA copies of gene-encoding mRNAs that have been cloned into plasmid or virus vectors) could be amplified using biotinylated primers which could then be bound to magnetic beads. These bound cDNAs could then be hybridized to shear genomic DNA and purified. The genomic DNA could then be released from the bound cDNAs and cloned for conventional sequencing or directly sequenced using 454 technology. A derivation of this technique has utilized microarrays to hybridize and trap sheared genomic DNA that is complimentary to the probes on the array. These trapped sequences can then be stripped off, amplified, and sequenced with any of the 2nd or third generation sequencing technologies [11].

Sequencing of transcribed regions of the genome is the most common approach for developing genomics data and tools for non-model plants [12]. This is usually accomplished by sequencing cDNAs either by standard Sanger sequencing methods or using newer 454 and Illumina technologies. Collections of the cDNA sequences are commonly referred to as EST (expressed sequence tags) databases. Because sequences of individual cDNAs can provide an indication of the function of the gene which encoded them, EST databases allow one to identify various genes likely to be involved in specific physiological or developmental processes. This so called "candidate gene approach" serves as a potential staring point to investigate the impact of a given treatment on expression of particular genes of interest [13].

Collections of sequenced cDNAs also allow the production of microarrays. Microarrays are glass slides on which thousands of individual cDNAs or fragments of cDNAs (referred to as probes) are spotted in a grid pattern [14]. Fluorescently labeled cDNAs produced from various tissues or following specific treatments can be hybridized to the microarrays. Because the amount of a specific cDNAs that hybridizes to it's complementary sequences on the array is directly relative to the expression of the gene which encoded them, the brightness of a given spot on the arrays is an indication of the expression level of the gene from which that probe was derived. Since the genes on the array have been characterized, it is possible to draw inferences about the physiological processes that are occurring in a given sample based on the probable function of the genes that are being expressed. However, since sequencing large numbers of ESTs is often not practical in weeds, in some cases, microarrays have been developed from unsequenced cDNAs. In most cases, some or all of the probes from these so called "anonymous arrays" are usually sequenced after they are determined to be of interest by differentially hybridization [15]. Another novel use of microarray technology that has been developed specifically for weedy and non-model systems is the use of commercially available microarrays developed for crop or model plant to follow gene expression in non-model organisms [16-18].

In cases where whole genome sequences are lacking, high density genetic and physical maps compliment and rely on other genomic resources. Simple sequence repeats (SSRs), microsatilites, amplified fragment length polymorphisms (ALFPs), and inter-simple sequence repeats (ISSRs) are polymerase chain reaction (PCR)- based genetic markers that can be developed without prior knowledge of the genomic sequences, and have been a mainstay for researchers working in non-model plants and weeds [19]. The development of high-density genetic maps based on these markers has proven to be of great value for identifying and cloning genes of interest when combined with BAC libraries [20]. Such high density maps can also be used to identify regions of chromosomes that are undergoing evolution in response to altered niches or environments [6].

Combining newer sequencing technologies with marker development also offers significant possibilities for identifying markers associated with chromosome regions undergoing evolution. Sequencing multiple individuals from a given population can identify thousands of single nucleotide polymorphisms (SNPs) or other small differences in the gene sequences. Comparison of SNP frequency between populations can identify cases where these alleles are not maintained in Hardy Weinberg equilibrium and are thus associated with particular populations or environments. Although the prospect of sequencing the 100s of individuals needed for a population analysis may seem daunting, the huge number of sequences produced in a single sequencing run using Illumina technologies allows for the possibility of sequencing numerous individuals simultaneously (a process referred to as multiplex sequencing), so long as the individuals can be distinguished from each other and there is enough iterations of each sequence produced to allow one to distinguish between sequence errors and real differences in the DNA between individuals. Distinguishing between individuals in a multiplexed sequence is easily done by adding short sequence "tags" to the end of each of the fragments produced from a specific individual or sample. However, even with technologies like Illumina where up to 9 Gb of sequence is produced in a single run, sequencing of even a single individual is unlikely to produce sufficient iterations of a given genomic fragment to allow identification of SNPs unless the genome size is less than a Gb in size.

For population level studies methods to reduce the complexity of genomes is required. Thus, techniques to create reduced representation libraries (RRLs) to limit the complexity of the DNA being sequenced have been developed so that multiplex sequencing will provide sufficient iterations of any given fragment to ensure any SNPs observed are not the result of sequencing error. One way to reduce the complexity is to just sequence the transcribed genes of the individuals. Most transcripts are about 2000 bases long, and there is an average of about 20,000 genes transcribed in most samples. Thus, a single lane of Illumina will produce over 200 sequences covering each transcript. Consequently, it is reasonable to multiplex 20-30 individuals providing the transcriptome libraries were normalized so that all transcripts were present in about the same molar ratios. However, coding sequences are generally well conserved and thus fewer SNPs or other sequence variations are likely to be found if transcriptomes are sequenced. A popular method of producing RRLs from genomic sequences which are far more likely to accumulate sequence variants is to only sequence DNA at specific restriction enzyme cleavage sites. This technique is known as Restriction site associated DNA sequencing or RAD sequencing. By chance alone, one would expect a 1 Gb genome to contain 250,000 sites from a 6 base cutting enzyme, and only 15,000 sites from an 8 base cutting enzyme. Thus, a single Illumina lane would provide 360X coverage of the 100 bases following each 6 base restriction site. Thus, as many as 36 individuals could be multiplexed on a single lane. Even more individuals could be multiplexed from species with smaller genomes (for example more than 100 individual arabidopsis plants could be sequenced simultaneously and still provide adequate representation from each locus). These technologies can identify 1000s of SNPs segregating within and between populations. This opens the possibility to use genome wide association mapping for non-model plants such as invasive weeds to find genomic regions that are undergoing selection in response to everything from herbicide pressure to adaptation to invaded environments.

2. CURRENT STATE OF GENOMICS RESEARCH IN WEEDS

2.1. Comparative Genomics

The use of genomic information for studies in weed science is still in its infancy in most cases. There are currently eight weeds with well developed genomic resources and significant research efforts. Three of these (wild rice, (*Oryza ssp.*), wild sunflowers (*Helianthus ssp.*), and perennial ryegrass (*Lolium perenne*)) have close crop relatives. *Amaranthus* ssp., are related to grain amaranth (*Amaranthus caudatus*) which is a minor crop species in Latin America and was an important crop in the pre-columbian period [21]. The other species leafy spurge (*Euphorbia esua*), dodder (*Cuscuta ssp.*), starthistle (*Centaurea solstitialis*), and spotted knapweed (*Centaurea solstitialis*) do not have close crop relatives.

2.2. Functional Genomics

As yet, the only weedy species for which significant amount of genomic sequence data is available is waterhemp (*Amaranthus tuberculatus*) [22]. 454 technology provided 42 Mbp of sequence data for this weed using only one half of the sequencing capacity of a single experiment. The sequences were aligned into more than 50,000 discrete fragments and included pieces of more than 10,000 different recognizable genes. The number of genes identified in these experiments is surprising since the genome of waterhemp is about 750 Mbps in size. Given the likelihood that most plants have less than 50,000 genes, this single experiment identified sequence fragments for more than a 5th of them. The experiment also generated sequence data for nearly the entire chloroplast genome of this species. Consequently numerous markers, and candidate genes, including genes for herbicide resistance, are now available for waterhemp. The information generated by this pilot study should augment the usefulness of a BAC library resources developed for a related grain amaranth species [23].

There are growing numbers of EST databases for numerous weeds including but not limited to leafy spurge, starthistle, and ryegrass, spotted knapweed, and the parasitic weed dodder [24-27]. These databases have provided vast resources for candidate genes used to analyze various physiological processes. For example, Horvath *et al.* 2002 [28], used the leafy spurge EST database to identify molecular markers for cell cycle transitions, and Anderson *et al.* 2005 [29], used the same EST resources to identify several likely sugar metabolism genes to investigate the suspected impact of environmental conditions resulting in dormancy transitions on sugar metabolism in leafy spurge.

As noted above, EST databases can be used to develop high density microarrays to assess global changes in gene expression (often referred to as transcriptome analysis). However, the leafy spurge is the only completely weedy species (ie a weed that is not also a crop or ornamental species) for which high density microarrays have been developed [30]. There are high density microarrays available for perennial ryegrass [26], but this species is a common forage crop that can be weedy in some ecosystems. These resources have already been used to provide insight into processes relevant to the biology and invasiveness of these weeds. In ryegrass, transcriptome analysis using microarrays have provided insight into vernalization processes required for perennial growth and disease responses [31, 32]. Both of these traits may influence the invasiveness and fitness of this weed. The leafy spurge microarrays have been used to identify signals and pathways important for dormancy transitions in underground buds, drought responses, and disease resistance [30] [Anderson personal communication] [Santana personal communication]. Anonymous arrays have been used to identify changes in gene expression associated with floral regulation following hybridization between *Senecio* ssps. [33].

The earliest reports of using non-weed microarrays to study weed transcriptomes described the use of arabidopsis (*Arabidopsis thaliana*) cDNA arrays and other genomic resources to follow changes in gene expression between shoots and mature leaves of wild oat and leafy spurge [16, 17]. The results from these experiments indicated that hybridization to heterologous arrays could provide useful information on gene expression in a non-model species. More recently, we used cotton microarrays to follow changes in gene expression in velvetleaf (*Abutilon theophrasti*) in response to crop competition [18]. These studies indicated that velvetleaf responded by induction of classic shade avoidance responses. This contrasted to earlier studies using microarray analysis to study how corn responded to velvetleaf [34] which indicated that corn did not exhibit the classic shade avoidance response when competing with velvetleaf. Likewise,

Lai *et al.* 2008 [35], used sunflower microarrays to follow differences in gene expression between wild and weedy sunflower species that identified potential changes in gene expression associated with invasiveness.

There are currently no mutant libraries, such as the T-DNA insertion collection developed for arabidopsis, developed for a weedy species. However, ecotilling techniques have been used in combination with a candidate gene approach to identify mutations in specific genes for a weedy species [45, 46]. In these studies, naturally occurring mutations in the acetolactate synthase gene which is the active site for ALS inhibiting herbicides were identified and catalogued from *Monochoria vaginalis*, a common paddy weed in Japan. This same group used ecotilling techniques to identify naturally occurring mutations in paralogous ALS genes in the same species. These two proof-of-concept papers highlight the potential for utilizing ecotilling to identify mutations in any gene that can be amplified by PCR, even those from polyploid plants.

The development of mutation libraries through the use of interfering RNAs [47], or insertional-mutagenesis using transposons [48] or agrobacterium derived T-DNA [49] is generally dependent on the ability to transform the target species. Likewise, good transformation systems are needed to confirm hypotheses derived from functional genomics studies. Currently there are only a small number of weeds that have been transformed. These include but are not limited to several weedy *Lolium* ssp. and a related grassy weed, *Agrostis stolonifera*, [50], dodder- *Cuscuta trifolii* [51], horseweed- *Conyza canadensis* [52], *Amaranthus hypochondriacus* [53] and leafy spurge (Chao, personal communications).

There are several well developed libraries available for several weedy species. BAC libraries have been produced for grain amaranth [23], and several weedy rice species [54], and have recently been developed for leafy spurge. Whole plant normalized libraries used for construction of the leafy spurge EST database, and a similar library developed from Canada thistle (*Cirsium arvense*) have been produced [24] (Anderson, personal communication). Likewise, we have produced a 2-hybrid capable cDNA library from growing underground buds of leafy spurge, and a lambda zap-based genomic library with average insert size of about 6 kb. The 2-hybird library could be used to identify protein-protein interactions for high throughput proteomics studies. Broz *et al.* 2007 [55], have produced a whole plant normalized cDNA library from spotted knapweed.

2.3. Structural Genomics

The use of high density maps and PCR-based or sequence-based markers has been extensively used in wild sunflowers to identify changes associated with interspecies hybridization [36], selection for stress resistance [37], and invasiveness [38]. Because cultivated rice (*Oryza sativa*) has been fully sequenced, and can cross-hybridize to numerous weedy rice species, molecular markers for cultivated rice have been used identify and in some cases clone genes for weedy characteristics such as shattering [39], dormancy [40], and seed color [41]. In addition to these complex studies that have used large numbers of well mapped markers to identify regions of the genome that may be under selection during invasion processes, there are numerous additional examples where a small number of ISSR, AFLP, and microsatellite markers have been used to compare the allelic diversity between invasive and non-invasive weed populations [42-44], among others.

One of the surprising findings from animal genomics projects is that there appears to be various sizable sections of genomes that have undergone duplication. These duplications have been associated with specific diseases or sub-populations of species, or even in specific tissues within an organism [56]. The discovery and function of the copy number variations (CNV) as they are called is currently a very hot topic in animal genomics. Although CNV has long been associated with transposable elements and complex gene families such as ribosomal RNA genes in plants, there has not been much focus on the type of structural gene amplifications associated with multi-drug resistant cancers, or speciation in mammals. However, recently, there have been as yet unpublished reports of CNV playing a role in herbicide resistance in weeds [57]. Specifically, amplification of the EPSPS gene which is that target site for the herbicide glyphosate appears to be the mechanism responsible for glyphosate resistance of Palmer amaranth (*Amaranthus palmeri*).Given the impact and ubiquity of CNV in animals development, physiology, and evolution, it seems likely that similar processes will be illuminated by genomic studies on these aspects of weeds as well.

3. POTENTIAL FOR GENOMICS RESEARCH IN WEEDS

Although there are limited resources for genomics research in weedy species, there is exceptional potential for utilizing genomic resources developed for crops such as potato (*Solanum tuberosum*), corn (*Zea maize*), and model species such as arabidopsis and poplar (*populus* ssp.). Since many genes are reasonably conserved, it is often possible to use sequence data from model and crop species to amplify and clone candidate genes from non-model and weedy species. Such approaches have been highly successful in identifying herbicide resistance genes from numerous weedy species. For example, the target sites of action for ALS inhibiting herbicides was fist cloned from arabidopsis [58]. These sequences were subsequently used to design primers for amplification and cloning homologous genes from many weedy species. In another example, the *SHOOTMERISTEMLESS* gene which is likely involved in production and growth of underground shoot buds on roots of leafy spurge was cloned by identifying conserved sequences from this gene and developing primers that would amplify a portion of the gene which was sued as a probe to isolate the full length gene form cDNA and genomic libraries [59]. Using available sequence databases should readily allow cloning of numerous genes of interest to weed biologists. Many genes involved in altered growth, development, response to stress, *etc.* have been well characterized in model systems, and are likely to play a role in invasiveness and competitiveness of weeds.

Fundamental questions of concern to weed scientists such as the evolution and nature of invasiveness and weediness, response to competition and control measures, and origin of weed populations has until recently been unanswerable through conventional studies and techniques. However, the introduction of genomic-based tools and technologies such as inexpensive large scale sequencing and genome wide association mapping could begin to illuminate the mechanisms and biological processes that make weeds so difficult to control. Plans are underway in several laboratories to use microarray analysis to compare transcriptomes of various invasive weeds in their native and introduced ranges and under common garden conditions to help identify gene expression differences related to invasiveness. Identification of the genes and physiological processes required for weediness and invasiveness should allow the development of methods to help control weeds. Additionally, it is possible that some of the genes that make weeds so competitive may provide targets or information needed to or modify crops and/or management practices to make crops more competitive and able to thrive in less than optimal growing conditions.

ACKNOWLEDGEMENTS

The author is supported by USDA Agricultural Research Service CRIS 5442-21220-026-00, and would like to thank all of the innovators and pioneers in the weed science community with special thanks to Drs. Patrick Tranel and Neil Stewart and my colleagues in the Sunflower and Plant Biology Research Unit for their support, encouragement, and efforts to develop and use genomics tools to answer fundamental questions in weed science.

CONFLICT OF INTEREST

The author confirm that this chapter contents have no conflict of interest.

DISCLOSURE

The chapter submitted for Advances in Genome Science, Vol. 3 entitled "Recent Genomic Advances for Weed Scienc" is an update of our article published in Current Genomics, Volume 11, Number 1, March 2010, pp. 47-51(5).

REFERENCES

[1] McKusick, V.A.; Ruddle, F.H. A new discipline, a new name, a new journal. *Genomics*, **1987**, 1, 1-2.

[2] Butte, A. The use and analysis of microarray data. *Nature Reviews Drug Discovery*, **2002**, 1, 951-960.

[3] Barkley, N.A.; Wang, M.L. Application of TILLING and EcoTILLING as reverse genetic approaches to elucidate the function of genes in plants and animals. *Current Genomics*, **2008**, 9, 212-226.

[4] Hardison, R.C. Comparative genomics. Plos Biology, 2003, 1, 156-160.

[5] Skolnick, J.; Fetrow, J.S.; Kolinski, A. Structural genomics and its importance for gene function analysis. *Nature Biotechnology*, **2000**, 18, 283-287.

[6] Luikart, G.; England, P.R.; Tallmon. D.; Jordan. S.; Taberlet, P. The power and promise of population genomics: From genotyping to genome typing. *Nature Reviews Genetics*, **2003**, 4, 981-994.

[7] Koonin, E.V. Bridging the gap between sequence and function. *Trends Genet*, **2000**, 16, 16.

[8] Sterky, F.; Lundeberg, J. Sequence analysis of genes and genomes. *Journal of Biotechnology*, **2000**, 76, 1-31.

[9] Palmer, L.E.; Rabinowicz, P.D.; O'Shaughnessy, A.L.; Balija, V.S.; Nascimento, L.U.; Dike, S.; de la Bastide, M.; Martienssen, R.A.; McCombie, W.R. Maize genome sequencing by methylation filtrations. *Science*, **2003**, 302, 2115-2117.

[10] Peterson, D.G.; Wessler, S.R.; Paterson, A.H. Efficient capture of unique sequences from eukaryotic genomes. *Trends in Genetics*, **2002**, 18, 547-550.

[11] Okou, D.T.; Steinberg, K.M.; Middle, C.; Cutler, D.J.; Albert, T.J.; Zwick, M.E. Microarray-based genomic selection for high-throughput resequencing. *Nature Methods*, **2007**, 4, 907-909.

[12] Hudson, M.E. Sequencing breakthroughs for genomic ecology and evolutionary biology. *Molecular Ecology Resources*, **2008**, 8, 3-17.

[13] Pflieger, S.; Lefebvre, V.; Causse, M. The candidate gene approach in plant genetics: a review. *Molecular Breeding*, **2001**, 7, 275-291.

[14] Schena, M.,; Shalon, D.; Davis, R.W.; Brown, P.O. Quantitative monitoring of gene-expression patterns with a complementary-DNA microarray. *Science*, **1995**, 270, 467-470.

[15] Hofmann, H.A. Functional genomics of neural and behavioral plasticity. *Journal of Neurobiology*, **2003**, 54, 272-282.

[16] Horvath, D.P.; Schaffer, R.; West, M.; Wisman, E. *Arabidopsis* microarrays identify conserved and differentially expressed genes involved in shoot growth and development from distantly related plant species. *Plant J*, **2003**, 34, 125-134.

[17] Horvath, D.P.; Schaffer, R.; Wisman, E. Identification of genes induced in emerging tillers of wild oat (*Avena fatua*) using Arabidopsis microarrays. *Weed Science*, **2003**, 51, 503-508.

[18] Horvath, D.P.; Llewellyn, D.; Clay, S.A. Heterologous hybridization of cotton microarrays with velvetleaf (*Abutilon theophrasti*) reveals physiological responses due to corn competition. *Weed Science*, **2007**, 55, 546-557.

[19] Baker, J.; Hidayat, I.; Preston, C. Molecular tools for understanding distribution and spread of weed genotypes. *Crop Protection*, **2007**, 26, 198-205.

[20] Kumar, L.S. DNA markers in plant improvement: An overview. *Biotechnology Advances*, **1999**, 17, 143-182.

[21] Pedersen, B.; Kalinowski, L.S.; Eggum, B.O. The Nutritive-Value of Amaranth Grain (*Amaranthus-Caudatus*).1. Protein and Minerals of Raw and Processed Grain. *Qualitas Plantarum-Plant Foods for Human Nutrition*, **1986**, 36, 309-324.

[22] Lee, R.M.; Thimmapuram, J.; Thinglum, K.A.; Gong, G.; Hernandez, A.G.; Wright, C.L.; Kim, R.W.; Mikel, M.A.; Tranel, P.J. Sampling the waterhemp (*Amaranthus tuberculatus*) genome using pyrosequencing technology. *Weed Science*, **In press**.

[23] Maughan, P.J.; Sisneros, N.; Luo, M.Z.; Kudrna, D.; Ammiraju, J.S.S.; Wing, R.A. Construction of an *Amaranthus hypochondriacus* bacterial artificial chromosome library and genomic sequencing of herbicide target genes. *Crop Science*, **2008**, 48, S85-S94.

[24] Anderson, J.V.; Horvath, D.P.; Chao, W.S.; Foley, M.E.; Hernandez, A.G.; Thimmapuram, J.; Liu, L.; Gong, G.L.; Band, M.; Kim, R.; Mikel, M.A. Characterization of an EST database for the perennial weed leafy spurge: An important resource for weed biology research. *Weed Science*, **2007**, 55, 193-203.

[25] Michelmore, R.W.; Knapp, S.; Rieseberg, L.; Kesseli, R.; Boore, J.; Kozik, A.; Matvienko, M.; Lavelle, D.; Chang, J.; Hierro, J. Yellow starthistle ESTs from The Compositae Genome Project. **2007**, unpublished.

[26] Sawbridge, T.; Ong, E.K.; Binnion, C.; Emmerling, M.; McInnes, R.; Meath, K.; Nguyen, N.; Nunan,K.; O'Neill, M.; O'Toole, F.; Rhodes, C.; Simmonds, J.; Pei, T.; Wearne, K.; Webster, T.; Winkworth, A.; Spangenberg, G. Generation and analysis of expressed sequence tags in perennial ryegrass (*Lolium perenne* L.). *Plant Science*, **2003**, 165, 1089-1100.

[27] Tomilov, A.A.; Tomilova, N.B.; Matvienko, M.; Yoder, J.I. Early haustorium development in hemiparasitic Orobanchaceae. **2007**, Unpublished

[28] Horvath, D.P.; Chao, W.S.; Anderson, J.V. Molecular analysis of signals controlling dormancy and growth in underground adventitious buds of leafy spurge. *Plant Physiol*, **2002**, 128, 1439-1446.

[29] Anderson, J.V.; Gesch, R.W.; Jia, Y.; Chao, W.S.; Horvath, D.P. Seasonal shifts in dormancy status, carbohydrate metabolism, and related gene expression in crown buds of leafy spurge. *Plant Cell and Environment*, **2005**, 28, 1567-1578.

[30] Horvath, D.P.; Chao, W.S.; Suttle, J.C.; Thimmapuram, J.; Anderson, J.A. Transcriptome analysis identifies novel responses and potential regulatory genes involved in seasonal dormancy transitions of leafy spurge (*Euphorbia esula* L.). *BMC Genomics*, **2008**, 9, 536.

[31] Ciannamea, S.; Busscher-Lange, J.; de Folter, S.; Angenent, G.C.; Immink, R.G.H. Characterization of the vernalization response in *Lolium perenne* by a cDNA microarray approach. *Plant and Cell Physiology*, **2006**, 47, 481-492.

[32] Dinkins, R.D.; Barnes, A.; Waters, W. Microarray and real-time PCR comparison of endophyte-infected and endophyte-free tall fescue gene expression. In Plant & Animal Genomes XVI Town & Country Convention Center, San Diego, CA. 2008

[33] Hegarty, M.J.; Jones, J.M.; Wilson, I.D.; Barker, G.L.; Coghill, J.A.; Sanchez-Baracaldo, P.; Liu, G.Q.; Buggs, R.J.A.; Abbott, R.J.; Edwards, K.J.; Hiscock, S.J. Development of anonymous cDNA microarrays to study changes to the Senecio floral transcriptome during hybrid speciation. *Molecular Ecology*, **2005**, 14, 2493-2510.

[34] Horvath, D.P.; Gulden, R.; Clay, S.A. Microarray analysis of late-season velvetleaf (*Abutilon theophrasti*) effect on corn. *Weed Science*, **2006**, 54, 983-994.

[35] Lai, Z.; Kane, N.C.; Zou, Y.; Rieseberg, L.H. Natural variation in gene expression between wild and weedy populations of *Helianthus annuus*. *Genetics*, **2008**, 179, 1881-1890.

[36] Rieseberg, L.H.; Baird, S.J.E.; Desrochers, A.M. Patterns of mating in wild sunflower hybrid zones. *Evolution*, **1998**, 52, 713-726.

[37] Lexer, C.; Welch, M.E.; Durphy, J.L.; Rieseberg, L.H. Natural selection for salt tolerance quantitative trait loci (QTLs) in wild sunflower hybrids: Implications for the origin of *Helianthus paradoxus*, a diploid hybrid species. *Molecular Ecology*, **2003**, 12, 1225-1235.

[38] Rieseberg, L.H.; Kim, S.C.; Randell, R.A.; Whitney, K.D.; Gross, B.L.; Lexer, C.; Clay, K. Hybridization and the colonization of novel habitats by annual sunflowers. *Genetica*, **2007**, 129, 149-165.

[39] Li, C.B.; Zhou, A.L.; Sang, T. Rice domestication by reducing shattering. *Science*, **2006**, 311 1936-1939.

[40] Gu, X.Y.; Kianian, S.F.; Hareland, G.A.; Hoffer, B.L.; Foley, M.E. Genetic analysis of adaptive syndromes interrelated with seed dormancy in weedy rice (Oryza sativa). *Theoretical and Applied Genetics*, **2005**, 110, 1108-1118.

[41] Sweeney, M.T.; Thomson, M.J.; Pfeil, B.E.; McCouch, S. Caught red-handed: *Rc* encodes a basic helix-loop-helix protein conditioning red pericarp in rice. *Plant Cell*, **2006**, 18, 283-294.

[42] Meekins, J.F.; Ballard, H.E.; McCarthy, B.C. Genetic variation and molecular biogeography of a North American invasive plant species (*Alliaria petiolata*, Brassicaceae). *International Journal of Plant Sciences*, **2001**, 162, 161-169.

[43] Maron, J.L.; Vila, M.; Bommarco, R.; Elmendorf, S.; Beardsley, P. Rapid evolution of an invasive plant. *Ecological Monographs*, **2004**, 74, 261-280.

[44] Durka, W.; Bossdorf, O.; Prati, D.; Auge, H. Molecular evidence for multiple introductions of garlic mustard (*Alliaria petiolata*, Brassicaceae) to North America. *Molecular Ecology*, **2005**, 14, 1697-1706.

[45] Wang, G.X.; Tan, M.K.; Suj, A.R.C.; Saitoh, H.; Terauchi, R.; Imaizumi, T.; Ohsako, T.; Tominaga, T. Discovery of single-nucleotide mutations in acetolactate synthase genes by Ecotilling. *Pesticide Biochemistry and Physiology*, **2007**, 88, 143-148.

[46] Wang, G.X.; Imaizumi, T.; Li, W., Saitoh, H.; Terauchi, R.; Ohsako, T.; Tominaga, T. Self-EcoTILLING to identify single-nucleotide mutations in multigene family. *Pesticide Biochemistry and Physiology*, **2008**, 92, 24-29.

[47] Waterhouse, P.M.; Helliwell, C.A. Exploring plant genomes by RNA-induced gene silencing. *Nature Reviews Genetics*, **2003**, 4, 29-38.

[48] Osborne, B.I.; Wirtz, U.; Baker, B. A system for insertional mutagenesis and chromosomal rearrangement using the Ds transposon and *Cre-Lox*. *Plant Journal*, **1995**, 7, 687-701.

[49] Krysan, P.J.; Young, J.C.; Sussman, M.R. T-DNA as an insertional mutagen in *Arabidopsis*. *Plant Cell*, **1999**, 11, 2283-2290.

[50] Dalton, S.J.; Bettany, A.J.E.; Timms, E.; Morris, P. Transgenic plants of *Lolium multiflorum*, *Lolium perenne*, *Festuca arundinacea* and *Agrostis stolonifera* by silicon carbide fibre-mediated transformation of cell suspension cultures. *Plant Science*, **1998**, 132, 31-43.

[51] Borsics, T.; Mihalka, V.; Oreifig, A.S.; Barany, I.; Lados, M.; Nagy, I.; Jenes, B.; Toldi, O. Methods for genetic transformation of the parasitic weed dodder (*Cuscuta trifolii* Bab. et Gibs) and for PCR-based detection of early transformation events. *Plant Science*, **2002**, 162, 193-199.

[52] Scheiber, P.A.; Tran, M.; Duncan, D.R. Tissue culture and transient transformation of Marestail (*Conyza canadensis* (L.) Cronquist). *Plant Cell Reports*, **2006**, 25, 507-512.

[53] Jofre-Garfias, A.E.; Villegas-Sepulveda, N.; Cabrera-Ponce. J.L.; Adame-Alvarez, R.M.; Herrera-Estrella, L.; Simpson, J. *Agrobacterium*-mediated transformation of *Amaranthus hypochondriacus*: light- and tissue-specific expression of a pea chlorophyll a/b-binding protein promoter. *Plant Cell Reports*, **1997**, 16, 847-852.

[54] Wing, R.A.; Ammiraju, J.S.S.; Luo, M.Z.X.; Kim, H.; Yu, Y.S.; Kudrna, D.; Goicoechea, J.L.; Wang, W.M.; Nelson, W.; Rao, K.; Brar, D.; Mackill, D.J.; Han, B.; Soderlund, C.; Stein, L.; SanMiguel, P.; Jackson, S. The *Oryza* Map Alignment Project: The golden path to unlocking the genetic potential of wild rice species. *Plant Molecular Biology*, **2005**, 59, 53-62.

[55] Broz, A.K.; Broeckling, C.D.; He, J.B.; Dai, X.; Zhao, P.X.; Vivanco, J.M. A first step in understanding an invasive weed through its genes: an EST analysis of invasive *Centaurea maculosa*. *BMC Plant Biology*, **2007**, 7, 25.

[56] Freeman, J.L.; Perry, G.H.; Feuk, L.; Redon, R.; McCarroll, S.A.; Altshuler, D.M.; Aburatani, H.; Jones, K.W.; Tyler-Smith, C.; Hurles, M.E.; Carter, N.P.; Scherer, S.W.; Lee, C. Copy number variation: New insights in genome diversity. *Genome Research*, **2006**, 16, 949-961.

[57] Gaines, T.; Preston, C.; Shaner, D.; Leach, J.; Chisholm, S.; Bukun, B.; Ward, S.; Culpepper, A.S.; Tranel, P.; Westra, P. A novel mechanism of resistance to glyphosate in Palmer amaranth (*Amaranthus palmeri*). In Annual Meeting of the Weed Science Society of America, Orlando FL, **2009**.
[58] Haughn, G.W.; Smith, J.; Mazur, B.; Somerville, C. Transformation with a mutant *Arabidopsis* acetolactate synthase gene renders tobacco resistant to sulfonylurea herbicides. *Molecular & General Genetics*, **1988**, 211, 266-271.
[59] Varanasi, V.; Slotta, T.; Horvath, D. Cloning and characterization of a critical meristem developmental gene (*EeSTM*) from leafy spurge (*Euphorbia esula*). *Weed Science*, **2008**, 56, 490-495.

Send Orders for Reprints to reprints@benthamscince.net
Advances in Genome Science, Vol. 3, **2014, 143-160**

CHAPTER 5

Oncogenomic Approaches in Exploring Gain of Function of Mutant p53

Sara Donzelli[1], Francesca Biagioni[1], Francesca Fausti[1], Sabrina Strano[2], Giulia Fontemaggi[1,3] and Giovanni Blandino[1,*]

[1]Translational Oncogenomics Unit, National Cancer Institute Regina Elena Rome, Italy; [2]Molecular Chemoprevention Group, Scientific Direction, National Cancer Institute Regina Elena Rome, Italy and [3]General Pathology Section, Department of Clinical and Experimental Medicine, Perugia University, Perugia, Italy

Abstract: Cancer is caused by the spatial and temporal accumulation of alterations in the genome of a given cell. This leads to the deregulation of key signaling pathways that play a pivotal role in the control of cell proliferation and cell fate. The p53 tumor suppressor gene is the most frequent target in genetic alterations in human cancers. The primary selective advantage of such mutations is the elimination of cellular wild type p53 activity. In addition, many evidences *in vitro* and *in vivo* have demonstrated that at least certain mutant forms of p53 may possess a gain of function, whereby they contribute positively to cancer progression. The fine mapping and deciphering of specific cancer phenotypes is taking advantage of molecular-profiling studies based on genome-wide approaches. Currently, high-throughput methods such as array-based comparative genomic hybridization (CGH array), single nucleotide polymorphism array (SNP array), expression arrays and ChIP-on-chip arrays are available to study mutant p53-associated alterations in human cancers. Here we will mainly focus on the integration of the results raised through oncogenomic platforms that aim to shed light on the molecular mechanisms underlying mutant p53 gain of function activities and to provide useful information on the molecular stratification of tumor patients.

Keywords: CGH arrays, ChIP-chip, gain of function, mutant p53, oncogenomic approaches, SNP arrays.

INTRODUCTION

The p53 tumor suppressor protein is a sequence-specific transcription factor that

***Address correspondence to Giovanni Blandino:** Translational Oncogenomics Unit, Regina Elena Cancer Institute, via Elio Chianesi, 53, 00144-Rome, Italy; Tel: +39-06-52662878; Fax: +39-06-52665523; E-mail: blandino@ifo.it

Christian Neri (Ed)

is present in a latent form under normal conditions and becomes activated by a variety of stress signals, such as DNA damage, oncogene activation and improper mitogenic stimulation. p53 acts as a potent transcriptional activator through the binding as a homo-tetramer to specific sequences, named p53 responsive elements (p53RE). Activated p53 modulates the expression of a large set of target genes involved in many cellular processes, including cell cycle arrest, DNA repair, apoptosis and senescence. Growth arrest or cell death prevents damaged DNA from being replicated; thus suggesting a role for p53 in the maintenance of genome integrity [1-5]. p53 is a member of a family of proteins that has recently been established through the identification of p73, p63 and their related isoforms, all heavily involved in tumor suppression and development [6]. Half of all human cancers bear p53 mutations [7]. The most prevalent type of p53 alterations are missense mutations, often within the conserved DNA binding core domain of the protein [8, 9]. In this area each residue has been found to be possibly mutated in human tumors, and the resulting proteins display a marked heterogeneity in terms of loss of structure and function. Several evidences demonstrate that a subset of p53 mutant proteins display gain of function activity [10], thereby actively participating to tumorigenesis (Fig. **1**). Structural studies have allowed the classification of p53 mutant proteins in two major categories [11,12]: a) DNA contact defective (His273, Trp248), that includes those mutants whose mutation impacts on the residues composing the DNA/protein interaction surface; b) defective structure (His175, His179), whose point mutation determines an important conformational alteration. At variance with the observations of several other tumor suppressor genes, cells with p53 mutations typically maintain expression of full-length mutant protein, often at markedly elevated levels. This observation suggests the possibility that some mutant p53 proteins gain additional functions over the mere loss of wild-type TP53 activity. This hypothesis has been validated by a combination of *in vitro* and *in vivo* studies that show how mutant p53 proteins confer transformed properties to cell cultures and increased tumorigenicity in mice [13, 14]. It has also been reported that conformational mutants, such as p53His175, but not DNA contact mutants (p53His273, p53Trp248), can increase cellular resistance to etoposide or can contribute to genomic instability by abrogating the mitotic spindle checkpoint, thereby facilitating the generation of aneuploid cells [15, 16]. p53-/- mice develop a different spectrum of tumours compared to that of knock-in mice expressing various p53 hot spot mutants [15, 17]. Indeed, mutant p53 knock-in mice have a higher frequency of solid tumours with high potential for metastasis, a feature not seen in knock-out animals. This peculiar spectrum of tumours is also observed in mice expressing one mutant p53 allele in a p53-null background, thus strongly suggesting a gain of function of mutant p53 proteins. Further analysis of mutant p53 knock-in models in mice defective in p63 and p73 suggests that alteration in the activity of the entire p53 family is an important feature in achieving function of mutant p53. Many questions regarding the molecular mechanisms at the basis of gain of function activities are still unsolved.

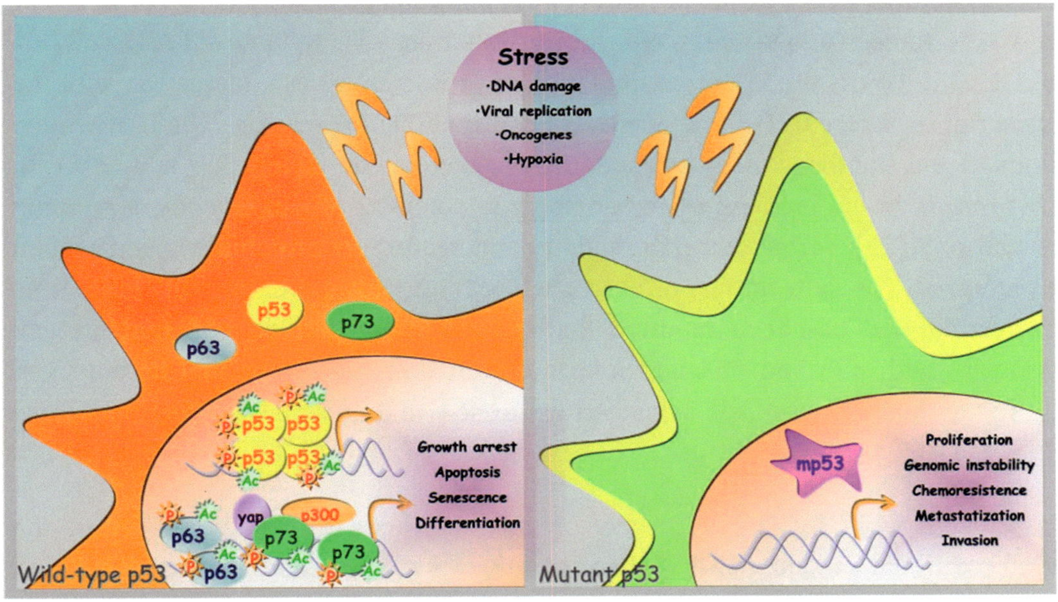

Figure 1: Wild type *versus* mutant p53 protein. In response to various types of cellular stress signals, wild-type p53 protein and its family members, p63 and p73, undergo different kinds of post-translational modifications, including phosphorylation and acetylation. These events cause p53 family members to stabilize and activate which results in growth arrest, apoptosis, senescence, and differentiation. However, cells carrying mutant p53 display increased cellular proliferation, genomic instability, chemoresistance, metastatization and invasion.

To date, we can depict the three following molecular scenarios to explain gain of function of mutant p53 proteins. **(a)** mutant p53 can bind to DNA through the association with DNA binding proteins and transcriptionally activate specific target genes using its functional transactivation domain (TAD). In support of this molecular mechanism, it has been reported that human tumor-derived p53 mutants, whose TAD was inactivated by site-directed mutagenesis, lost the ability to increase tumorigenicity *in vitro* and *in vivo* [13, 18-21]; **(b)** Di Agostino *et al.* have shown that mutant p53 proteins physically interact *in vivo* with the transcription factor NF-Y, whose DNA binding consensus are present in the regulatory regions of many key genes involved in the regulation of the cell cycle. Mutant p53 can be recruited *in vivo* onto the promoter of NF-Y target genes, as a member of a large transcriptional complex that also includes histone acetyltransferases (HAT). Of note, mutant p53 favours the selective recruitment of HAT in response to DNA damage, thereby dictating the transcriptional activation of its target genes [22]. Recently, Fontemaggi *et al.* have identified ID4 as a new transcriptional target of gain-of-function p53 mutants R175H, R273H and R280K : mutant p53 is recruited on ID4 promoter, by the interaction with the transcription factor E2F-1 and positively controls ID4 expression. The ID4 protein binds to and stabilizes mRNAs encoding pro-angiogenic factors IL8 and GRO- α. This results in the increase of the angiogenic potential of cancer cells expressing mutant p53 [23]; **(c)** mutant p53 binds to and sequesters proteins whose function is required for anti-tumor effects such as apoptosis or growth inhibition. Interestingly, it has been reported that human tumor-derived p53 mutants can associate with p73 and p63 and interfere with their transcriptional activity and ability to induce apoptosis when co-expressed in transient transfection assays (Fig. **2**) [19, 24-27].

In a recent work by Donzelli *et al.*, it has also been shown the ability of the conformational mutant p53-175H to modulate the expression of microRNAs, in particular the miR-128-2, whose expression is increased by mutant p53 in lung cancer cells [28]. Up-regulation of miR-128 was achieved by transactivating the expression of its host gene, ARPP21, and inhibits apoptosis and confers increased resistance to chemoterapeutic agents. The anti-apoptotic effect of miR-128-2 is mediated by its ability to suppress E2F5, which in turn induces p21. The induced p21 is predominantly cytoplasmic and inhibits apoptosis by preventing pro-caspase 3 cleavage.

Many of the studies exploring the molecular mechanisms that underlie gain of function of mutant p53 proteins have mainly characterized the activities of p53 hot spot mutations. These mutants display an oncogenic potential by themselves, as demonstrated by transformation assays *in vitro* and knock-in models *in vivo*. However, as mentioned above, mutations can occur within the entire DBD of the protein. The activity of a large number of p53 mutants has been analyzed by Kato *et al.* [29]. This analysis demonstrated that hot spot mutations lost their transactivation activity completely, while the rest of the mutants retained some ability to transactivate target genes. In parallel to these observations, additional evidences have shown that the clinical penetrance of p53 mutations in cancer is strongly influenced by the genetic background of the individual [13, 17, 30] and by the presence of other cancer-associated somatic mutations [31].

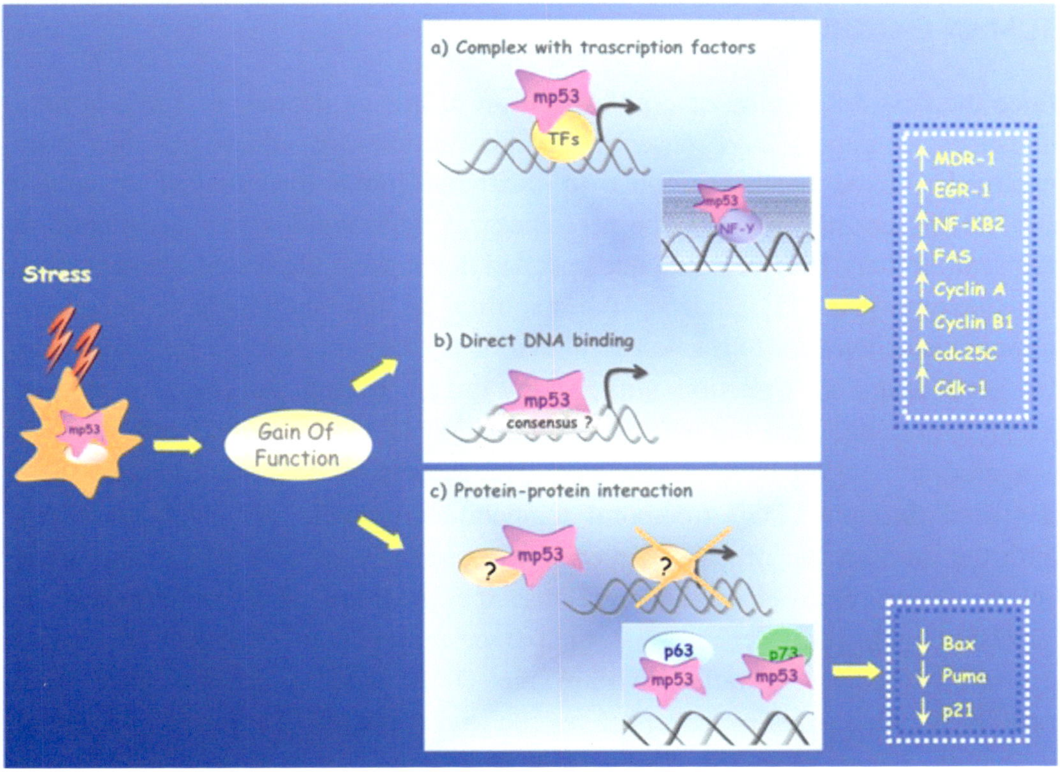

Figure 2: Molecular mechanisms underlying gain of function of mutant p53 protein. *DNA binding.* Mutant p53 protein can be recruited on the promoters of its target genes by the interaction with different transcription factors, as has been demonstrated for the transcription factor NF-Y in response to DNA damage (a), or by direct binding to specific consensus sequences that are still to be identified (b). *Protein-protein interaction.* (c) Mutant p53 protein binds to and sequesters proteins (p73 and p63) whose activities are closely connected to anti-tumoral effects such as apoptosis or growth inhibition.

Since p53 mutations of the DBD are selected in cancer, it is reasonable to hypothesize that hot spot mutations are directly selected because they represent the driving force during the neoplastic transformation. The remaining p53 mutations are selected following a second mutational event that might be permissive for mutant p53 gain of function activities. Cell-type specificity for gain of function of mutant p53 might explain why certain p53 mutants do not display any activity when exogenously overexpressed in cell systems different from those where the mutation was originated.

p53 MUTATIONS AS PREDICTIVE/PROGNOSTIC MARKERS IN HUMAN CANCERS

Since the first description of a p53 mutation in human cancers in 1989, several thousands of papers have described clinical studies in which p53 has been tentatively linked to the response to treatment (predictive power) or patient survival (prognostic power). The usefulness of these studies was questioned in terms of strategy (too few patients enrolled for correct statistical analysis) and methodology [32, 33]. Indeed, the majority of the studies have relied on immunohistochemistry to assess p53 alterations. This approach is a poor surrogate for gene mutation detection since many mutations do not lead to protein accumulation, and also because an accumulation of wild type p53 might occur. Langerod *et al.* [34] have shown how IHC detects 50% of the p53 mutations identified through TTGE (temporal temperature gradient gel electrophoresis). With the introduction of gene sequencing to precisely identify p53 mutations in cancer, it has been clearly defined that the presence of a p53 mutation is correlated with a shorter survival or a poor response to treatment in several cancers. Moreover, mutations within the DNA binding domain have been associated with a worst prognosis compared to that related to other p53 mutations in breast cancer [35-38]. Now that the power of mutant p53 as a predictive/prognostic marker in breast cancer has been established, the complete characterization of mutant p53 contribution to tumorigenesis will require a combination of oncogenomic approaches, for example copy number analyses, mutational studies and expression profiling (protein coding genes and microRNAs) on a large number of patients.

MUTANT p53 AND CHROMOSOMAL INSTABILITY

The genome of cancer cells is thought to be unstable as they often harbour many chromosomal abnormalities. Due to the role of TP53 in maintaining genomic stability, its inactivation profoundly impairs repair of DNA damage and could promote genomic alterations. Several research groups have reported a correlation between genomic instability and TP53 mutations in breast cancer tissues. Using fluorescence *in situ* hybridization, Sigurdsson *et al.* [39] have shown that breast cancer cells with abnormal p53 display a higher copy number of chromosome 17 than cells without p53 staining. This points to a relationship between TP53 alterations and chromosomal instability. Jain *et al.* [40] combined information on TP53 mutation status in breast carcinomas with their genomic complexity, as assessed by chromosomal comparative genomic hybridization (CGH), a technique capable of detecting copy number alterations at chromosomal level [41]. They found that gain at 8q24 and loss at 5q15-5q21 are linked to mutant p53. Subsequently, Jong YJ *et al.* [42] and Kleivi K *et al.* [43] performed CGH analysis in relation to TP53 status on breast cancer specimens and confirmed a higher degree of abnormalities in patients carrying TP53 mutation, compared to those with wild type p53. These abnormalities appeared to cluster in certain chromosomal regions at significantly higher frequencies. Of note, the result of 8q gain associated with the TP53 mutated status was shared by all three studies. Identification of putative cancer-related genes that reside within these genomic loci, such as c-myc (at 8q24), and their potential interactions with mutant p53 might represent an exciting new area for breast cancer research.

Currently, high throughput methods such as array-based comparative genomic hybridization (CGH array) and single nucleotide polymorphism array (SNP array) are available to study mutant p53-associated alterations in a genome-wide fashion. Array-CGH is an established high-resolution method to study the whole genome for chromosomal amplifications and deletions [44]. One limitation of the array-CGH method is its lack of genotype information. Subsequently, it fails to provide information about regions of LOH without copy number alteration (CNA) such as mitotic recombination or gene conversion events. Recently, high-density oligonucleotide-based single polymorphism arrays (SNP arrays) have been used to identify copy number and LOH of chromosomal regions. The advantage of a combined SNP-CGH approach is the identification of allele specific gain/loss by SNP array and the robust copy number detection by array CGH. In addition to the copy number detection, SNP array would also lead to an increase in information about the genetic background of the analysed patients and about additional cancer-associated mutations. For example, this method would reveal the presence of specific SNP variants in components of p53-interacting pathways. The integration of these two platforms (CGH, SNP) with expression arrays will provide detailed and relevant information about the genomic alterations associated with the mutated TP53 status and about the consequences determined by these alterations at the expression level. It is still being debated as to whether copy number alterations lead to changes in gene expression. So far, few studies have indicated that a correlation exists between chromosomal instability and gene expression in colorectal carcinoma and other cancer types [45, 46]. Moreover, if particular associations between specific p53 mutations and other genomic alterations will be shown to be related to the clinical outcome of given tumors, it is highly possible that these combined analyses would reinforce the value of mutant p53 as a molecular biomarker in specific genetic contexts (Fig. **3**).

IDENTIFICATION OF MTp53-ASSOCIATED GENE-EXPRESSION SIGNATURES

Two different approaches have been followed to identify mutant p53-associated gene-expression signatures: **a**) cell lines overexpressing different mutant forms of p53 have been generated to analyze the impact of mutant p53 on cell transcriptome; **b**) tumor samples with characterized p53 status (mutated or wild-type) have been examined by microarray analysis for their global gene expression. The latter has been compared with that of normal counterparts to define (mainly through hierarchical clustering or uninvariate analysis by SAM) clusters of genes associated with the mutated status of p53.

A number of studies employed microarrays to evaluate global gene expression in H1299 lung cancer cells expressing p53-175H, p53-273H and p53-281G [47-49], p53-deficient HCT116 colorectal cancer cells overexpressing p53-138P and p53-175H [50], osteosarcoma U2OS cells expressing p53-157F, p53-

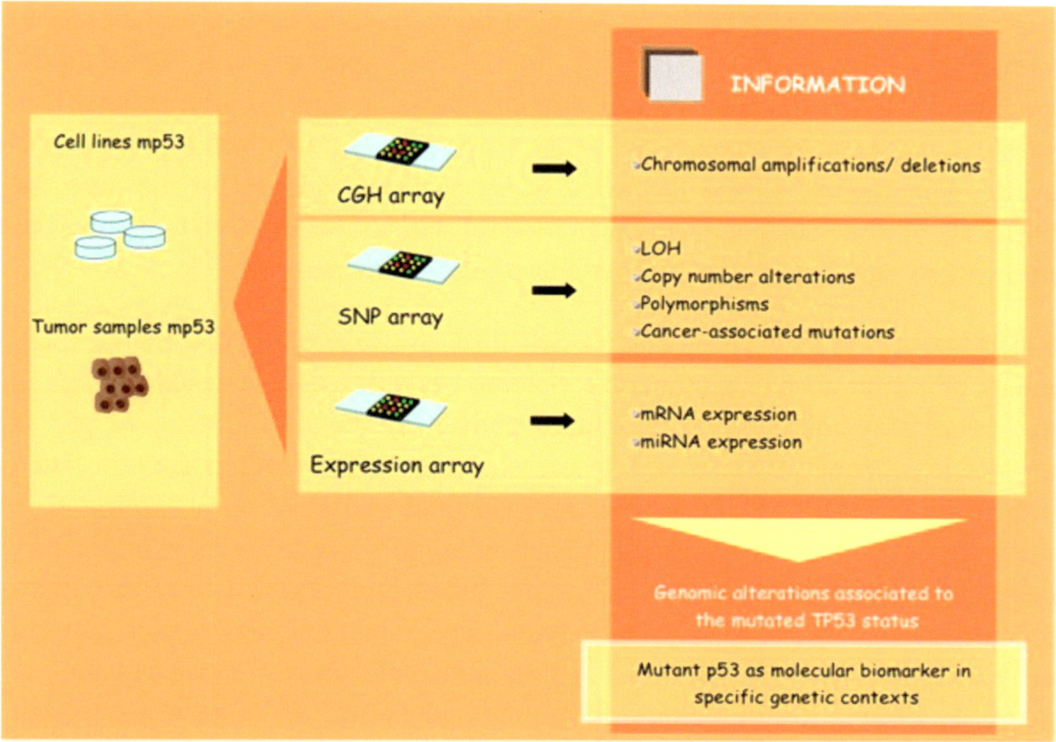

Figure 3: Oncogenomic approaches. Different kinds of oncogenomic approaches can be used to analyze genomic alterations in cell lines or tumor samples carrying mutant p53 proteins: *CGH arrays*, which allow the identification of chromosomal amplifications and deletions, and *SNP arrays,* by which it is possible to analyze the presence of LOH, copy number alterations, polymorphisms and cancer-associated mutations.

175H and p53-248Q [51], LNCaP prostate cancer cells expressing p53-245S, p53-248W, p53-273C, p53-273H [52] as well as in Li Fraumeni syndrome-derived fibroblasts expressing p53-175H [51] have been performed. The main limitation of these studies is based on the fact that, except for the O'Farrell study [53], they failed to analyze the impact of wt p53 expression on the same cell system. For this reason, they were unable to assess whether some of the transcriptional modifications observed would have been found, to a different or opposite extent, upon expression of wild-type p53. Indeed, O'Farrell *et al.* report that the most obvious difference related to the ectopic expression of two diverse p53 mutants (R175H and A138P) is that R175H retained less wild-type transcriptional regulatory events than A138P. A long list of mutant p53-regulated genes has been obtained from microarray analyses, suggesting the involvement of mutant p53 in several cellular processes ranging from transcriptional and translational regulation, signal transduction, cell motility, invasion and metastatization. This list obviously includes some of the genes previously characterized as mutant p53 targets. Importantly, all the identified genes provide evidence for the potential disclosure of new molecular targets and for substantial insights into the molecular mechanisms underlying gain of function activities of mutant p53.

In addition to the *in vitro* models of cell lines overexpressing p53 mutants, the impact of a mutated p53 on the transcriptome has also been evaluated through *in vivo* microarray analyses. As mentioned above, the mutational status of p53 is prognostic in many malignancies. In breast cancer, p53 mutations are associated with worse overall and disease-free survival rates and have been implicated in resistance to anticancer therapies. Miller *et al.* [54] reported on a microarray analysis performed on 251 primary invasive breast tumors with known p53 status (58 carrying mutant p53 and 193 wild-type). They found that tumors with mutated and wild-type p53 are distinguished by pervasive molecular differences (mainly in genes controlling proliferation). They also identified a 32-gene expression signature (as optimal classifier) that distinguishes mutant *versus* wild-type p53 tumors of different histological types. Furthermore, it also enables the prediction of the p53 status when tested on two publicly available microarray data sets (breast cancer [55], and liver cancer [56]) where p53 mutational status is known. The identified p53 signature predicts outcome better than p53 mutation status alone, also independently from the therapy-specific data sets.

Despite the efficacy of these *in vivo* analyses in identifying gene expression signatures associated with wild type or mutant p53, unfortunately these studies are unable to assess whether the characterized genes are really dependent on the presence of mutant p53 or whether they merely represent the result of a loss of wild-type p53 activity. The data sets derived from microarray analyses on cell lines overexpressing p53 mutants, enclosing the putative genes responsible for gain of function, need to be matched with those derived from *in vivo* analyses to define which of the transcripts are truly expressed in cancer tissues carrying p53 mutation. Up until now few p53 mutations have been analyzed *in vitro* for their ability to transcriptionally affect gene expression. Further microarray analyses employing additional p53 mutants to be compared with those performed on tumor specimens with known p53 status might unveil important molecular details on the gain of function properties of the different mutant p53 proteins present in human cancers.

ChIP-chip AS A TOOL TO IDENTIFY MUT-p53 TRANSCRIPTIONAL SIGNATURES IN HUMAN CANCERS

Little is known about the mechanisms through which mutant p53 proteins, that have lost the ability to bind DNA in a sequence-specific manner, can achieve the specificity in gene regulation. Chromatin Immunoprecipitation experiments have shown that mutant p53 can be found onto the regulatory regions of putative target genes, but despite extensive efforts, the identification of the specific DNA binding consensus of mutant p53 has not yet been found. It has been proposed that mutant p53 can be recruited to its putative target promoters through the physical interaction with sequence specific transcription factors whose specific DNA binding consensus are present on those regulatory regions. Sp1 and Ets, which were shown to cooperate with mutant p53 in transcriptional regulation [57-59], are paradigmatic of the proposed molecular mechanism. The notion that mutant p53 is recruited to transcriptional regulatory regions of peculiar genes *via* interaction with other transcription factors is strongly supported by the work of Di Agostino *et al.* [22] which has shown the existence of a transcriptional competent protein complex involving mutant p53 and the transcription factor NF-Y. Along this transcriptional pathway mutant p53 favours the recruitment of the coactivator p300 onto NF-Y target promoters, leading to the transcriptional activation of its target genes. Our group has also found that the transcription factors E2F-1 and p65 (RelA) are both concomitantly present with mutant p53 on the promoters of its target gene ID4 [23].

Due to the absence of a canonical consensus sequence for mutant p53, the only approach that seems truly promising for the precise deciphering of its transcriptional code is the use of whole-genome tiling arrays, designed to interrogate an entire genome in an unbiased fashion, for ChIP-chip analysis. ChIP-chip (Chromatin Immunoprecipitation-chip) permits a simultaneous analysis of the occupancy of a specific transcription factor on thousands of its target genes.

Briefly, crosslinked chromatin (DNA/protein) complexes are extracted from the analyzed cell culture or from a tissue and sheared down to relatively short fragments (500-1000 bp). The chromatin fragments containing the transcription factor of interest are then immunoprecipitated using a specific antibody and the target sequences are subsequently identified through hybridization to DNA microarrays. At a high probe tiling resolution multiple overlapping probes may contain the actual transcription factor binding motif and thus enable a fine mapping of the binding site to a resolution of less than 25 bp. Apart from the most recent generated tiling arrays, different kinds of microarrays have been generated for ChIP-chip analysis. Initial studies were performed using slides carrying spotted oligonucleotides or PCR products from selected predicted promoters. However these arrays are unable to interrogate the entire genome and are unable to address the possibility that transcription factors might bind to other locations.

In a recent work by Dell'Orso *et al.* the authors designed and produced a low-density slide (mutp53-chip) carrying oligonucleotides that correspond to the promoter regions of 154 putative and proven target genes of mutant p53 [60]. By using the chromatin obtained from mutant p53-175H -immunoprecipitation in proliferating SKBr3 breast cancer cells, they found that mutant p53 binds to 40 of the 154 promoters analyzed. Two-thirds of the mutant p53- bound promoters were also engaged by either p300 or PCAF acetyl-transferases, strongly indicating the presence of transcriptionally active complexes. Moreover they confirmed the presence of p65 NF-κB on 27 of those promoters.

Recent studies using tiling arrays interrogating entire chromosomes show that a large fraction of *in vivo* binding sites are outside the predicted promoter regions of given genes. It has been shown that the binding locations of NFkB, cMyc, Sp1, p53, CREB are found within both coding and non-coding regions. More binding sites than those predicted were found and of those only a relatively small fraction of the sites occurred in genomic regions that would typically be considered "promoters". The use of tiling arrays in the study of mutant p53 transcriptional activity will overcome the difficulties existing in the prediction of the binding sites for this oncogenic transcription factor. In addition, mutant p53 has been reported to bind a wide range of DNA secondary structures. For example, mutant p53 was shown to preferentially bind matrix attachment regions (MARs) [61].

MARs are involved in the anchoring of chromatin fibres to the nuclear matrix, generating chromatin domains that may enhance or repress transcription. Further studies from this group suggest that the binding to non-B DNA rather than a specific sequence represents the basis of the interaction between mutant p53 and MARs. If the binding of mutant p53 to MARs leads to changes in chromatin accessibility and activity it becomes more necessary to identify the genes that will sense the effects of this non-sequence-specific binding. This information will likely be available after a hybridization of whole-genome tiling arrays with mutant p53-bound chromatin.

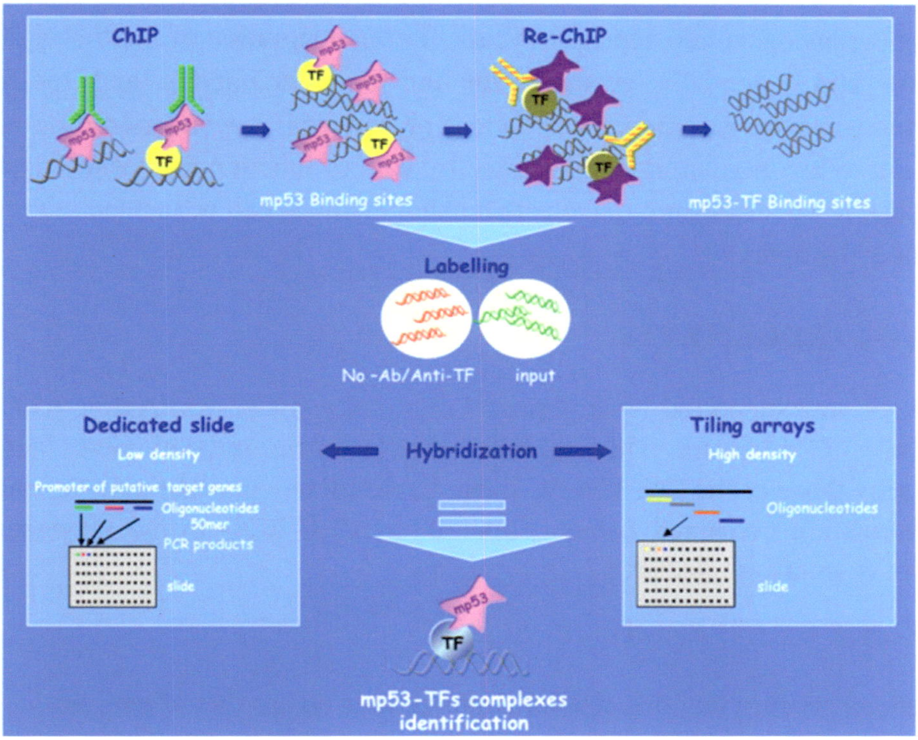

Figure 4: ChIP-chip approach. ChIP-chip technique allows for the simultaneous analysis of the presence of a specific transcription factor on the regulatory regions of thousands of its putative target genes. ChIP-chip analysis of chromatin immunoprecipitated with an antibody against p53 and subsequently re-immunoprecipitated with an antibody against a specific transcription factor (Re-ChIP), in a mutant p53 cellular context, allows for the identification of a transcriptional protein complex that might control the activation or repression of a target gene. Two different types of slides can be used for the ChIP-chip analysis: *dedicated slides* containing oligonucleotide probes specific for different regions of known and putative target gene promoters; *tiling slides* containing oligonucleotide probes representing the entire genomic sequences.

The ChIP-chip analysis of chromatin initially immunoprecipitated with antibodies against mutant p53 and successively with antibodies directed to the various transcription factors that mediate mutant p53's binding to its target genes (Re-ChIP) is likely to lead to the identification of subgroups of target genes controlled by different mutant p53-TF protein complexes (Fig. **4**).

CONCLUSIONS

Genome wide analyses are turning out to be powerful instruments in the molecular stratification of cancer patients. The complexity of the data raised through the oncogenomic approaches calls for the integration of multidisciplinary expertise and potentially accounts for the fine deciphering and molecular characterization of specific cancer patient phenotypes. An understanding of the molecular events regulating e/o governed by gain of function mutant p53 proteins might lead to the disclosure of cancer signaling pathways to be tackled with novel anticancer therapies.

ACKNOWLEDGEMENTS

This work was supported by European Community (EC) FP7 "Active p53" and "Mutant p53" consortia. This publication reflects the authors' views and not necessarily those of the EC. The EC is not liable for any use that may be made of the information contained herein. Support by AIRC-ROC to the oncogenomic platform is greatly appreciated.

CONFLICT OF INTEREST

The authors confirm that this chapter contents have no conflict of interest.

DISCLOSURE

The chapter submitted for Advances in Genome Science, Vol. 3 entitled **"Oncogenomic approaches in exploring gain of function of mutant p53 (review) S. Donzelli"** is an update of our article published in Curr. Genomics, 2008 May; 9(3):200-7.

REFERENCES

[1] Harris, S.L., and Levine, A.J. The p53 pathway: positive and negative feedback loops. *Oncogene* **2005**, *24*: 2899-908.

[2] Oren, M. Decision making by p53: life, death and cancer. *Cell Death Differ.* **2003**, *10*: 431-42.

[3] Toledo, F., and Wahl G. M. Regulating the p53 pathway: *in vitro* hypoteses, *in vivo* veritas. *Nat. Rev. Cancer.* **2006**, *6*: 909-23.

[4] Vousden, K.H., and Prives, C. P53 and prognosis: new insights and further complexity. *Cell* **2005**, *120*: 7-10.

[5] Beinz, B., Zacut-Houri, R., Givol, D., Oren, M. Analysis of the gene coding for the murine cellular tumor antigen p53. *EMBO J.* **1984**, *3*: 2179-83.

[6] Kaelin, Jr. W.G. The emerging p53 gene family. *J. Natl. Cancer. Inst.* **1999**, *91*: 594-98.

[7] Kern, S.E., Kinzler, K.W., Baker, S.J., Nigro, J.M., Rotter, V., Levine, A.J., *et al.* Mutations in the p53 gene occour in overal human tumor types. *Nature* **1989**, *342*: 705-8.

[8] Hollestein, M., Sidransky, D., Vogelstain, B., Harris, C.C. P53 mutations in human cancers. *Science* **1991**, *253*: 49-53.

[9] Olivier, M., Hussain, S.P., Cardon, dF., Hainaut, P., Harris, C.C. The IARC TP53 databese: new online mutation analysis and recommedations to users. *Hum. Mutat.* **2002**, *19*: 607-614.

[10] Haley, O., Michaloviz, D., Oren, M. Different tumor-derived p53 mutants exhibit distinct biological activities. *Science* **1990**, *250*: 113-6.

[11] Bullock, A., Fersht, A.R. Rescuing the function of mutant p53. *Nature Cancer. Rev.* **2001**, *1*: 68-76.

[12] Cho, Y.J., Gorina, S., Jeffrey, P.D., Pavletich, N.P. Crystal structure of a p53 tumor suppressor DNA complex: understanding tumorigenic mutations. *Science* **1994**, *265*: 346-355.

[13] Olive, K.P., Tuveson, D.A., Ruhe, Z.C., Yin, B., Willis, N.A., Bronson, R.T., *et al.* Mutant p53 gain of function in two mouse models of Li-Fraumeni syndrome. *Cell* **2004**, *119*: 847-60.

[14] Jacks, T., Remington, L., Williams, B.O., Schmitt, E.M., Halachmi, S., Bronson, R.T., *et al.* Tumor spectrum analysis in p53-mutant mice. *Curr. Biol.* **1994**, *4*: 1-7.

[15] Prives, C., and Hall, P.A. The p53 pathway. *J. Pathol.* **1999**, *187*: 112-26.

[16] Sigal, A., Rotter, V. Oncogenic mutations of the p53 tumor suppressor: the demons of the guardian of the genome. *Cancer Res.* **2000**, *60*: 6788-93.

[17] Lang, G.A., Iwakuma, T., Suh, Y.A., Liu, G., Rao, V.A., Parant, J.M., *et al.* Gain of function of a p53 hot spot mutation in a mouse model of Li-Fraumeni syndrome. *Cell* **2004**, *119*: 861-72.

[18] Blandino, G., Levine, A.J., Oren, M. Mutant p53 gain of function: differential effects of different p53 mutants on resistance of cultured cells to chemotherapy. *Oncogene* **1999**, *18*: 477-85.

[19] Gualberto, A., Aldape, K., Kozakiewicz, K., Tlsty, T.D. An oncogenic form of p53 confers a dominant, gain-of-function phenotype that disrupts spindle checkpoint control. *Proc. Natl. Acad. Sci. U.S.A.* **1998**, *95*: 5166-71.

[20] Frazier, M.W., He, X., Wang, J., Gu, Z., Cleveland, J.L., Zambetti, G.P. Activation of c-myc gene expression by tumor-derived p53 mutants requires a discrete c-terminal domain. *Mol. Cell. Biol.* **1998**, *18*: 3735-43.

[21] Lin, J., Teresky, A.K., Levine, A.J. Two critical hydrofobic amino acids in the N-terminal domain of the p53 protein are required for the gain of function phenotypes pf human p53 mutants. *Oncogene* **1995**, *10*: 2387-90.

[22] Di Agostino, S., Strano, S., Emiliozzi, V., Zerbini, V., Mottolese, M., Sacchi, *et al.* Gain of function of mutant p53: the mutant p53/NF-Y protein complex reveals an aberrant transcriptional mechanism of cell cycle regulation. *Cancer Cell* **2006**, *10*: 191-202.

[23] Fontemaggi, G., Dell'Orso, S., Trisciuoglio, D., Shay, T., Melucci, E., Fazi, F., *et al.* The execution of the transcriptional axis mutnt p53, E2F1 and ID4 promotes tumor neo-angiogenesis. *Nat. Struct. Mol. Biol.* **2009,** *16*: 1086-93

[24] Strano, S., Munarriz, E., Rossi, M., Cristofanelli, B., Shaul, Y., Castagnoli, L., *et al.* Physical and functional interaction between p53 mutants and different isoforms of p73. *J. Biol. Chem.* **2000**, *275*: 29503-12.

[25] Strano, S., Fontemaggi, G., Costanzo, A., Rizzo, M.G., Monti, O., Baccarini, A., *et al.* Physical interaction with human tumor-derived p53 mutants inhibits p63 activities. *J. Biol. Chem.* **2002**, *277*: 18817-26.

[26] Strano, S., Blandino, G. p73-mediated chemosensitivity: a preferential target of oncogenic mutant p53. *Cell Cycle* **2003**, *2*: 348-9.

[27] Di Como, C.J., Gaiddon, C., Prives, C. p73 function is inhibited by tumor-derived p53 mutants in mammalian cells. *Mol. Cell. Biol.* **1999**, *19*: 1438-49.

[28] Donzelli, S., Fontemaggi, G., Fazi, F., Di Agostino, S., Padula, F., Biagioni, F., *et al.* MicroRNA-128-2 targets the transcriptional repressor E2F5 enhancing mutant p53 gain of function. *Cell Death and Diff.* **2011**.

[29] Kato, S., Han, S.Y., Liu, W., Otsuka, K., Shibata, H., Kanamaru, R., *et al.* Understanding the function-structure and function-mutation relationships of p53 tumor suppressor protein by high-resolution missense mutation analysis. *Proc. Natl. Acad. Sci. U.S.A.* **2003**, *100*: 8424-9.

[30] Marin, M.C., Jost, C.A., Brooks, L.A., Irwin, M.S., O'Nions, J., Tidy, J.A., *et al.* A common polymorphism act as an intrgenic modifier of mutant p53 behaviour. *Nat. Genet.* **2000**, *25*: 47-54.

[31] Soussi, T., Wiman, K.G. Shaping genetic alterations in human cancer: the p53 mutation paradigm. *Cancer Cell* **2007**, *12*: 303-12.

[32] Hall, P.A., Lane, D.P. p53 in tumour pathology: can we trust immunohistochemistry?--Revisited! *J. Pathol.* **1994**, *172*: 1-4.

[33] Soussi, T., Beroud, C. Assessing TP53 status in human tumours to evaluate clinical outcome. *Nat. Rev. Cancer*. **2001**, *1*: 233-40.

[34] Langerod, A., Zhao, H., Borgan, O., Nesland, J.M., Bukholm, I.R., Ikdahl, T., *et al.* TP53 mutation status and gene expression profiles are powerful prognostic markers of breast cancer. *Breast Cancer Res*. **2007**, *9*: R30.

[35] Borresen, A.L., Andersen, T.I., Eyfjord, J.E., Cornelis, R.S., Thorlacius, S., Borg, A., *et al.* TP53 mutations and breast cancer prognosis: particularly poor survival rates for cases with mutations in the zinc-binding domains. *Genes Chromosomes Cancer* **1995**, *14*: 71-5.

[36] Temam, S., Flahault, A., Perie, S., Monceaux, G., Coulet, F., Callard, P., *et al.* p53 gene status as a predictor of tumor response to induction chemotherapy of patients with locoregionally advanced squamous cell carcinomas of the head and neck. J. *Clin. Oncol.* **2000**, *18*: 385-94.

[37] Olivier, M., Langerod, A., Carrieri, P., Bergh, J., Klaar, S., Eyfjord, J., *et al.* The clinical value of somatic TP53 gene mutations in 1,794 patients with breast cancer. *Clin. Cancer Res.* **2006**, *12*: 1157-67.

[38] Geisler, S., Lnning, P.E., Aas, T., Johnsen, H., Haugen, D.F. TP53 gene mutations predict the response to neoadjuvant treatment with 5-fluorouracil and mitomycin in locally advanced breast cancer. *Clin. Cancer Res.* **2003**, *9*: 5582-8.

[39] Sigurdsson, S., Bodvarsdottir, S.K., Anamthawat-Jonsson, K., Steinarsdottir, M., Jonasson, J.G., Ogmundsdottir, H.M., *et al.* p53 abnormality and chromosomal instability in the same breast tumor cells. *Cancer Genet. Cytogenet.* **2000**, *121*: 150-5.

[40] Jain, A.N., Chin, K., Borresen-Dale, A.L., Erikstein, B.K., Eynstein Lonning, P., Kaaresen, R., *et al.* Quantitative analysis of chromosomal CGH in human breast tumors associates copy number abnormalities with p53 status and patient survival. *Proc. Natl. Acad. Sci. U.S.A.* **2001**, *98*: 7952-7.

[41] Kallioniemi, A., Kallioniemi, O.P., Sudar, D., Rutovitz, D., Gray, J.W., Waldman, F., *et al.* Comparative genomic hybridization for molecular cytogenetic analysis of solid tumors. *Science* **1992**, *258*: 818-21.

[42] Jong, Y.J., Li, L.H., Tsou, M.H., Chen, Y.J., Cheng, S.H., Wang-Wuu, S., *et al.* Chromosomal comparative genomic hybridization abnormalities in early- and late-onset human breast cancers: correlation with disease progression and TP53 mutations. *Cancer Genet. Cytogenet.* **2004**, *148*: 55-65.

[43] Kleivi, K., Diep, C.B., Pandis, N., Heim, S., Teixeira, M.R., Lothe, R.A. TP53 mutations are associated with a particular pattern of genomic imbalances in breast carcinomas. *J. Pathol.* **2005**, *207*: 14-9.

[44] Oostlander, A.E., Meijer, G.A., Ylstra, B. Microarray-based comparative genomic hybridization and its applications in human genetics. *Clin. Genet.* **2004**, *66*: 488-95.

[45] Tsafrir, D., Bacolod, M., Selvanayagam, Z., Tsafrir, I., Shia, J., Zeng, Z., *et al.* Relationship of gene expression and chromosomal abnormalities in colorectal cancer. *Cancer Res.* **2006**, *66*: 2129-37.

[46] Carter, S.L., Eklund, A.C., Kohane, I.S., Harris, L.N., Szallasi, Z. A signature of chromosomal instability inferred from gene expression profiles predicts clinical outcome in multiple human cancers. *Nat. Genet.* **2006**, *38*: 1043-8.

[47] Weisz, L., Zalcenstein, A., Stambolsky, P., Cohen, Y., Goldfinger, N., Oren, M., *et al.* Transactivation of the EGR1 gene contributes to mutant p53 gain of function. *Cancer Res.* **2004**, *64*: 8318-27.

[48] Scian, M.J., Stagliano, K.E., Deb, D., Ellis, M.A., Carchman, E.H., Das, A., *et al.* Tumor-derived p53 mutants induce oncogenesis by transactivating growth-promoting genes. *Oncogene* **2004**, *23*: 4430-43.

[49] Scian, M.J., Stagliano, K.E., Ellis, M.A., Hassan, S., Bowman, M., Miles, M.F., *et al.* Modulation of gene expression by tumor-derived p53 mutants. *Cancer Res.* **2004**, *64*: 7447-54.

[50] O'Farrell, T.J., Ghosh, P., Dobashi, N., Sasaki, C.Y., Longo, D.L. Comparison of the effect of mutant and wild-type p53 on global gene expression. *Cancer Res.* **2004**, *64*: 8199-207.

[51] Mizuarai, S., Yamanaka, K., Kotani, H. Mutant p53 induces the GEF-H1 oncogene, a guanine nucleotide exchange factor-H1 for RhoA, resulting in accelerated cell proliferation in tumor cells. *Cancer Res.* **2006**, *66*: 6319-26.

[52] Tepper, C.G., Gregg, J.P., Shi, X.B., Vinall, R.L., Baron, C.A., Ryan, P.E., *et al.* Profiling of gene expression change caused by p53 gain of function mutant alleles in prostate cancer cells. *Prostate* **2005**, *65*: 375-89.

[53] Knaup, K.X., Roemer, K. Cell type-specific regulation of calmodulin 2 expression by mutant p53. *FEBS Lett.* **2004**, *569*: 70-4.

[54] Miller, L.D., Smeds, J., George, J., Vega, V.B., Vergara, L., Ploner, A., *et al.* An expression signature for p53 status in human breast cancer predicts mutation status, transcriptional effects, and patient survival. *Proc. Natl. Acad. Sci. U.S.A.* **2005**, *102*: 13550-5.

[55] Sorlie, T., Perou, C.M., Tibshirani, R., Aas, T., Geisler, S., Johnsen, H., *et al.* Gene expression patterns of breast carcinomas distinguish tumor subclasses with clinical implications. *Proc. Natl. Acad. Sci. U.S.A.* **2001**, *98*: 10869-74.

[56] Chen, X., Cheung, S.T., So, S., Fan, S.T., Barry, C., Higgins, J., *et al.* Gene expression patterns in human liver cancers. *Mol. Biol. Cell.* **2002**, *13*: 1929-39.

[57] Gualberto, A., Baldwin, A.S.Jr. p53 and Sp1 interact and cooperate in the tumor necrosis factor-induced transcriptional activation of the HIV-1 long terminal repeat. *J. Biol. Chem.* **1995**, *270*: 19680-3.

[58] Chicas, A., Molina, P., Bargonetti, J. Mutant p53 forms a complex with Sp1 on HIV-LTR DNA. *Biochem. Biophys Res. Commun.* **2000**, *279*: 383-90.

[59] Sampath, J., Sun, D., Kidd, V.J., Grenet, J., Gandhi, A., Shapiro, L.H., *et al.* Mutant p53 cooperates with ETS and selectively up-regulates human MDR1 not MRP1. *J. Biol. Chem.* **2001**, *276*: 39359-67.

[60] Dell'Orso, S., Fontemaggi, G., Stambolsky, P., Goeman, F., Voellenkle, C., Levrero, M., *et al.* ChIP-on-Chip analysis of *in vivo* mutant p53 binding to selected gene promoters. *OMICS.* **2011**, *15*: 305-12.

[61] Gohler, T., Jager, S., Warnecke, G., Yasuda, H., Kim, E., Deppert, W. Mutant p53 proteins bind DNA in a DNA structure-selective mode. *Nucleic Acid Res.* **2005**, *33*: 1087-100.

Advances in Genome Science, Vol. 3, 2014, 161-188

CHAPTER 6

DNA Instability at Chromosomal Fragile Sites in Cancer

Laura W. Dillon*, Allison B. Weckerle† and Yuh-Hwa Wang*

Department of Biochemistry, Wake Forest School of Medicine, Medical Center Boulevard, Winston-Salem, NC 27157-1016, USA

Abstract: Human chromosomal fragile sites are specific genomic regions which exhibit gaps or breaks on metaphase chromosomes following conditions of partial replication stress. Fragile sites often coincide with genes that are frequently rearranged or deleted in human cancers, with over half of cancer-specific translocations containing breakpoints within fragile sites. But until recently, little direct evidence existed linking fragile site breakage to the formation of cancer-causing chromosomal aberrations. Studies have revealed that DNA breakage at fragile sites can induce formation of *RET/PTC* rearrangements, and deletions within the *FHIT* gene, resembling those observed in human tumors. These findings demonstrate the important role of fragile sites in cancer development, suggesting that a better understanding of the molecular basis of fragile site instability is crucial to insights in carcinogenesis. It is hypothesized that under conditions of replication stress, stable secondary structures form at fragile sites and stall replication fork progress, ultimately resulting in DNA breaks. A study examining an FRA16B fragment confirmed the formation of secondary structure and DNA polymerase stalling within this sequence *in vitro*, as well as reduced replication efficiency and increased instability in human cells. Polymerase stalling during synthesis of FRA16D has also been demonstrated. A recent study of endogenous FRA16C in human cells showed that replication fork stalling occurs at AT-rich sequences, and under mild replication stress, the frequency of stalling is increased. The ATR DNA damage checkpoint pathway plays a critical role in maintaining stability at fragile sites. Recent findings have confirmed binding of the ATR protein to three regions of FRA3B under conditions of mild replication stress. This review will discuss recent advances made in understanding the role and mechanism of fragile sites in cancer development.

Keywords: Aphidicolin, ATR pathway, cancer-specific chromosomal translocation, carcinogenesis, CCDC6, cell-cycle checkpoint protein, chromosomal rearrangement, DNA damage sensor protein, DNA repair, DNA secondary structure, environmental mutagen, Fragile Site, genome instability, M-fold, NCOA4, occupational exposure, papillary thyroid carcinomas, RET, RET/PTC rearrangement, stalled replication fork.

***Address correspondence to Yuh-Hwa Wang:** Department of Biochemistry, Wake Forest School of Medicine, Medical Center Boulevard, Winston-Salem, NC 27157-1016, USA; Tel: (336) 716-6186; Fax: (336) 716-7671; E-mail: ywang@wakehealth.edu

†These two authors contributed equally to the work.

Christian Neri (Ed)

INTRODUCTION

Genomic instability is a common cause of chromosomal aberrations in many types of tumor cells. Cells can acquire DNA damage through various extrinsic or intrinsic factors. DNA damage commonly arises following exposure to exogenous factors such as UV radiation, ionizing radiation, chemotherapy, and endogenous factors such as reactive oxygen species [1]. Chromosomal fragile sites, which are especially susceptible to DNA breakage, have also been suggested as contributory to the formation of cancer-specific chromosomal aberrations [2].

Chromosomal fragile sites are defined as regions of the genome which exhibit gaps or breaks on metaphase chromosomes under conditions of partial replication stress [3]. Many genes identified as tumor suppressors or oncogenes are located at or within fragile sites [4]. The deletion of tumor suppressors and the amplification of oncogenes are frequently consequences of breakage at these sites. Studies have also demonstrated a significant association between sites of breakage in cancer-specific chromosomal rearrangements and the location of fragile sites [5-7]. Several oncogenic viruses target and preferentially integrate at these chromosomal regions [4]. Additionally, mutational signatures of some unexplained homozygous deletions observed in cancer cell lines match those at fragile sites [8]. Based on this information, it has been proposed that fragile sites may directly contribute to cancer development, although their direct role remains unclear. In the study by Gandhi *et al.*, DNA breakage at fragile sites directly contributed to the formation of a cancer-specific chromosomal translocation found in human papillary thyroid carcinoma [9].

Fragile sites are divided into two major classes, based on their frequency in the population, and are further divided according to their mode of induction in cultured cells. Common fragile sites have been observed in all individuals and are therefore believed to represent a normal component of chromosome architecture [10]. Most common fragile sites are induced by low doses of aphidicolin (APH), an inhibitor of DNA polymerases α, δ and ε [11, 12]. Other common fragile sites can occur following treatment with bromodeoxyuridine (BrdU) or 5-azacytidine (5-aza). In contrast, rare fragile sites are found in less than 5% of the population and are inherited in a Mendelian manner [13, 14]. Most of these sites are expressed under folate-deficient conditions, whereas others are induced by chemicals that bind AT-rich DNA, such as distamycin-A or berenil. Recently, Mrasek *et al.* found that APH can induce all types of common and rare fragile sites, suggesting that their expression is less dependent on their currently defined mode of induction, and instead, a classification of fragile sites based on their frequency is more appropriate [15]. In addition, various dietary and environmental factors can significantly increase fragile site breakage, including caffeine, ethanol, pesticides, and cigarette smoke [3].

While the molecular basis remains elusive, several factors may contribute to fragile site breakage. Although a consensus sequence has not yet been identified among all fragile sites, all common fragile DNAs examined to date are comprised of AT-rich flexibility islands, with the potential of forming secondary structures that are much more stable than other genomic regions [16, 17]. These regions can extend over megabases of DNA, with gaps or breaks occurring throughout [18]. Fragility at rare sites is attributed to the expansion of either a CGG repeat or an AT-rich minisatellite, which can also form stable secondary structures. All fragile sites studied so far are late-replicating regions of the genome, and their replication can be further delayed in the presence of induction agents, with some fragile site alleles remaining unreplicated in late G_2 [19-23]. In addition, the ATR (Ataxia-telangiectasia and Rad3 Related)-dependent DNA damage checkpoint pathway is crucial for fragile site maintenance, since a deficiency of proteins including ATR [24, 25] and its downstream targets, such as BRCA1 [26] and CHK1 [27], result in a dramatic increase in fragile site breakage. Therefore, it is hypothesized that at fragile sites, under conditions of replication stress, the replicative polymerases may uncouple from the helicase, resulting in long regions of single-stranded DNA and thus promoting the formation of stable secondary structures. Consequently, these structures may stall replication fork progression, triggering an ATR pathway response. Defective ATR pathway signaling could result in DNA breakage at these sites, and thus lead to cancer-specific chromosomal aberrations [6]. Other mechanisms, in addition to replication fork stalling, have been recently suggested to contribute to fragile site instability. The lack of replication initiation at fragile sites was shown to force the origins of the flanking regions to initiate DNA replication, resulting in fragile sites incompletely replicated under replication stress [28]. Also, the conflict between DNA replication and transcription of large genes at fragile sites promoting the formation of stable R-loop structures was shown to ultimately lead to common fragile site breakage [29].

Here, we will review recent advances made in understanding the role and mechanism of fragile sites in cancer development. We will discuss the importance of fragile sites, direct evidence of their involvement in cancer, sequence characteristics of fragile sites that contribute to their instability, and the role of the ATR-dependent DNA damage checkpoint pathway in their breakage.

IMPORTANCE OF FRAGILE SITES IN CANCER

Fragile sites are normally stable in cultured cells. However, these regions are highly susceptible to chromosomal deletions, rearrangements, and sister chromatid exchanges following their induction with replication inhibitors [30, 31]. Most rare fragile sites can be induced by removal of folic acid or treatment with fluorodeoxyuridine (FUdR), an inhibitor of folate metabolism. Other rare sites are expressed upon treatment with minor groove binders, such as distamycin-A or berenil, as well as BrdU, a nucleoside analog of thymidine that is incorporated into newly synthesized DNA. BrdU is also an inducer of several common fragile sites, along with 5-aza, whereas most sites are induced by APH, an inhibitor of DNA polymerases α, δ and ε. Common fragile sites can also be induced by various environmental agents and chemicals, including caffeine, ethanol, and cigarette smoke (Table **1**).

Table 1: Environmental, Dietary and Medicinal Inducers/Enhancers of Fragile Sites.

Chemical/Condition	Uses	References
5-azacytidine	chemotherapeutic agent	[43]
actinomycin D	chemotherapeutic agent	[43]
atenolol	hypertension drug	[44]
benzene	found in cigarette smoke, gasoline fumes	[43]
bleomycin	chemotherapeutic agent	[43]
busulfan	chemotherapeutic agent	[43]
caffeine	dietary agent	[5, 32]
carbon tetrachloride	found in refrigerants, pesticides	[43]

Table 1: contd…

chlorambucil	chemotherapeutic agent	[43]
cigarette smoke	dietary and environmental agent	[36, 37]
cytosine arabinoside	chemotherapeutic agent	[43]
diethylnitrosamine	found in cigarette smoke, pesticides, cured meat, whiskey	[43]
dimethyl sulfate	found in dyes, drugs, perfumes, pesticides	[43]
ethanol	dietary agent	[33, 34]
FUdR	chemotherapeutic agent	[43]
hypoxia	low oxygen; found in tumor microenvironment	[35]
methotrexate	chemotherapeutic agent	[43]
pesticides	environmental agent	[38, 39, 42]

Caffeine, an inhibitor of phosphoinositide 3-kinase related kinases, including ATR and ATM (Ataxia-telangiectasia mutated), significantly increases common fragile site breakage in conjunction with FUdR and APH [5, 32]. Ethanol treatment combined with APH also significantly increases fragile site breakage [33]. Interestingly, cells from chronic alcohol users show a significantly higher frequency of fragile site and chromosomal breakage compared to control individuals, suggesting long term alcohol use alone can induce fragile site expression [34]. Hypoxic conditions induced fragile site breakage in GMA32 Chinese hamster cells, with or without the addition of other fragile site-inducing chemicals [35]. Kao-Shan *et al.* found that peripheral blood lymphocytes from cigarette smokers show significantly greater fragile site breakage compared to non-smokers [36]. Stein *et al.* found that treatment of peripheral blood lymphocytes with low-doses of APH results in increased fragile site breakage in active smokers compared to non-smokers and patients with small cell lung cancer who stopped smoking [37]. These results suggest that active exposure to cigarette smoke increases the potential of breakage at fragile sites, and that this risk is reversible.

Exposure to pesticides also results in an increased susceptibility to fragile site breakage. Blood lymphocytes from pesticide sprayers and flower collectors working in greenhouses show greater fragile site breakage than normal individuals following treatment with APH, with these results being reproducible a year later [38, 39]. An association between pesticide exposure and an increased risk of hematopoietic tumors has been observed [40, 41]. The APH-induced damage was enhanced at fragile sites containing breakpoints involved in leukemias and non-Hodgkin's lymphoma, supporting a role for pesticide-associated fragile site breakage in development of these cancers. Like cigarette smoke exposure, the effect of pesticide exposure on fragile site breakage is also transient. APH-induced fragile site breakage in blood cells from farmers increased significantly post-exposure to organophosphate-based pesticides compared to pre-exposure samples, where the farmers had not been exposed to pesticides for at least one month [42].

Several mutagens and carcinogens can also induce fragile site breakage. Yunis *et al.* screened various mutagens for their ability to induce fragile site breakage [43]. The mutagens found to induce breakage were diverse and included benzene, a component found in cigarette smoke and gasoline fumes; carbon tetrachloride, formerly widely used in refrigerants and pesticides; diethylnitrosamine, found in cigarette smoke, pesticides, cured meat, and whiskey; and dimethyl sulfate, used in the manufacturing of dyes, drugs, perfumes, and pesticides. Several chemotherapeutic agents were also identified, including actinomycin D, bleomycin, busulfan, chlorambucil, cytosine arabinoside, 5-aza, FUdR, and methotrexate. The action of these mutagens was enhanced 3- to 8-fold with the addition of caffeine.

Peripheral blood lymphocytes from hypertensive patients taking atenolol, a common β-blocker, have more frequent chromatid and chromosome breaks than normal individuals, and these breaks are preferentially located at fragile sites [44]. While antihypertensive drugs have been examined for genotoxicity and carcinogenicity, a recent review suggests that studies to date may lack sufficient evidence to evaluate the potential risks of these drugs in humans [45]. One genotoxicity study of atenolol revealed a statistically significant increase in the number of micronuclei in patents taking atenolol compared to control [46]. Combined, these data suggest that more long-term research needs to be performed to assess the risk of these drugs on carcinogenesis, and that fragile site breakage may be involved in the carcinogenesis process.

Variability of fragile site breakage has been observed within individuals [47]. This phenomenon may reflect exposure to such mutagens/carcinogens, with high levels being associated with cancer patients [48]. Tumor cells also demonstrate an association between sites of breakage in recurrent chromosome abnormalities and the position of fragile sites. The two most highly breakable fragile sites, FRA3B and FRA16D, lie within the tumor suppressor genes *FHIT* and *WWOX*, respectively. The *FHIT* gene is often involved in deletions which map specifically to the fragile site region in various types of cancer [49-54], and deletion of *WWOX* is also frequently observed in tumor cells [55-60]. It has been proposed that fragile sites may also play a role in the breakage-fusion-bridge model of gene amplification [61, 62], since proto-oncogenes such as *MYC* [63] and *MET* [64] are located at fragile sites FRA8C and FRA7G, respectively. In addition to environmental mutagens, several oncogenic viruses including human papilloma virus [65], Hepatitis B [66], and Epstein-Barr [67], have been shown to target and preferentially integrate at fragile sites.

While strong correlations exist between fragile site locations and sites of breakage in other chromosome rearrangements, such as translocations, until recently there was no systematic demonstration of the location of fragile sites relative to all reported translocations in tumor cells. After the examination of all known cancer-specific recurrent translocations, it was determined that over half (52%) of translocation breakpoints in participating gene sets correspond to fragile site positions [68]. Furthermore, 65% of the breakpoints were located within common fragile sites, as opposed to rare sites, and most cancers associated with the translocations examined had little or no genetic component. These results suggest that exposure to fragile site-inducing chemicals may confer a risk for the formation of cancer-specific rearrangements. This study focused on simple translocations involving two genes, or deletions which lead to fusion transcripts, and excluded single gene deletions (the most common aberrations associated with fragile sites) and chromosomal insertions or more complex translocations involving multiple genes. Therefore, the association between translocation breakpoints and fragile site locations revealed in this study is likely to be an underestimation.

Because the evidence strongly supports a link between fragile sites and cancer development, it is essential to investigate the mechanism of fragile site breakage, and to examine the consequences of breakage at these sites in order to demonstrate a causative role in tumorigenesis. Also, it will be important to identify factors that contribute to chromosomal fragility, such as DNA sequence, proteins and environmental/dietary agents, since fragile sites can be induced by a variety of environmental and chemical agents.

DIRECT EVIDENCE OF FRAGILE SITE INVOLVEMENT IN CANCER DEVELOPMENT

Despite a long-established connection between fragile sites and the formation of cancer-specific chromosomal aberrations in many different studies [69], no studies have demonstrated direct evidence of fragile site breakage leading to cancer-causing chromosomal aberrations. Most experiments testing fragile site induction through exposure to chemicals examined DNA breakage on the cytogenetic level. However, two recent studies examining fragile site breakage at the nucleotide level, have revealed that DNA breakage at fragile sites can lead to the formation of *RET/PTC* rearrangements [9], as well as deletions within the *FHIT* gene [70], resembling those observed in human tumors.

FRA3B, the most frequently expressed fragile site in the genome, is located within the tumor supressor gene *FHIT* [71]. Deletions within *FHIT* have been associated with various human cancers including breast, lung, cervical, and esophageal [69]. Durkin *et al.* observed submicroscopic deletions within *FHIT*/FRA3B following treatment of human-mouse chromosome 3 somatic hybrid cells with low doses of APH [70]. The APH-induced deletions spanned ~200-600 Kb within *FHIT* and were centered on exon 5, within the breakpoint cluster of FRA3B. The location and size of APH-induced *FHIT* deletions were consistent with those observed in esophageal cancer cell lines, small-cell and non-small cell lung carcinomas, and breast cancers. Sequence analysis of the APH-induced deletion breakpoints showed no sequence homology, suggesting nonhomologous end-joining (NHEJ)-mediated repair. Interestingly, chromosomes from clones containing APH-induced FRA3B deletions exhibited a significant decrease in FRA3B breakage following additional APH treatment. While these studies show directly that fragile site breakage can lead to mutations like those seen in human tumors, these experiments were carried out in mouse hybrid cells, which may not respond similarly to human cells.

Genes participating in the two major types of *RET/PTC* rearrangements, *RET/PTC1* and *RET/PTC3*, are all located within known common fragile sites [68]. *RET/PTC* rearrangements, a common cause of papillary thyroid carcinoma, result in the fusion of the *RET* tyrosine kinase domain to the 5' portion of various constitutively expressed genes [72]. *RET* (rearranged during transfection) encodes for a cell membrane receptor tyrosine kinase protein whose ligands belong to the glial cell line-neurotropic factor (GDNF) family [73]. In the thyroid gland, RET is highly expressed in neural crest derived C-cells but not in follicular cells, where it can be activated through the formation of *RET/PTC* rearrangements [72]. In *RET/PTC1*, *RET* is rearranged with *CCDC6*, while in *RET/PTC3* is rearranged with *NCOA4* [74]. Both *CCDC6* and *NCOA4* are located closer than expected to *RET* in interphase nuclei of normal human thyroid cells, potentiating rearrangement formation [75, 76]. *RET* and *NCOA4* are both located within the APH-induced common fragile site FRA10G, and *CCDC6* is located within the BrdU-induced common fragile site FRA10C. Recently, Gandhi *et al.* found that exposure of human thyroid epithelial cells to the fragile site-inducing chemicals APH, BrdU, and 2-AP results in the formation of *RET/PTC1* rearrangements like those observed in patients [9]. DNA breakage was observed within *RET*, *CCDC6*, and to a lesser extent *NCOA4* using fluorescence *in situ* hybridization, following exposure to fragile site-inducing chemicals, consistent with the mode of induction for the fragile site where each gene is located.

APH-induced DNA breakage within *RET* was identified to be located within intron 11 [9], the major breakpoint cluster region within patients [77]. Interestingly, these breakpoints were located within 2-15 basepairs of breakpoints identified from human papillary thyroid carcinomas containing *RET/PTC* rearrangements. The breakpoints isolated in patient samples were identified post-rearrangement, while the breakpoints induced by APH were identified pre-rearrangement. In most patient tumors, small insertions or deletions ranging from 1-18 nucleotides surround the fusion points, suggesting a possible break repair mechanism after the initial breaks that form the re-arrangement. Schwartz *et al.* found that down-regulation of Rad51, DNA-PKcs, and Ligase IV, key components of the homologous recombination (HR) and NHEJ repair pathways, significantly increases fragile site breakage with APH treatment, and that γH2AX and phosphorylated DNA-PKcs foci were located at expressed fragile sites [78]. Together, these data suggest a role for both the HR and NHEJ repair pathways in the repair of fragile site breakage. However, more research is needed to elucidate the contribution of these pathways in the formation of *RET/PTC* rearrangements as a result of fragile site breakage.

RET/PTC rearrangements are commonly associated with radiation exposure, especially in children. However, most adult tumors containing *RET/PTC* rearrangements are sporadic and patients lack a history of radiation exposure [79]. Most importantly, the observation that fragile site-inducing chemicals can cause the formation of *RET/PTC1* rearrangements, and significant breakage at *RET* and *CCDC6* but not *NCOA4* [9], suggests a role for fragile sites in the formation of sporadic *RET/PTC1* tumors. Interestingly, an increasing prevalence of *RET/PTC1* over *RET/PTC3* rearrangements has been observed in sporadic papillary thyroid carcinomas [80], which is consistent with the experimental data and further supports fragile site involvement in sporadic papillary thyroid carcinomas. Overall, these experiments show direct involvement of fragile sites in the formation of cancer-causing chromosomal translocations within human cells.

According to the National Cancer Institute, thyroid cancer is now the fastest growing cancer among both men and women, increasing at a rate of 6.5 percent a year from 1997-2006 [81]. Of all types of thyroid cancer, papillary thyroid carcinoma is the only subtype in which the rate of incidence increased consistently regardless of race or ethnicity [82]. Interestingly, in a study examining cancer incidence in U.S. Air Force active duty personnel between 1989-2002, thyroid cancer was the third most frequent invasive cancer in women and the fifth most frequent in men [83]. Compared to the U.S. general population, thyroid cancer in women was over four times more prevalent in the U.S. Air Force. More importantly, overall cancer incidence in the U.S. Air Force were significantly reduced compared to the general population, suggesting that the increase in thyroid cancer may partly be the result of occupational exposure. Several components of unique occupational conditions for active duty Air Force personnel, including jet fuel, napalm, and high altitudes, have been shown to induce fragile site breakage. While the cause of the increased thyroid cancer incidence is unknown, exposure to various environmental and dietary fragile site-inducing chemicals and conditions is a potential causal factor.

DNA SECONDARY STRUCTURE AND REPLICATION FORK STALLING AT FRAGILE SITES

Understanding the molecular basis of fragile site breakage is critical for dissecting the role of fragile sites in cancer. Several intrinsic factors may contribute to their expression. Replication timing experiments have demonstrated that all fragile sites examined so far, including FRA1H [23], FRA2G [23], FRA3B [20], FRA7H [21], FRA10B [22], FRA16B [22], and FRAXA [19] exhibit late replication, which can be further delayed by the addition of replication inhibitors, with some fragile site alleles remaining unreplicated in the late G_2 phase [20, 21]. Although there is no consensus sequence among fragile sites, most fragile DNAs studied to date can form highly stable secondary structures [16, 17]. Evidence for secondary structure formation at fragile sites has largely been generated by the Mfold program [84], which predicts secondary structure formation of a single-stranded DNA. The first evidence of fragile DNA forming a secondary structure *in vitro* came from the ability of the CGG repeat, which underlies the basis of fragility at rare, folate-sensitive fragile sites, to form both quadruplex [85] and hairpin structures [86]. Furthermore, the formation of such structures by the CGG repeat presents a significant block to replication both *in vitro* [87] and *in vivo* [88]. Additionally, a polymorphic AT-rich sequence within the common fragile site FRA16D is predicted to form a cruciform that can block replication and increase chromosomal instability in yeast [89]. A recent study examining the rare fragile site FRA16B demonstrated the ability of an AT-rich site, which comprises the majority of fragile sites, to form a secondary structure *in vitro* [90].

Several studies have demonstrated a significant effect of *cis*-acting factors, including replication orientation and distance relative to the origin on the instability of fragile site DNA sequences. This effect may be due to the ability of an unstable DNA to form a stable secondary structure, and/or location of an alternative DNA structure within a region that maintains single-strandedness during replication, such as the Okazaki initiation zone (OIZ), which could promote the formation of such structures. Reports examining replication of expanded trinucleotide repeats in mammalian cells, such as CTG [91], GAA [92], and CGG [93], indicate that when the more structure-prone strand serves as the lagging strand template (a situation dictated by replication orientation), mutation events increase. The types of mutations (*i.e.* insertions and deletions) also differed based on the orientation of the DNA. When both strands can form an alternative structure (as with FRA16B), a similar orientation effect is also seen, likely due to the difference in the structures, as evidenced by the different electrophoretic mobilities produced by the two strands in native polyacrylamide gels [90]. The mutation rate for the clone in which the more stable structure served as the lagging strand template was statistically significant higher than that of the clone in opposite orientation [90]. These studies also revealed a significant effect on the distance from the repeat to the replication origin. Placing a repetitive fragment of DNA close to the origin of replication would cause the unstable sequence to occupy the first OIZ, potentially out-competing the binding of RPA protein. Repeats located further away from the origin would occupy different OIZs, and could be trumped by the initial binding of RPA in the first OIZ, which would propagate into the second, third, and so on [90]. In addition to mutation rate, replication orientation and distance relative to the origin also significantly affect replication efficiency of CGG and FRA16B-containing constructs. Edamura *et al.* found that constructs with expanded CGG repeats located further from the origin of replication replicated less efficiently compared to constructs with the same orientation located closer to the origin; plasmids in which CGG repeats served as the lagging strand template demonstrated a higher replication rate relative to plasmids in which CCG repeats served as the lagging strand template [93]. In a study of FRA16B replication in human cells, Burrow *et al.* found that various FRA16B-containing constructs replicated less efficiently compared to a control without fragile DNA, and that both orientation and distance from the origin significantly affected its replication rate [90].

A decrease in replication rate by a stalled fork could be responsible for the late replication observed at fragile sites. To date, synthesis of fragile site DNA has been investigated using primer extension assays both *in vitro* and in human cell-free extracts, and analyzing synthesis intermediates isolated for 2D gel electrophoresis in yeast. Examination of replication intermediates from cells containing AT-rich sequences within common fragile site FRA16D in *S. cerevisiae* showed site-specific replication fork stalling, depending on the length of the AT repeat [89]. Furthermore, correlation with secondary structure predictions suggests that the structure formed by this repeat is responsible for the fork stalling. Using an *in vitro* primer extension assay, synthesis of the same fragile site by human replicative polymerases δ and α confirmed polymerase stalling at sites predicted to form inhibitory DNA structures, which was alleviated by the addition of WRN protein [94]. The same study examined primer extension using HeLa cell-free extracts, and obtained similar results. Interestingly, FRA16D regions with increased DNA flexibility and accompanying high A/T content were not sufficient to inhibit DNA synthesis, but sequences with the propensity to form secondary structures were significantly more inhibitory to replication. Fragile site FRA16B exhibits characteristics of both common and rare sites, and showed strong polymerase pause sites specifically within the fragile DNA sequence *in vitro*, likely due to its ability to form an alternative DNA structure [90]. Ozeri-Galai *et al.* elegantly demonstrated that replication fork stalling occurs at endogenous AT-rich sequences within the common fragile site FRA16C in human lymphoblastoid cells, and the stalling is enhanced by APH treatment [95].

These data strongly suggest that the secondary-structure forming ability of fragile site DNAs greatly contributes to their instability by inhibiting replication. Even without replication stress, fragile sites show replication blockage during synthesis. While fragile site instability primarily arises from DNA polymerase stalling, visualization of FRA16B replication fork constructs by electron microscopy showed a high propensity for stalled forks to spontaneously regress during synthesis, a previously unidentified mechanism of instability at these sites [90]. In addition, measurements of the non-regressed FRA16B tracts confirmed synthesis products that were much shorter than predicted, consistent with the deletion mutants produced from replication of FRA16B in HEK293T cells. These results suggest yet another mechanism of fragile site instability in which polymerase bypass may occur at regions of secondary structure formation. Based on these results, fragile DNAs likely contain intrinsic features that make them difficult to replicate, and in the presence of replication stress, which can be produced by the addition of chemicals such as APH, replication is further delayed. The addition of

APH or other fragile site-inducing chemicals will likely create an accumulation of long, single-stranded DNA regions caused by the functional uncoupling of replicative DNA polymerase and helicase activities. These regions are then highly prone to forming stable secondary structures, which could further stall progression of the replication fork, triggering activation of ATR and its downstream products. A small fraction of fragile sites escapes the replication checkpoint, which would lead to breakage at fragile sites.

In addition to replication fork stalling, paucity of replication initiation in fragile site regions [28] and the presence of transcription-derived R-loops during DNA replication of fragile sites [29, 96-98], are also suggested to be involved in the mechanism of fragility. However, it is not clear whether the potential of fragile site sequences to form highly stable secondary structures participates in the latter two mechanisms. Interestingly, the formation of R-loops promotes trinucleotide repeat instability. The ability of trinucleotide repeats to form stable secondary structures may stabilize the presence of R-loops by adopting hairpin structure on the non-template DNA strand, therefore favoring hybrid formation between RNA transcripts and the DNA template strand [99]. Further investigation of the specific role of secondary structure-forming ability of fragile sites in the mechanism of fragility is needed.

REGULATION OF COMMON FRAGILE SITE STABILITY BY THE ATR PATHWAY

ATR kinase is a DNA damage sensor protein that has a major role in regulating stability at common fragile sites. ATR works with downstream target proteins to respond to stalled and collapsed replication forks, resulting in a block in further replication and mitosis progression and the promotion of DNA repair, recombination, or apoptosis [100, 101]. The loss of functional ATR in cells results in a defective DNA damage response to agents which block replication fork progression, including APH and hydroxyurea [102-104], and conditions of hypoxia [105]. Casper *et al.* found that cells deficient in ATR, but not ATM, display up to a 20-fold increase in fragile site breakage following treatment with low doses of APH compared to control cells [24]. Also, a deficiency in ATR alone is enough to induce fragile site breakage in cells without treatment with replication inhibitors. Cells from patients with Seckel syndrome, who express low levels of ATR protein due to a hypomorphic mutation in the *ATR* gene, exhibit an increase in chromosomal breakage at common fragile sites compared to unaffected individuals [25]. Furthermore, mice hypomorphic for ATR also display an increase in common fragile site breakage and a significant delay in checkpoint induction [106]. While the loss of ATM alone does not cause increased common fragile site breakage [24], it is involved in maintaining fragile site stability in the absence of ATR. Ozeri-Galai *et al.* found that a loss of both ATR and ATM significantly increases APH-induced common fragile site breakage compared to the loss of ATR alone. Also, ATM is activated and forms nuclear foci with γH2AX following treatment with low doses of APH [107]. These findings indicate that ATR is the major pathway responsible for maintaining fragile site stability, but that ATM also plays a secondary role, perhaps through a downstream response to double strand breaks that form as a result of ATR deficiency.

Other downstream targets of the ATR-mediated pathway involved in maintaining fragile site stability include BRCA1 [26], CHK1 [27], SMC1 [108], FANCD2 [109], HUS1 [110], WRN [111], and Claspin [112] (Table **2**). BRCA1 is a primary target of both ATR and ATM phosphorylation in response to DNA damage. Cells lacking BRCA1 show significantly more fragile site breakage after treatment with APH compared to control cells [26]. Also, cells expressing mutant BRCA1 exhibit elevated levels of fragile site breakage but lack the G_2/M checkpoint, suggesting BRCA1 regulates fragile site stability through its role at this checkpoint.

Table 2: DNA Damage Checkpoint Proteins Shown to Regulate Common Fragile Site Stability.

Protein	Function	References
ATM	Kinase, maintains fragile site stability in the absense of ATR	[107]
ATR	Kinase, binds to fragile DNA in response to replication stress, phosphorylates downstream targets to activate checkpoint response	[24, 119]
BRCA1	Phosphorylated by ATR, major downstream target of ATR, necessary for G_2/M checkpoint activation following replication stress	[26]
CHK1	Kinase, phosphorylated by ATR in response to replication stress, central regulator of ATR pathway	[27]
Claspin	Phosphorylated and interacts with CHK1 in response to replication stress	[112]

Table 2: contd….

FANCD2	Fanconi Anemia pathway protein, phosphorylated by ATR leading to activation by mono-Ub, activated by replication stress	[109]
HUS1	Member of the 9-1-1 complex, promotes phosphorylation of ATR substrates	[110]
SMC1	Chromosomal structural maintenance protein, member of the cohesion complex	[108]
WRN	ATP-dependent 3'-5' helicase, 3'-5' exonuclease	[94, 111]

CHK1 kinase is the major downstream target of ATR and serves as the central regulator of the ATR checkpoint pathway. Loss of CHK1, but not the ATM regulated CHK2, in cells was found to result in a significant increase in fragile site breakage after treatment with APH [27]. Also, it was found that both ATR and ATM phosphorylate CHK1 following treatment with low doses of APH [107]. These data suggest that the role of ATM in fragile site maintenance may be to activate the ATR pathway through phosphorylation of CHK1, when ATR is missing or fails to properly respond to damage. HUS1 is a member of the PCNA-related 9-1-1 complex which promotes the phosphorylation of ATR substrates like CHK1 and helps aid in DNA repair through association with multiple factors. A significant increase in DNA breakage at common fragile sites was observed after inactivation of HUS1 [110]. SMC1 is a chromosomal structural maintenance protein that belongs to the cohesin complex, which is necessary for sister chromatid cohesion and DNA repair and acts to hold DNA strands in place. After treatment with APH, cells exhibit an ATR-dependent, ATM-independent, phosphorylation of SMC1 and increased fragile site breakage after SMC1 inhibition [108]. Claspin is another member of the ATR pathway that is required for ATR-mediated phosphorylation of CHK1 in response to replication stress. Inhibition of claspin expression increases fragile site expression, with or without APH treatment [112].

Several studies have focused on the Fanconi anemia pathway, which responds to DNA cross-linking damage and chromosomal instability through a yet unknown mechanism involving interactions with BRCA1 and RAD51 and recruitment of BRCA2, in regulation of fragile site stability [109]. Fanconi anemia is an autosomal recessive disease associated with an increase in cancer susceptibility, and is most often a result of mutations in FA genes (subtypes A, B, C, D1, D2, E, F, G, I, J, L, M, N) [113]. Chromosomal breaks in blood lymphocytes of FA patients are preferentially located at fragile sites [114], and FANCD2 and FANCI specifically associate with common fragile site loci under conditions of replication stress [115]. Also, ATR phosphorylates the FA protein, FANCD2, and is required for its mono-ubiquitination [116], which is necessary for its activation during S-phase and subsequent colocalization with BRCA1 and RAD51 [117], in response to replication stress. Treatment of both FANCD2 knockdown cells and FA-patient cells with APH results in increased fragile site breakage [109]. Interestingly, cigarette smoke, a known induction agent of fragile site expression, suppresses FANCD2 expression in airway epithelial cells [118]. It will be intriguing to investigate the connection between the cancer susceptibility of FA patients and fragile site-mediated cancer-causing rearrangements.

WRN is an ATP-dependent 3'-5' helicase and 3'-5' exonuclease that is targeted by ATR and interacts with ATR-pathway proteins. Increased fragile site breakage is seen in cells of patients with Werner syndrome (a premature aging disease associated with a greater susceptibility to cancer development), and in WRN knockdown cells after treatment with APH [111]. In addition, double knockdown of WRN and ATR did not result in increased chromosomal damage compared to WRN or ATR knockdown alone, suggesting these proteins work in a common pathway. The activity of WRN in fragile site maintenance still remains unclear. Pirzio *et al.* presented data suggesting that WRN helicase, not exonuclease activity, plays the main role in stabilizing fragile sites [111]. In contrast, Shah *et al.* found that neither WRN helicase or exonuclease activity was necessary for polymerase δ progression past stalled replication forks within various FRA16D sequences *in vitro* [114].

While the importance of the ATR pathway in fragile site maintenance has been established, the mechanism is not fully understood. Recently, Wan *et al.* found that ATR binds (directly or through complexes) to fragile site FRA3B preferentially compared to non-fragile regions under conditions of mild replication stress [119]. This binding increases in a dose-dependent manner, peaking at 0.4μM APH, and decreases at higher APH concentrations. While the level of ATR binding to FRA3B changes with treatment, the cellular levels of ATR, phospho-ATR (Ser 428), and ATR interacting proteins ATRIP and TopBP1 remain unchanged. This suggests that ATR binding to the fragile site is guided initially by the level of replication stress signals generated at FRA3B due to APH treatment, and then sequestered from FRA3B regions by successive signals from other non-fragile site regions, which are produced at the higher concentrations of APH. Furthermore, the kinase activity of ATR was required for ATR binding to FRA3B in response to APH treatment. While ATR kinase activity is known to be necessary for phosphorylation of downstream targets to activate the checkpoint signaling cascade [101], these data indicate that the kinase activity of ATR is also necessary for ATR interaction to fragile site regions, most likely through phosphorylation of ATRIP and TopBP1 to stabilize the interaction between these three proteins and the fragile DNA. Two models which are not mutually exclusive have been proposed to explain how ATR helps to maintain fragile site stability [120]. The first model states that a loss of ATR can lead to a bypass of stalled replication forks at fragile sites, ultimately resulting in a failure of checkpoint pathways to prevent entry into mitosis, thus leaving DNA breakage at the unreplicated DNA. The second model states that a loss of ATR leads to replication fork collapse at fragile sites and improper resolution of these structures by ATR leads to DNA breaks. The current information about the involvement of ATR at fragile sites supports a combination of both models. The preferential binding of ATR protein to FRA3B fragile DNAs following APH treatment [119] suggests that ATR plays a possible local role in stabilizing stalled replication forks at fragile regions. Also, this binding and increased fragile site breakage following the inhibition of various members of the ATR pathway suggest that ATR response to fragile sites under conditions of replication stress can activate the ATR-dependent pathway. Finally, decreased ATR binding to FRA3B at higher concentrations of APH [119], which induce more chromosomal gaps or genomic breaks, supports the idea that DNA breakage at fragile sites is due to a failure of ATR to stabilize replication forks and to signal a checkpoint response.

ATR and many downstream proteins (*e.g.* ATM, BRCA1, CHK1, HUS1, SMC1, FANCD2, WRN, and claspin) are important in maintaining fragile site stability. However, in response to replication stress, whether these proteins act in the ATR pathway and their specific role as adaptor, transducer or effector molecules, and/or participate in an alternative pathway, remains unclear. More research is necessary to better understand the mechanism of the ATR pathway at fragile sites and its effect on cancer development.

PERSPECTIVES

Herein we present a model for fragile site instability in the formation of cancer-specific chromosomal rearrangements in the context of *RET/PTC1* rearrangement formation, since fragile site breakage at *RET* and *CCDC6* directly results in the formation of translocations like those observed in patient tumors (Fig. **1**). *RET* and *CCDC6* genes, located within the fragile sites FRA10G and FRA10C respectively, are in close proximity during interphase in normal thyroid cells [75, 76], thus promoting the formation of rearrangement. Under conditions of replication stress, such as those environmental and dietary agents known to induce fragile sites, replicative DNA polymerases become uncoupled from the helicase/topoisomerase complex, resulting in long stretches on single-stranded DNA. Regions that maintain single-strandedness, such as the OIZ, may promote the formation of stable secondary structures due to the intrinsic features of fragile DNA. These structures can cause significant difficulties during replication, resulting in a stalled replication fork. The ATR-dependent DNA damage checkpoint pathway, which responds to stalled or collapsed replication forks, is then triggered. In some cases, a stalled replication fork may spontaneously regress during synthesis of fragile DNA, generating a Holliday junction-like intermediate which could lead to breakage through cleavage by resolvase enzymes [121], or polymerase bypass might occur at regions of structure formation resulting in chromosome deletions. For repair of stalled forks, ATR, the main sensor of the pathway, binds to the fragile DNA either directly or through complexes and activates a downstream signaling cascade through phosphorylation of various targets, including central regulator CHK1. Other components of the ATR pathway, including BRCA1, FANCD2, WRN, Claspin, HUS1, and SMC1, are crucial for fragile site maintenance, although their direct role remains unclear. If the ATR pathway properly responds, the replication fork will be repaired and DNA replication will resume normally. However, a loss, deficiency, or defect in ATR pathway proteins could lead to checkpoint failure and/or replication fork collapse resulting in DNA breakage, the initiating event of translocation formation. Therefore, breakage at *RET* and *CCDC6* can lead to the formation of *RET/PTC1* translocations in sporadic tumors through a fragile site-mediated mechanism, resulting in the development of sporadic papillary thyroid carcinomas.

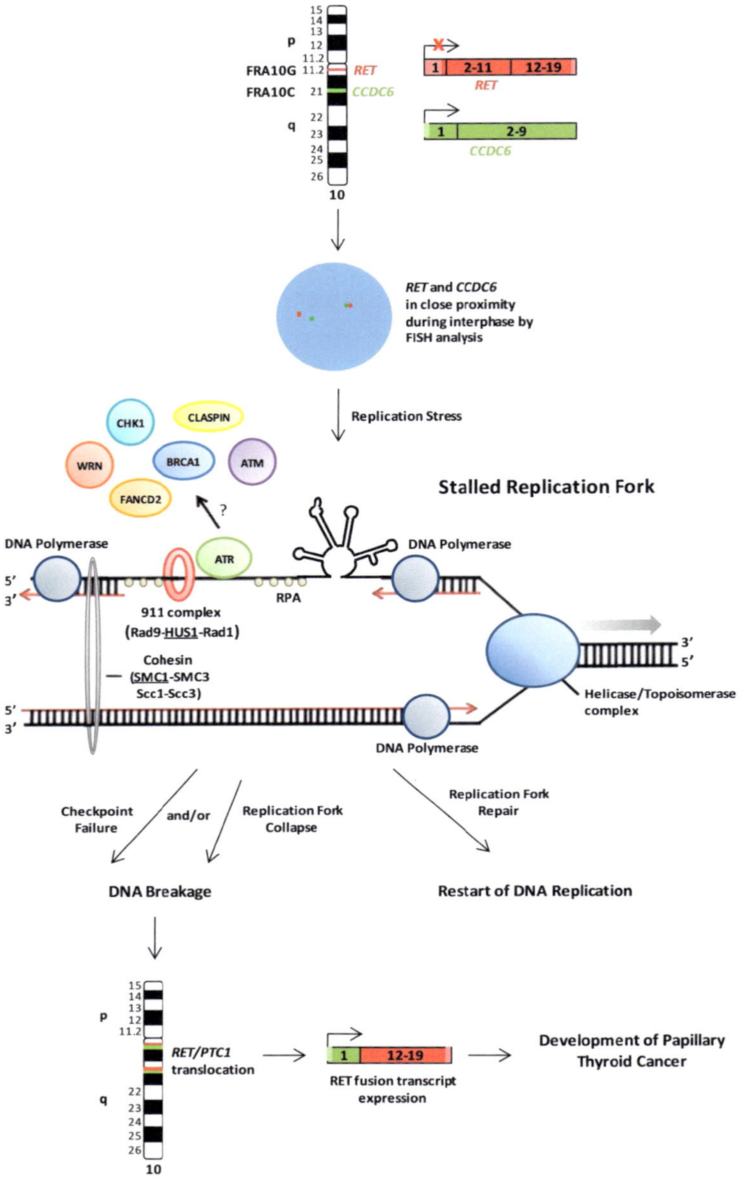

Figure 1: Model of Fragile Site Instability in the Formation of Cancer-Specific Chromosomal Rearrangements. *RET* and *CCDC6* genes, located on chromosome 10 within the fragile sites FRA10G and FRA10C respectively, are closer than expected during interphase in normal thyroid cells. In the normal thyroid gland, *RET* is not expressed in follicular cells, while *CCDC6* is constitutively expressed. Under conditions of replication stress, replicative DNA polymerases α, δ, and ε become uncoupled from the helicase/topoisomerase complex, resulting in long stretches on single-stranded DNA susceptible to the formation of stable secondary structures.

These structures can cause replication fork stalling, triggering the ATR-dependent DNA damage checkpoint pathway. Fragile sites may also be susceptible to spontaneous fork reversal or polymerase skipping at regions of secondary structure. For repair of stalled forks, ATR binds to the fragile DNA either directly or through complexes, and activates a downstream signaling cascade with other proteins, including CHK1, BRCA1, FANCD2, WRN, Claspin, HUS1, ATM, and SMC1. If the ATR pathway properly responds, the replication fork will be repaired and DNA replication will resume normally. A loss, deficiency, or defect in ATR pathway proteins could lead to checkpoint failure and/or replication fork collapse resulting in DNA breakage at *RET* and *CCDC6*. DNA breakage at these sites can lead to the formation of *RET/PTC1* translocations, the expression of oncogenic RET protein, and the development of papillary thyroid carcinoma.

Several studies have shown a strong correlation between the location of fragile sites and sites of breakage in a few cancer-specific chromosomal aberrations. We showed that over half of breakpoints in all reported recurrent cancer-specific chromosomal translocations are within known fragile sites, but that this is an underestimation, since only simple translocations were investigated [68]. There may be other fragile sites in the genome as yet unidentified. A recent report by Mrasek *et al.* found 61 fragile sites previously not observed and 52 previously not verified, all of which are APH-inducible [15]. Intriguingly, all classes of rare and common fragile sites were induced by APH. These results suggest that fragile sites may have an even greater role in carcinogenesis than previously thought. Since certain dietary and environmental factors can induce or enhance breakage of various APH-induced common fragile sites, it is conceivable that the newly identified sites are also targeted by these agents, thus increasing the risk of cancer development. Futhermore, these results strongly suggest a role for fragile sites in the development of sporadic cancer. Future studies are needed to elucidate the mechanism of fragile site instability in the formation of cancer-specific chromosomal aberrations and the contribution of dietary and environmental factors in this process.

ACKNOWLEDGEMENTS

This work was supported by the National Cancer Institute (R01 CA85826 and R01 CA113863 to Y.-H. W).

CONFLICT OF INTEREST

The authors confirm that this chapter contents have no conflict of interest.

DISCLOSURE

The chapter submitted for Advances in Genome Science, Vol. 3 entitled **"DNA Instability at Chromosomal Fragile Sites in Cancer (review) L. W. Dillon, *et al.*"** is an update of our article published in Current Genomics, 2010, 11: 326-337.

REFERENCES

[1] Jackson SP. Sensing and repairing DNA double-strand breaks. *Carcinogenesis*, **2002**, *23*, 687-96.

[2] O'Keefe LV, Richards RI. Common chromosomal fragile sites and cancer: focus on FRA16D. *Cancer Lett*, **2006**, *232*, 37-47.

[3] Richards RI. Fragile and unstable chromosomes in cancer: causes and consequences. *Trends Genet*, **2001**, *17*, 339-45.

[4] Popescu NC. Genetic alterations in cancer as a result of breakage at fragile sites. *Cancer Letters*, **2003**, *192*, 1-17.

[5] Yunis JJ, Soreng AL. Constitutive fragile sites and cancer. *Science*, **1984**, *226*, 1199-204.

[6] Durkin SG, Glover TW. Chromosome fragile sites. *Annu Rev Genet*, **2007**, *41*, 169-92.

[7] Re A, Cora D, Puliti AM, Caselle M, Sbrana I. Correlated fragile site expression allows the identification of candidate fragile genes involved in immunity and associated with carcinogenesis. *BMC Bioinformatics*, **2006**, *7*, 413.

[8] Bignell GR, Greenman CD, Davies H, Butler AP, Edkins S, Andrews JM, *et al.* Signatures of mutation and selection in the cancer genome. *Nature*, **2010**, *463*, 893-8.

[9] Gandhi M, Dillon LW, Pramanik S, Nikiforov YE, Wang YH. DNA breaks at fragile sites generate oncogenic RET/PTC rearrangements in human thyroid cells. *Oncogene*, **2010**, *29*, 2272-80.

[10] Glover TW. Common fragile sites. *Cancer Lett*, **2006**, *232*, 4-12.

[11] Glover TW, Berger C, Coyle J, Echo B. DNA polymerase alpha inhibition by aphidicolin induces gaps and breaks at common fragile sites in human chromosomes. *Hum Genet*, **1984**, *67*, 136-42.

[12] Cheng CH, Kuchta RD. DNA polymerase epsilon: aphidicolin inhibition and the relationship between polymerase and exonuclease activity. *Biochemistry*, **1993**, *32*, 8568-74.

[13] Kremer EJ, Pritchard M, Lynch M, Yu S, Holman K, Baker E, *et al.* Mapping of DNA instability at the fragile X to a trinucleotide repeat sequence p(CCG)n. *Science*, **1991**, *252*, 1711-4.

[14] Sutherland GR, Baker E, Richards RI. Fragile sites still breaking. *Trends Genet*, **1998**, *14*, 501-6.

[15] Mrasek K, Schoder C, Teichmann AC, Behr K, Franze B, Wilhelm K, *et al.* Global screening and extended nomenclature for 230 aphidicolin-inducible fragile sites, including 61 yet unreported ones. *Int J Oncol*, **2010**, *36*, 929-40.

[16] Mishmar D, Rahat A, Scherer SW, Nyakatura G, Hinzmann B, Kohwi Y, *et al.* Molecular characterization of a common fragile site (FRA7H) on human chromosome 7 by the cloning of a simian virus 40 integration site. *Proc Natl Acad Sci U S A*, **1998**, *95*, 8141-6.

[17] Zlotorynski E, Rahat A, Skaug J, Ben-Porat N, Ozeri E, Hershberg R, *et al.* Molecular basis for expression of common and rare fragile sites. *Mol Cell Biol*, **2003**, *23*, 7143-51.

[18] McAvoy S, Ganapathiraju SC, Ducharme-Smith AL, Pritchett JR, Kosari F, Perez DS, *et al.* Non-random inactivation of large common fragile site genes in different cancers. *Cytogenet Genome Res*, **2007**, *118*, 260-9.

[19] Hansen RS, Canfield TK, Fjeld AD, Mumm S, Laird CD, Gartler SM. A variable domain of delayed replication in FRAXA fragile X chromosomes: X inactivation-like spread of late replication. *Proc Natl Acad Sci U S A*, **1997**, *94*, 4587-92.

[20] Le Beau MM, Rassool FV, Neilly ME, Espinosa R, 3rd, Glover TW, Smith DI, *et al.* Replication of a common fragile site, FRA3B, occurs late in S phase and is delayed further upon induction: implications for the mechanism of fragile site induction. *Hum Mol Genet*, **1998**, *7*, 755-61.

[21] Hellman A, Rahat A, Scherer SW, Darvasi A, Tsui LC, Kerem B. Replication delay along FRA7H, a common fragile site on human chromosome 7, leads to chromosomal instability. *Mol Cell Biol*, **2000**, *20*, 4420-7.

[22] Handt O, Baker E, Dayan S, Gartler SM, Woollatt E, Richards RI, *et al.* Analysis of replication timing at the FRA10B and FRA16B fragile site loci. *Chromosome Res*, **2000**, *8*, 677-88.

[23] Pelliccia F, Bosco N, Curatolo A, Rocchi A. Replication timing of two human common fragile sites: FRA1H and FRA2G. *Cytogenet Genome Res*, **2008**, *121*, 196-200.

[24] Casper AM, Nghiem P, Arlt MF, Glover TW. ATR regulates fragile site stability. *Cell*, **2002**, *111*, 779-89.

[25] Casper AM, Durkin SG, Arlt MF, Glover TW. Chromosomal instability at common fragile sites in Seckel syndrome. *Am J Hum Genet*, **2004**, *75*, 654-60.

[26] Arlt MF, Xu B, Durkin SG, Casper AM, Kastan MB, Glover TW. BRCA1 is required for common-fragile-site stability *via* its G2/M checkpoint function. *Mol Cell Biol*, **2004**, *24*, 6701-9.

[27] Durkin SG, Arlt MF, Howlett NG, Glover TW. Depletion of CHK1, but not CHK2, induces chromosomal instability and breaks at common fragile sites. *Oncogene*, **2006**, *25*, 4381-8.

[28] Letessier A, Millot GA, Koundrioukoff S, Lachages AM, Vogt N, Hansen RS, *et al.* Cell-type-specific replication initiation programs set fragility of the FRA3B fragile site. *Nature*, **2011**, *470*, 120-3.

[29] Helmrich A, Ballarino M, Tora L. Collisions between replication and transcription complexes cause common fragile site instability at the longest human genes. *Mol Cell*, **2011**, *44*, 966-77.

[30] Glover TW, Stein CK. Induction of sister chromatid exchanges at common fragile sites. *Am J Hum Genet*, **1987**, *41*, 882-90.

[31] Glover TW, Stein CK. Chromosome breakage and recombination at fragile sites. *Am J Hum Genet*, **1988**, *43*, 265-73.

[32] Glover TW, Coyle-Morris J, Morgan R. Fragile sites: overview, occurrence in acute nonlymphocytic leukemia and effects of caffeine on expression. *Cancer Genet Cytogenet*, **1986**, *19*, 141-50.

[33] Kuwano A, Kajii T. Synergistic effect of aphidicolin and ethanol on the induction of common fragile sites. *Hum Genet*, **1987**, *75*, 75-8.

[34] Demirhan O, Tastemir D. Cytogenetic effects of ethanol on chronic alcohol users. *Alcohol Alcohol*, **2008**, *43*, 127-36.

[35] Coquelle A, Toledo F, Stern S, Bieth A, Debatisse M. A new role for hypoxia in tumor progression: induction of fragile site triggering genomic rearrangements and formation of complex DMs and HSRs. *Mol Cell*, **1998**, *2*, 259-65.

[36] Kao-Shan CS, Fine RL, Whang-Peng J, Lee EC, Chabner BA. Increased fragile sites and sister chromatid exchanges in bone marrow and peripheral blood of young cigarette smokers. *Cancer Res*, **1987**, *47*, 6278-82.

[37] Stein CK, Glover TW, Palmer JL, Glisson BS. Direct correlation between FRA3B expression and cigarette smoking. *Genes Chromosomes Cancer*, **2002**, *34*, 333-40.

[38] Musio A, Sbrana I. Aphidicolin-sensitive specific common fragile sites: a biomarker of exposure to pesticides. *Environ Mol Mutagen*, **1997**, *29*, 250-5.

[39] Sbrana I, Musio A. Enhanced expression of common fragile site with occupational exposure to pesticides. *Cancer Genet Cytogenet*, **1995**, *82*, 123-7.

[40] Blair A, Zahm SH. Cancer among farmers. *Occup Med*, **1991**, *6*, 335-54.

[41] Brown LM, Blair A, Gibson R, Everett GD, Cantor KP, Schuman LM, *et al.* Pesticide exposures and other agricultural risk factors for leukemia among men in Iowa and Minnesota. *Cancer Res*, **1990**, *50*, 6585-91.

[42] Webster LR, McKenzie GH, Moriarty HT. Organophosphate-based pesticides and genetic damage implicated in bladder cancer. *Cancer Genet Cytogenet*, **2002**, *133*, 112-7.

[43] Yunis JJ, Soreng AL, Bowe AE. Fragile sites are targets of diverse mutagens and carcinogens. *Oncogene*, **1987**, *1*, 59-69.

[44] Telez M, Ortiz-Lastra E, Gonzalez AJ, Flores P, Huerta I, Ramirez JM, *et al.* Assessment of the genotoxicity of atenolol in human peripheral blood lymphocytes: correlation between chromosomal fragility and content of micronuclei. *Mutat Res*, **2010**, *695*, 46-54.

[45] Brambilla G, Martelli A. Genotoxicity and carcinogenicity studies of antihypertensive agents. *Mutat Res*, **2006**, *612*, 115-49.

[46] Telez M, Martinez B, Criado B, Lostao CM, Penagarikano O, Ortega B, *et al. In vitro* and *in vivo* evaluation of the antihypertensive drug atenolol in cultured human lymphocytes: effects of long-term therapy. *Mutagenesis*, **2000**, *15*, 195-202.

[47] Craig-Holmes AP, Strong LC, Goodacre A, Pathak S. Variation in the expression of aphidicolin-induced fragile sites in human lymphocyte cultures. *Hum Genet*, **1987**, *76*, 134-7.

[48] Tunca B, Egeli U, Zorluoglu A, Yilmazlar T, Yerci O, Kizil A. The expression of fragile sites in lymphocytes of patients with rectum cancer and their first-degree relatives. *Cancer Lett*, **2000**, *152*, 201-9.

[49] Kameoka Y, Tagawa H, Tsuzuki S, Karnan S, Ota A, Suguro M, *et al.* Contig array CGH at 3p14.2 points to the FRA3B/FHIT common fragile region as the target gene in diffuse large B-cell lymphoma. *Oncogene*, **2004**, *23*, 9148-54.

[50] Lee SH, Kim WH, Kim HK, Woo KM, Nam HS, Kim HS, *et al.* Altered expression of the fragile histidine triad gene in primary gastric adenocarcinomas. *Biochem Biophys Res Commun*, **2001**, *284*, 850-5.

[51] Michael D, Beer DG, Wilke CW, Miller DE, Glover TW. Frequent deletions of FHIT and FRA3B in Barrett's metaplasia and esophageal adenocarcinomas. *Oncogene*, **1997**, *15*, 1653-9.

[52] Sozzi G, Veronese ML, Negrini M, Baffa R, Cotticelli MG, Inoue H, *et al.* The FHIT gene 3p14.2 is abnormal in lung cancer. *Cell*, **1996**, *85*, 17-26.

[53] Virgilio L, Shuster M, Gollin SM, Veronese ML, Ohta M, Huebner K, *et al.* FHIT gene alterations in head and neck squamous cell carcinomas. *Proc Natl Acad Sci U S A*, **1996**, *93*, 9770-5.

[54] Huebner K, Croce CM. FRA3B and other common fragile sites: the weakest links. *Nat Rev Cancer*, **2001**, *1*, 214-21.

[55] Driouch K, Prydz H, Monese R, Johansen H, Lidereau R, Frengen E. Alternative transcripts of the candidate tumor suppressor gene, WWOX, are expressed at high levels in human breast tumors. *Oncogene*, **2002**, *21*, 1832-40.

[56] Kuroki T, Trapasso F, Shiraishi T, Alder H, Mimori K, Mori M, *et al.* Genetic alterations of the tumor suppressor gene WWOX in esophageal squamous cell carcinoma. *Cancer Res*, **2002**, *62*, 2258-60.

[57] Kuroki T, Yendamuri S, Trapasso F, Matsuyama A, Aqeilan RI, Alder H, *et al.* The tumor suppressor gene WWOX at FRA16D is involved in pancreatic carcinogenesis. *Clin Cancer Res*, **2004**, *10*, 2459-65.

[58] Paige AJ, Taylor KJ, Taylor C, Hillier SG, Farrington S, Scott D, *et al.* WWOX: a candidate tumor suppressor gene involved in multiple tumor types. *Proc Natl Acad Sci U S A*, **2001**, *98*, 11417-22.

[59] Ried K, Finnis M, Hobson L, Mangelsdorf M, Dayan S, Nancarrow JK, *et al.* Common chromosomal fragile site FRA16D sequence: identification of the FOR gene spanning FRA16D and homozygous deletions and translocation breakpoints in cancer cells. *Hum Mol Genet*, **2000**, *9*, 1651-63.

[60] Yendamuri S, Kuroki T, Trapasso F, Henry AC, Dumon KR, Huebner K, *et al.* WW domain containing oxidoreductase gene expression is altered in non-small cell lung cancer. *Cancer Res*, **2003**, *63*, 878-81.

[61] McClintock B. Chromosome organization and genic expression. *Cold Spring Harb Symp Quant Biol*, **1951**, *16*, 13-47.

[62] Coquelle A, Pipiras E, Toledo F, Buttin G, Debatisse M. Expression of fragile sites triggers intrachromosomal mammalian gene amplification and sets boundaries to early amplicons. *Cell*, **1997**, *89*, 215-25.

[63] Zimonjic DB, Keck-Waggoner CL, Yuan BZ, Kraus MH, Popescu NC. Profile of genetic alterations and tumorigenicity of human breast cancer cells. *Int J Oncol*, **2000**, *16*, 221-30.

[64] Hellman A, Zlotorynski E, Scherer SW, Cheung J, Vincent JB, Smith DI, Trakhtenbrot L, Kerem B. A role for common fragile site induction in amplification of human oncogenes. *Cancer Cell*, **2002**, *1*, 89-97.

[65] Popescu NC, DiPaolo JA, Amsbaugh SC. Integration sites of human papillomavirus 18 DNA sequences on HeLa cell chromosomes. *Cytogenet Cell Genet*, **1987**, *44*, 58-62.

[66] Feitelson MA, Lee J. Hepatitis B virus integration, fragile sites, and hepatocarcinogenesis. *Cancer Lett*, **2007**, *252*, 157-70.

[67] Popescu NC, Chen MC, Simpson S, Solinas S, DiPaolo JA. A Burkitt lymphoma cell line with integrated Epstein-Barr virus at a stable chromosome modification site. *Virology*, **1993**, *195*, 248-51.

[68] Burrow AA, Williams LE, Pierce LC, Wang YH. Over half of breakpoints in gene pairs involved in cancer-specific recurrent translocations are mapped to human chromosomal fragile sites. *BMC Genomics*, **2009**, *10*, 59.

[69] Arlt MF, Durkin SG, Ragland RL, Glover TW. Common fragile sites as targets for chromosome rearrangements. *DNA Repair (Amst)*, **2006**, *5*, 1126-35.

[70] Durkin SG, Ragland RL, Arlt MF, Mulle JG, Warren ST, Glover TW. Replication stress induces tumor-like microdeletions in FHIT/FRA3B. *Proc Natl Acad Sci U S A*, **2008**, *105*, 246-51.

[71] Ohta M, Inoue H, Cotticelli MG, Kastury K, Baffa R, Palazzo J, *et al.* The FHIT gene, spanning the chromosome 3p14.2 fragile site and renal carcinoma-associated t(3;8) breakpoint, is abnormal in digestive tract cancers. *Cell*, **1996**, *84*, 587-97.

[72] Nikiforov YE. Thyroid carcinoma: molecular pathways and therapeutic targets. *Mod Pathol*, **2008**, *21 Suppl 2*, S37-43.

[73] Arighi E, Borrello MG, Sariola H. RET tyrosine kinase signaling in development and cancer. *Cytokine Growth Factor Rev*, **2005**, *16*, 441-67.

[74] Santoro M, Melillo RM, Fusco A. RET/PTC activation in papillary thyroid carcinoma: European Journal of Endocrinology Prize Lecture. *Eur J Endocrinol*, **2006**, *155*, 645-53.

[75] Gandhi M, Medvedovic M, Stringer JR, Nikiforov YE. Interphase chromosome folding determines spatial proximity of genes participating in carcinogenic RET/PTC rearrangements. *Oncogene*, **2006**, *25*, 2360-6.

[76] Nikiforova MN, Stringer JR, Blough R, Medvedovic M, Fagin JA, Nikiforov YE. Proximity of chromosomal loci that participate in radiation-induced rearrangements in human cells. *Science*, **2000**, *290*, 138-41.

[77] Smanik PA, Furminger TL, Mazzaferri EL, Jhiang SM. Breakpoint characterization of the ret/PTC oncogene in human papillary thyroid carcinoma. *Hum Mol Genet*, **1995**, *4*, 2313-8.

[78] Schwartz M, Zlotorynski E, Goldberg M, Ozeri E, Rahat A, le Sage C, *et al.* Homologous recombination and nonhomologous end-joining repair pathways regulate fragile site stability. *Genes Dev*, **2005**, *19*, 2715-26.

[79] Nikiforova MN, Nikiforov YE. Molecular genetics of thyroid cancer: implications for diagnosis, treatment and prognosis. *Expert Rev Mol Diagn*, **2008**, *8*, 83-95.

[80] Fenton CL, Lukes Y, Nicholson D, Dinauer CA, Francis GL, Tuttle RM. The ret/PTC mutations are common in sporadic papillary thyroid carcinoma of children and young adults. *J Clin Endocrinol Metab*, **2000**, *85*, 1170-5.

[81] Horner MJ RL, Krapcho M, Neyman N, Aminou R, Howlader N, Altekruse SF, *et al.* (eds). SEER Cancer Statistics Review, 1975-2006. National Cancer Institute. Bethesda, MD. http://seer.cancer.gov/csr/1975_2006/, based on November 2008 SEER data submission, posted to the SEER web site, 2009.

[82] Enewold L, Zhu K, Ron E, Marrogi AJ, Stojadinovic A, Peoples GE, *et al.* Rising thyroid cancer incidence in the United States by demographic and tumor characteristics, 1980-2005. *Cancer Epidemiol Biomarkers Prev*, **2009**, *18*, 784-91.

[83] Yamane GK. Cancer incidence in the U.S. Air Force: 1989-2002. *Aviat Space Environ Med*, **2006**, *77*, 789-94.

[84] Zuker M. Mfold web server for nucleic acid folding and hybridization prediction. *Nucleic Acids Res*, **2003**, *31*, 3406-15.

[85] Fry M, Loeb LA. The fragile X syndrome d(CGG)n nucleotide repeats form a stable tetrahelical structure. *Proc Natl Acad Sci U S A*, **1994**, *91*, 4950-4.

[86] Gacy AM, Goellner G, Juranic N, Macura S, McMurray CT. Trinucleotide repeats that expand in human disease form hairpin structures *in vitro*. *Cell*, **1995**, *81*, 533-40.

[87] Usdin K, Woodford KJ. CGG repeats associated with DNA instability and chromosome fragility form structures that block DNA synthesis *in vitro*. *Nucleic Acids Res*, **1995**, *23*, 4202-9.

[88] Samadashwily GM, Raca G, Mirkin SM. Trinucleotide repeats affect DNA replication *in vivo*. *Nat Genet*, **1997**, *17*, 298-304.
[89] Zhang H, Freudenreich CH. An AT-rich sequence in human common fragile site FRA16D causes fork stalling and chromosome breakage in S. cervisiae. *Molecular Cell*, **2007**, *27*, 367-379.
[90] Burrow AA, Marullo A, Holder LR, Wang YH. Secondary structure formation and DNA instability at fragile site FRA16B. *Nucleic Acids Res*, **2010**, *38*, 2865-77.
[91] Cleary JD, Nichol K, Wang YH, Pearson CE. Evidence of cis-acting factors in replication-mediated trinucleotide repeat instability in primate cells. *Nat Genet*, **2002**, *31*, 37-46.
[92] Rindler PM, Clark, R.M., Pollard, L.M., De Biase, I., Bidichandani, S.I. Replication in mammalian cells recapitulates the locus-specific differences in somatic instability of genomic GAA triplet-repeats. *Nucleic Acids Res*, **2006**, *34*, 6352-6361.
[93] Edamura KN, Leonard, M.R., and Pearson, C.E. Role of replication and CpG methylation in fragile X syndrome CGG deletions in primate cells. *Am J Hum Genet*, **2005**, *76*, 302-311.
[94] Shah SN, Opresko PL, Meng X, Lee MY, Eckert KA. DNA structure and the Werner protein modulate human DNA polymerase delta-dependent replication dynamics within the common fragile site FRA16D. *Nucleic Acids Res*, **2009**.
[95] Ozeri-Galai E, Lebofsky R, Rahat A, Bester AC, Bensimon A, Kerem B. Failure of origin activation in response to fork stalling leads to chromosomal instability at fragile sites. *Mol Cell*, **2011**, *43*, 122-31.
[96] Lin Y, Dent SY, Wilson JH, Wells RD, Napierala M. R loops stimulate genetic instability of CTG.CAG repeats. *Proc Natl Acad Sci U S A*, **2010**, *107*, 692-7.
[97] Reddy K, Tam M, Bowater RP, Barber M, Tomlinson M, Nichol Edamura K, *et al.* Determinants of R-loop formation at convergent bidirectionally transcribed trinucleotide repeats. *Nucleic Acids Res*, **2011**, *39*, 1749-62.
[98] Lin Y, Wilson JH. Transcription-induced DNA toxicity at trinucleotide repeats: double bubble is trouble. *Cell Cycle*, **2011**, *10*, 611-8.
[99] Grabczyk E, Mancuso M, Sammarco MC. A persistent RNA.DNA hybrid formed by transcription of the Friedreich ataxia triplet repeat in live bacteria, and by T7 RNAP *in vitro*. *Nucleic Acids Res*, **2007**, *35*, 5351-9.
[100] Abraham RT. Cell cycle checkpoint signaling through the ATM and ATR kinases. *Genes Dev*, **2001**, *15*, 2177-96.
[101] Cimprich KA, Cortez D. ATR: an essential regulator of genome integrity. *Nat Rev Mol Cell Biol*, **2008**, *9*, 616-27.
[102] Cliby WA, Roberts CJ, Cimprich KA, Stringer CM, Lamb JR, Schreiber SL, Friend SH. Overexpression of a kinase-inactive ATR protein causes sensitivity to DNA-damaging agents and defects in cell cycle checkpoints. *EMBO J*, **1998**, *17*, 159-69.
[103] Cortez D, Guntuku S, Qin J, Elledge SJ. ATR and ATRIP: partners in checkpoint signaling. *Science*, **2001**, *294*, 1713-6.
[104] Nghiem P, Park PK, Kim YS, Desai BN, Schreiber SL. ATR is not required for p53 activation but synergizes with p53 in the replication checkpoint. *J Biol Chem*, **2002**, *277*, 4428-34.
[105] Hammond EM, Denko NC, Dorie MJ, Abraham RT, Giaccia AJ. Hypoxia links ATR and p53 through replication arrest. *Mol Cell Biol*, **2002**, *22*, 1834-43.

[106] Ragland RL, Arlt MF, Hughes ED, Saunders TL, Glover TW. Mice hypomorphic for Atr have increased DNA damage and abnormal checkpoint response. *Mamm Genome*, **2009**, *20*, 375-85.

[107] Ozeri-Galai E, Schwartz M, Rahat A, Kerem B. Interplay between ATM and ATR in the regulation of common fragile site stability. *Oncogene*, **2008**, *27*, 2109-17.

[108] Musio A, Montagna C, Mariani T, Tilenni M, Focarelli ML, Brait L, *et al.* SMC1 involvement in fragile site expression. *Hum Mol Genet*, **2005**, *14*, 525-33.

[109] Howlett NG, Taniguchi T, Durkin SG, D'Andrea AD, Glover TW. The Fanconi anemia pathway is required for the DNA replication stress response and for the regulation of common fragile site stability. *Hum Mol Genet*, **2005**, *14*, 693-701.

[110] Zhu M, Weiss RS. Increased common fragile site expression, cell proliferation defects, and apoptosis following conditional inactivation of mouse Hus1 in primary cultured cells. *Mol Biol Cell*, **2007**, *18*, 1044-55.

[111] Pirzio LM, Pichierri P, Bignami M, Franchitto A. Werner syndrome helicase activity is essential in maintaining fragile site stability. *J Cell Biol*, **2008**, *180*, 305-14.

[112] Focarelli ML, Soza S, Mannini L, Paulis M, Montecucco A, Musio A. Claspin inhibition leads to fragile site expression. *Genes Chromosomes Cancer*, **2009**, *48*, 1083-90.

[113] Moldovan GL, D'Andrea AD. How the fanconi anemia pathway guards the genome. *Annu Rev Genet*, **2009**, *43*, 223-49.

[114] Schoder C, Liehr T, Velleuer E, Wilhelm K, Blaurock N, Weise A, *et al.* New aspects on chromosomal instability: chromosomal break-points in Fanconi anemia patients co-localize on the molecular level with fragile sites. *Int J Oncol*, **2010**, *36*, 307-12.

[115] Chan KL, Palmai-Pallag T, Ying S, Hickson ID. Replication stress induces sister-chromatid bridging at fragile site loci in mitosis. *Nat Cell Biol*, **2009**, *11*, 753-60.

[116] Andreassen PR, D'Andrea AD, Taniguchi T. ATR couples FANCD2 monoubiquitination to the DNA-damage response. *Genes Dev*, **2004**, *18*, 1958-63.

[117] Taniguchi T, D'Andrea AD. The Fanconi anemia protein, FANCE, promotes the nuclear accumulation of FANCC. *Blood*, **2002**, *100*, 2457-62.

[118] Hays LE, Zodrow DM, Yates JE, Deffebach ME, Jacoby DB, Olson SB, *et al.* Cigarette smoke induces genetic instability in airway epithelial cells by suppressing FANCD2 expression. *Br J Cancer*, **2008**, *98*, 1653-61.

[119] Wan C, Kulkarni A, Wang YH. ATR preferentially interacts with common fragile site FRA3B and the binding requires its kinase activity in response to aphidicolin treatment. *Mutat Res*, **2010**, *686*, 39-46.

[120] Cimprich KA. Fragile sites: breaking up over a slowdown. *Curr Biol*, **2003**, *13*, R231-3.

[121] Sharples GJ. The X philes: structure-specific endonucleases that resolve Holliday junctions. *Mol Microbiol*, **2001**, *39*, 823-34.

Send Orders for Reprints to reprints@benthamscincе.net
Advances in Genome Science, Vol. 3, 2014, 189-212

CHAPTER 7

Unraveling the Origin of Aneuploidy: Role of Epigenetic Marks

Diddier Prada, Marco A. Andonegui, and Luis A. Herrera[*]

Unidad de Investigación Biomédica en Cáncer-Instituto Nacional de Cancerología-Instituto de Investigaciones Biomédicas, Universidad Nacional Autónoma de México (UNAM), Mexico City, Mexico

Abstract: Theodore Boveri, eminent German biologist, embryologist and pathologist, observed aneuploidy in cancer cells more than a century ago and suggested that cancer cells derived from a single progenitor cell that acquires the potential for uncontrolled continuous proliferation. Currently, it is well known that aneuploidy is observed in virtually all cancers. Gain and loss of chromosomal material in neoplastic cells is considered a process of diversification that leads to survival of the fittest clones. According to Darwin's theory of evolution, the environment determines the grounds upon which selection takes place and the genetic characteristics necessary for better adaptation. This concept can be applied to the carcinogenesis process, connecting the ability of cancer cells to adapt to different environments and to resist chemotherapy, genomic instability being the driving force of tumor development and progression. What causes this genome instability? Mutations have been recognized for a long time as the major source of genome instability in cancer cells. Nevertheless, an alternative hypothesis suggests that aneuploidy is a primary cause of genome instability rather than solely a simple consequence of the malignant transformation process. Whether genome instability results from mutations or from aneuploidy is not a matter of discussion in this review. It is most likely both phenomena are intimately related; however, we will focus on the mechanisms involved in aneuploidy formation and more specifically on the epigenetic origin of aneuploid cells. Epigenetic inheritance is defined as cellular information—other than the DNA sequence itself—that is heritable during cell division. DNA methylation and histone modifications comprise two of the main epigenetic modifications that are important for many physiological and pathological conditions, including cancer. Aberrant DNA methylation is the most common molecular cancer-cell lesion, even more frequent than gene mutations; global hypomethylation and aberrant local hypermethylation are perhaps the most frequent epigenetic modifications in cancer cells. Epigenetic characteristics of cells may be modified by several factors including environmental exposure, certain nutrient deficiencies, radiation, *etc.* Some of these alterations have been correlated with the formation of aneuploid cells *in vivo*. A growing body of evidence suggests that aneuploidy is produced and caused by chromosome instability. We propose and support in this manuscript that not only

***Address correspondence to Luis A. Herrera:** Instituto de Investigaciones Biomédicas, UNAM, P.O. Box 70-228, Ciudad Universitaria, 04510 México, D.F., México; Tel: (+52) (55) 5628-0425; E-mail: herreram@biomedicas.unam.mx

Christian Neri (Ed)

genetics but also epigenetics, and specifically alterations in DNA methylation, contribute in a major fashion to aneuploid cell formation.

Keywords: Aneuploidy, arsenic, cadmium, cancer, centromere, centrosome, chromatin, chromosome instability, chromosomes, DNA methylation, dnmt, epigenetics, heterochromatin, histone marks, hypomethylation, methyltransferases, mitosis, pericentromeric chromatin, subtelomeric regions, telomeres.

INTRODUCTION

Chromosome number is transmitted with a high rate of fidelity to daughter cells in each cell division. Deviation from normal chromosome number leading to an unbalanced chromosome complement or to any chromosome number that is not an exact multiple of the haploid number is referred to as aneuploidy [1]. In humans, aneuploidy has different clinical consequences depending on the cell type affected. If germinal cells are affected, aneuploidy is responsible for nearly 20% of spontaneous abortions, most frequently within the first trimester, affecting mainly women aged >35 years [2]. Nonetheless, aneuploidy affecting certain combinations of sex chromosomes and small chromosomes such as 13, 18, and 21 may be present in products that survive to gestation, although with birth defects, abnormal development, infertility, mental retardation, and early death in some cases [3]. If aneuploidy occurs in somatic cells, it could result in apoptosis or chromosome instability, the latter strongly associated with generation, aggressiveness, and resistance of cancer cells [4]. Recent large-scale DNA copy number analyses confirmed the common recurrent presence of aneuploidy in human cancers [5]. Janssen *et al.* reported an interesting association between chromosome missegregation and structural aberrations in cancer cells, showing that errors in mitosis can activate DNA double-strand damage responses, involving the ATM, Chk2 and p53 genes, generating unbalanced translocations in daughter cells [6]. On the other hand, aneuploidy has also been proposed as a contributor to cognitive impairment and aging neuropathies [7].

Aneuploidy is also present in normal tissues, as it has been reported both for human and for rodent livers, but the physiological or pathological significance of this finding is unknown [8]. However, Cleveland *et al.* have shown that aneuploidy may also behave as a tumor suppressor phenomena which makes controversial its association with cancer diseases [9].

The causes of aneuploidy and chromosome instability remain largely unknown. Many experiments support the hypothesis of a genetic cause of aneuploidy and chromosome instability [10]; notwithstanding this, it is important to point out that the human genome is contained in chromosomes, whose structure is more than a simple array of genes, and one that determines chromosomal behavior during mitosis and meiosis, distribution of genes to daughter cells a direct consequence of this behavior. In fact, the human genome contains a large proportion of DNA that forms highly repeated DNA sequences incapable of coding for proteins; this may play a role in the stabilization and normal functioning of chromosomes that can be implicated in the normal segregation process and in prevention of genomic instability.

It is now accepted that gene expression and chromosome biology are largely affected by epigenetic marks within chromosomes. These marks occur in the chromatin of eukaryotic genomes, are present in both DNA and the associated histones, and are relatively stable throughout rounds of cell division but do not involve changes in the organism's underlying DNA sequence. Epigenetic changes play a role in the cellular differentiation process, allowing cells to maintain differential gene expression stably despite containing the same genomic material. Epigenetic information is also essential for the formation of heterochromatin, which is highly compacted, not so accessible to transcription and recombination machinery as are other chromosome regions, and forms structured nucleosome arrays. These characteristics of heterochromatin depend on epigenetic marks that include post-translational modifications of histones and DNA methylation. Other epigenetic marks, such as 5-hidroxymethylcytosines, have been reported which can be affected in tumors compared with normal tissues; however it is not known its impact on chromosome stability [11]. Although for some time the repetitive DNA contained in heterochromatin was considered as "junk", evidences indicate that heterochromatin can also play important roles during chromosome segregation [12]. Therefore, epigenetics is not only important for genome expression, but also for its correct segregation during mitosis and meiosis.

Here, we review the role of DNA methylation in chromosome stability and the impact of modifications on this epigenetic mark in correct chromosome segregation, with special emphasis on cancer cells. This is not only because the majority of experiments have been carried out either in tumor cells or in cancer cell lines, but also due to the importance of aneuploidy and epigenetic modifications in carcinogenesis. Currently, it is well known that aneuploidy is observed in virtually all cancers. Indeed, gains and losses of chromosomal material in neoplastic cells is considered a process of diversification that leads to survival of the fittest clones [13].

From our point of view, and based on published evidences, DNA methylation plays a major role in chromosome stability, as well as in the organization of chromatin and in gene regulation. DNA methylation can be affected by genetic, endogenous and environmental factors, affecting chromatin organization and generating chromosome missegregation and aneuploidy.

THE CLASSIC CONCEPT OF THE ORIGIN OF ANEUPLOIDY

Induction of aneuploidy has been associated with defects in chromosome segregation resulting from mitotic spindle alterations such as merotelic attachments and chromosome cohesion defects, centrosome amplifications, cell-cycle checkpoint defects, and telomere stability, among others. A common of view is that these alterations are induced mainly by mutations in some genes that control chromosome segregation. More than 100 genes are expected to cause chromosome instability when mutated in eukaryotic cells, including genes involved in telomere metabolism, chromatid cohesion, spindle assembly and dynamics, cell-cycle regulation, DNA repair, and checkpoint controls [14]. A recent finding supports this hypothesis. Solomon *et al.* found, by integrative genomic analysis, that STAG2, a gene encoding a subunit of cohesin complex located on X chromosome, is absent in many tumors, and its inactivation can induce chromatin cohesion defects and aneuploidy; furthermore, its targeted correction is associated with enhanced chromosomal stability [10]. However, some important genes are not mutated in aneuploid cells, but can be expressed aberrantly [15].

Other groups have proposed that aneuploidy and chromosome instability in cancer cells could result from inappropriate union between kinetochores and microtubules, which does not activate spindle mitotic checkpoint and allows chromatids separation to daughter cells [16]; however, recent evidences have shown that abnormal unions between kinetochores and microtubules, such as merotelic attachments, are associated with lagging chromosomes but at the end of mitosis, these chromosomes are segregated accurately [17].

Centrosome proteins such as Aurora kinases may contribute to aneuploidy and usually are overexpressed in cancer cell lines. Aurora A overexpression is associated with centrosome amplification and generation of polyploid cells containing multiple centrosomes. Aurora A overexpression also overrides the spindle assembly checkpoint, resulting in arrested mitosis with incomplete cytokinesis and leading to taxol resistance-associated multinucleation [18]. Studies with pre-invasive lesions found a high frequency of centrosome defects in addition to spindle abnormalities, correlated with a high rate of progression to cancer [19]. In the same way, low prevalence of centrosome defects was associated with low-grade lesions. Also, centrosome amplifications have been detected in *in situ*-stage breast carcinoma cells, indicating that these may develop early in the neoplastic process [19]. Centrosome amplification can be associated with diverse genes. For instance, centrosome duplication is controlled by Cdk2/cyclin E complex, which is inhibited by p21. Thus, overexpression of cyclin E or p21 inhibition results in centrosome amplification; moreover, dysfunction of p53, a positive p21 regulator, has the same outcome [20]. Plk1, other protein required for centrosome maturation, is overexpressed in many tumors, and correlates with poor prognosis [21].

Failure in the DNA damage response and double-strand break repair can lead to genetic alteration or chromosome instability. Similarly, defects in cellular response to double-strand breakage result in genetic mutations, gene amplification, and chromosomal aberrations, and are associated with cancer [22]. For example, ATR activation culminates in cell-cycle arrest, apoptosis, and in DNA repair alteration. ATR has been shown to be a critical factor in the maintenance of chromosomal integrity, and its inhibition leads to chromosome instability and over expression of fragile sites. ATM and ATR function as key molecules in the DNA damage response, with a strong influence on the control of cell-cycle checkpoints, DNA repair, telomere maintenance, and apoptosis [23]. Many other specific DNA damage response genes and hereditary mutations in genes such as histone H2AX [24], mre11A, CHEK1, BRCA1, and -2 have been associated with chromosome instability, and have been shown to cause tumor predisposition by initiating chromosome instability [25].

Some experiments suggest that short telomeres could possess a key role in chromosome instability, as observed in telomerase-negative immortalized cells, which tend to develop a tetraploid cell population [26]. Polyploidization is an event occurring concurrently with gradual loss of individual chromosome copies. Some preneoplastic lesions, such as Barrett's esophagus [27] and ulcerative colitis [28], have a high frequency of tetraploid cells. In fact, it has been suggested that some aneuploid cancer cells develop through a tetraploid intermediate [29].

The function of the spindle checkpoint is to ensure that all chromosomes are correctly aligned in metaphase cells and properly attached to the mitotic spindle before chromosome separation can proceed. Chromosomally instable cancer cells in tissue culture tend to override the mitotic arrest in metaphase when incubated with microtubule-disrupting agents, most likely due to an abnormal mitotic checkpoint. Spindle checkpoint defects also have been demonstrated in head and neck cancer cell lines [30]. Spindle checkpoint gene mutations were initially reported in colon and pancreatic cancer, as well as in a single breast cancer cell line. However, further analyses revealed that mutations in these genes are rather uncommon [31]; therefore, chromosome instability in tumors is due to other factors. Subsequent investigations have demonstrated aberrant expression of spindle checkpoint genes (over- and sub-expression) rather than chromosome instability-related mutations [32]. Mutations and control of chromosome segregation genes such as hZW10, -ILCH, and hROD have also been identified as playing a role in the stabilization of this mechanism, and their dysfunction might contribute to the aneuploidy process [25b]. Other studies have suggested that lagging chromosomes could be efficiently aligned, and chromosome segregation can be achieved after misalignments are repaired, suggesting a functional spindle mitotic checkpoint in cancer cells [16].

DNA METHYLATION AND INDUCTION OF CHROMOSOME INSTABILITY

Although there are several important elements involved in the generation and maintenance of epigenetic marks, we will focus on DNA methylation and its contribution to chromosome stability. Excellent reviews on the elements that make up the DNA methylation machinery have been published recently [33]. Nevertheless, we consider it important to describe briefly some of the most relevant characteristics of this process.

DNA methylation is the covalent attachment of a methyl group from S-adenosyl-methionine (SAM) to carbon 5 of the cytosine ring [34, 35]. It is one of the best studied factors involved in cellular epigenetics and is considered an important gear in the organization of chromatin in mammalian genomes. In constitutive heterochromatin, DNA methylation is probably directed by post-translational modifications of histones, such as H3K36me3 [36] and stabilized by proteins, such as methyl-CpG-binding proteins, which are characterized by a chromodomain that binds specifically to methylated DNA and recruits enzymatic complexes, *e.g.*, histone deacetilases or histone methyltransferases to create a compacted chromatin [37]. This model has been extensively reviewed elsewhere [38] and is currently used to explain the perpetuation of epigenetic modifications that derive in constitutive heterochromatin region formation. In facultative heterochromatin and in maintenance methylation, the epigenetic signal lies in the parental strand and corresponds to a methylated cytosine [39].

Feinberg *et al.* have proposed a model of DNA methylation based on chromatin organization of large regions, denominated DMRs (differentially methylated regions), and some small regions called sDMRs, which contribute to organization of repetitive and unique sequences in human genomes. This group has found that genes located in DMRs are regulated in healthy tissues by DNA methylation and may be affected by global hypomethylation, which is usually present in various types of cancer [40].

The enzymes directly involved in DNA methylation are known as DNA-methyltransferases, or Dnmts. Mammals in general, express three families of

Dnmts: Dnmt1; -2, and -3 [41, 42]. All mammalian Dnmts known possess a common catalytic domain that is characterized by 10 conserved amino acid motifs implicated in the catalytic function. Dnmt1 and -3 enzymes contain a large N-terminal regulatory domain responsible for the differential function between them [33]. Dnmt1 is highly expressed in somatic differentiated cells and is responsible for generation of methylation patterns in many daughter cells after differentiation. In other words, this enzyme is responsible for DNA methylation perpetuation in promoter regions of tissue-specific genes, juxtacentromeric satellites, and imprinting control regions, although its work in maintenance methylation may be complemented by the remaining Dnmts [43]. This enzyme recognizes an asymmetric specific methylated sequence for transferring the methyl group to the newly incorporated cytosine. Hemimethylation is an essential process to maintain DNA methylation patterns during cell proliferation, because any mistake—including alterations in methylation machinery and/or methyl group bioavailability—could generate important scarring in the epigenome. It has been shown that, with the use of demethylating agents, hypomethylating stress may induce changes in the methylation pattern that can be conserved even in absence of the substance [44]. Dnmt3a, -b, and -L establish the methylation patterns during embryonic stages, contributes to cell differentiation and participates in methylation of repetitive and imprinted sequences [33, 45].

It is important to emphasize some genome structures that are normally DNA methylated, and participate in the process of chromosome segregation and genome stabilization in somatic cells. It is known that modifications in DNA methylation patterns on these structures are related to missegregation and aneuploidy [12].

DNA Methylation and Chromosome Instability

Cells with reduced DNA methylation levels appear to be more susceptible to undergoing chromosomal loss, gain, or rearrangement, probably because hypomethylation reduces chromosomal stability [46]. Experiments performed with embryonic stem cells show that cells suffered global DNA hypomethylation after Dnmt3b inactivation, generating chromosome instability characterized by aneuploidy, polyploidy, and some forms of chromosomal aberrations such as chromosomal breaks and fusions. In this model, DNA hypomethylation was also associated with premature senescence or spontaneous immortalization [47].

Global hypomethylation in male germ cells may also result in meiotic abnormalities [48]. DNA demethylating drugs such as 5-azacytidine, a cytidine analogue [49], and its derived 5-aza-2' deoxycytidine, induce progressive increase in micronuclei formation. These demethylating substances lead to chromosomal aberrations such as deletions, chromosome breaks, isochromosome formation, and translocations [50]. In some cases, these substances induce undercondensation and somatic pairing among the constitutive heterocromatin of chromosomes 1, 9, and 16, 90% of these involving pericentromeric regions [51], figures denominated multibranched chromosomes, and delay in the centromere separation sequence [49], generating preferential exclusion of chromosomes 1, 9, and 16 in micronuclei [52, 53] as well as uncoiling and recombination in classical satellite-containing constitutive heterochromatin, but not in alpha satellite [54].

5-azacytidine, ethionine, and 9-b-D-arabinofuranosyl-adenine-induced hypomethylation increases sister chromatid exchange (SCE) formation in mammalian cells [55]. SCEs occur only after a second cell cycle following the demethylating pulse, indicating that the key event comprises demethylation of the old parental DNA strand. SCE persistence for many cell cycles after removal of the demethylating pulse may be due to methylation maintenance mechanisms [56].

In our laboratory, we have found that in cells exposed to aneugenic compounds, the chromosomal pattern of methylation changes, with an important reduction of the pericentromeric normal methylation that has been correlated with aneuploid cell formation (unpublished data). Changes in the methylation pattern of chromosomes has been previously described in Epstein-Barr-virus-transformed monocytes [57] and in sodium arsenite-treated immortalized cells [44]; nonetheless, its relationship with aneuploidy has not been evaluated.

DNA Methylation in Subtelomeric and Telomeric Regions

Telomeres are nucleoproteic structures at the end of chromosomes that consist of tandem repeats of the TTAGGG sequence, which is bound by associated proteins, and their alterations have been associated with chromosome instability in cancer [58]. Telomeres' length is regulated by telomerase, a reverse transcriptase, by alternative telomere lengthening and by histone modifications and subtelomeric methylation [59]. Telomere function is also controlled by proteins such as TRF1, -2, POT1, TIN2, and Rap1, a mammalian telomeric core complex that forms and protects the telomere [60]. Mammalian telomeres contain histone modifications such as the H3K9 di- and trimethylation produced by histone-methyltransferases Suv39h1 and -2, [61] trimethylation of H4K20 by Suv420h1 and -2, and the retinoblastoma family of proteins, as well as the binding of CBX1, -3, and -5 [62].

Subtelomeric regions are rich in repetitive DNA, have a high density of CpG sequences, and are methylated in human somatic cells [63]. The short arms of acrocentric chromosomes 13, 14, 15, 21, and 22 correspond to high concentrations of methylated CCGG repeats at the end of the chromosomes [64]. When epigenetic changes are lost, telomere elongation takes place, suggesting that a compacted chromatin state is fundamental for controlling telomere length. Lack of Dnmts increases telomeric recombination, indicating that DNA methylation protects the genome from illegitimate recombination between repetitive sequences of telomeres. Loss of methyl groups in telomeres by knock-out of the Dnmt3b has also been correlated with the presence of SCE in stem cells [59]. The importance of DNA methylation on subtelomeric regions and its role on prevention of shortening or elongation of telomeres has been recently confirmed in various types of chromosomes [65].

Recent reviews analyze the mechanisms implicated in telomere dysfunction-associated chromosome instability formation [60] and the importance of this phenomenon in the stability of the genome; nonetheless, the significance of methylation patterns in subtelomeric regions in chromosome instability induction has not been studied yet in depth. Nevertheless, in experiments with ICF cell lines the model of functional depletion of one of the *de novo* enzymes, Dnmt3b, demonstrate telomeric associations and clonal telomeric rearrangements between chromosomes and anaphase bridges. These are increased by the presence of short dysfunctional telomeres among various chromosomes [66], suggesting that subtelomeric methylation may be mediated by Dnmt3b.

Sodium arsenite, a substance associated with DNA demethylation, generates nucleoplasmic bridges and Breakage/Fusion/Bridge (B/F/B) cycles, as well as telomeric associations and dicentric chromosomes, conferring a selective advantage on proliferation associated with an aneuploid state [44]. However, the relationship between subtelomeric demethylation and chromosome instability in sodium arsenite-exposed models has not been analyzed.

Cancer cells commonly possess telomere instability [66, 67], and tumor cell lines have lost telomeres at a high rate (10^{-6} events per cell per generation). In tumor cell lines, telomere loss often results in sister chromatid fusion followed by B/F/B cycles [68] lasting for many generations. This inability to terminate B/F/B cycles is likely to contribute to the chromosome instability resulting from telomere loss in human tumor cells. Also, telomere function loss and cell-cycle control disruption are capable of triggering extensive chromosome instability in colon cancer cells [69]. Recently, Martinez-A *et al.* have discussed this commonly referred hypothesis. They found that most of carcinomas (the most frequent type of cancer), show frequent chromosome breakage at pericentromeric regions, which could be captured by telomeres, and that only a few percentage of chromosome fusions could be attributable to telomeric erosions or to random associations between chromosomes [70].

DNA Methylation at Pericentromeric Regions

Pericentromeric regions, the areas surrounding the centromere, are characterized by highly repetitive DNA segments termed classical satellites 2 and 3, which are mainly non-transcribed and highly methylated. These regions contain large amounts of methylated constitutive heterochromatin and are located at metacentric and submetacentric chromosomes 1, 9, and 16, and in the short arms of some acrocentric chromosomes such as 13, 14, 15, 21, and 22. Based on ICF syndrome findings, Ehrlich *et al.* proposed a relationship between satellite 2 hypomethylation and induction of centromeric decondensation [71]. Pericentromeric region hypomethylation is associated with the induction of a significant reorganization of constitutive pericentromeric heterochromatin [72] and can be clearly observed in cells from patients with ICF syndrome [73]. Somatic cells from patients with ICF exhibit hypomethylated pericentromeric regions associated with chromosomal rearrangements, centromere under-condensation, and the formation of micronuclei preferentially containing chromosomes 1, 9, and 16 [46]. We have recently reported that human somatic cells exposed to demethylating agent 5-azacytidine show missegregation of chromosomes with methylated pericentromeric regions [12]. Defects in pericentromeric epigenetic heterochromatin modifications initiate a dynamic HP1-dependent response that rescues pericentromeric heterochromatin function and is essential for viable progression through mitosis [74].

Recently, it has been described that Down Syndrome, a congenital syndrome characterized by trisomy of chromosome 21, is associated with a specific recombination during meiosis II in oocytes in women of advanced age [75]. It is currently known that aging is associated with DNA demethylation [76]. However, in this work the authors associated the presence of pericentromeric regions with missegregation of chromosome 21 during development of germ cells. Nevertheless, the effect on pericentromeric methylation could not be established [75].

Efficient DNA methylation of pericentromeric sequences requires previous trimethylation H3K9 (histone 3 lysine 9) by Suv39h histone methyltransferases [77]. Experimentally induced Suv39h1 overexpression increases trimethylation at H3K9, and is associated with defects in mitotic progression and chromosome segregation [78]. However, mice lacking Suv39h do not exhibit trimethylation at H3K9 within pericentromeric heterochromatin, but rather present increased genomic instability and cancer predisposition [79]. Other epigenetic marks, such as H3K36me3 have also been described as required for the accurate recruitment of DNA methyltransferases -3a and -3b to specific regions [36].

Chromosome Instability and Aneuploidy Induced by Factors Modifying Epigenetic Marks

One characteristic of epigenetic marks is that they can be modified by genetic and environmental factors, which also induce chromosome instability and aneuploidy. For instance, the ICF syndrome (Immunodeficiency, Centromeric instability, Facial abnormalities), is a rare genetic disease in which some mutations in the

Dnmt3b gene have been detected. These mutations not only affect the activity of the enzyme, but also reduce affinity by Dnmt3L [80, 81] Cells from patients with ICF syndrome present demethylation in repetitive Alu sequences [82], satellites 2 and 3, preferentially at pericentromeric regions [83], and hypomethylation of the CpG island of genes located at the inactive X chromosome and of non-satellite DNA repeats D4z4 and NBL2 [84]. These cells also present multiple chromosomal aberrations such as isochromosomes, multibranched chromosomes, and breaks [73]. These patients die at early age principally due to infections secondary to immunosupression [85]. Only few patients have been reported affected by cancer diseases (hematological [86] and angiosarcomas [87, 88]).

On the other hand, mutations have also been reported in Dnmt3a enzymes in patients with acute myeloid leukemia [89] and myelodisplastic syndrome [90], associated with worst prognosis as an independent factor, probably affecting expression in developmental-associated genes [91]. However, the impact of these mutations on pericentromeric or subtelomeric methylated regions has not been evaluated yet. Other genetic changes, such as polymorphisms could play some role in DNA methylation, as has been reported for the methylenetetrahydrofolate reductase (MTHFR). In fact, the polymorphism MTHFR 1298A>C has been associated with numeric chromosome alterations in spontaneous aborts; however, its association with DNA demethylation was not established [92].

Chemical substances such as cadmium, nickel, and sodium arsenite have been categorized as human carcinogens [93], affecting epigenetic marks by different mechanisms. Cadmium chloride induces aneuploidy in MRC-5 human cells [94, 95] as well as in other animal models [96, 97]. This substance also inhibits both mammalian and bacterial Dnmts *in vitro* and *ex vivo* [98, 99] in a non-competitive manner, indicating an interaction with the DNA binding domain, probably at the cysteine residue in the active center of Dnmts [100]. This induces DNA hypomethylation early in the cell cycle. Prolonged exposure to cadmium induces signs of cell transformation such as altered cell morphology, increased invasiveness, and increased growth rate accompanied by DNA hypermethylation and enhanced Dnmt activity [99].

Exposure to inorganic arsenic, which interferes with normal methyl- group metabolism, represses expression of DNA methyltransferase genes Dnmt1 and -3a [101]. Sodium arsenite generates hypomethylation and changes in chromosome methylation patterns (after acute exposure), that are maintained throughout several cell cycles without the substance [44]. This damage increased the frequency of aneuploid cells, up to the 50th generation. Other chromosome rearrangements can be observed even at the 90th cell generation [102]. In fact, arsenic-induced c-myc overexpression in TRL 1215 cells is probably associated with many other changes, such as hypomethylation, which is strongly correlated with malignant capacity [103]. Studies in exposed humans found significant DNA hypermethylation of p53 and -16 promoter regions in arsenic-exposed persons [104].

Nickel compounds are potent human and rodent carcinogens [105] and produce chromosome instability and aneuploidy in mammalian cells [106]. The carcinogenic potential of nickel compounds is thought to involve oxidative stress, genomic DNA damage, and epigenetic effects, including gene silencing. Nickel is a DNA methyltransferase activity inhibitor *in vitro* and *in vivo*, and induces an initial DNA methylation decrease but with a rebound elevation of total DNA methylation [107]. Water-insoluble nickel compounds induce gene silencing by DNA methylation as a result of *de novo* methylation, being one of the first models of possible epigenetic carcinogenesis [108]. Genes such as RAR-β2, RASSF1 and CDKN2A have been also found hypermethylated in response to nickel compounds [109]. Short-term exposure of cells to crystalline nickel particles (1–3 days) silences epigenetically, target genes placed near heterochromatin. A similar effect was found in yeast cells, in which nickel was able to silence the URA-3 gene while it is placed near a telomere silencing element [110, 111]. Recently, it was discovered that this change was associated with a decrease in histone H3 and H4 acetylation, as well as an increase in histone H3K9 dimethylation, and that nickel decreased histone demethylase activity [103].

Folates are important methyl-group donors for DNA methylation. Nearly two decades ago, it was observed that cells from patients with folate and B_{12} vitamin deficiencies have chromosomal abnormalities that persisted up to 12 months after hematological remission [112]. A study with postmenopausal women in whom moderate folate depletion was induced with a folate-deficient diet demonstrated increased chromosome missegregation frequency in their peripheral blood lymphocytes associated with a decrease on DNA methylation [113]. In contrast, a 3-month 700-μg folate and 7-μg vitamin B_{12} supplementation was associated with a lower frequency of micronuclei in young persons [114]. It has been suggested that folic acid deficiency may increase not only micronuclei, but also the frequency of nuclear budding and nucleoplasmic bridges, suggesting that folate depletion plays a role in B/F/B cycle initiation [115, 116]. Indeed, some authors have pointed out that folate depletion may be a factor associated with chromosome instability induction more importantly than mutations in genes such as BRCA1 or -2 [117]. Nevertheless, other group, working with human erythrocytes, reported that folate supplementation did not reduce chromosome missegregation frequency [118].

CONCLUSIONS

The molecular basis of heritable epigenetics and its effects on gene expression have been studied in a variety of organisms and associated with several human diseases. Although it is known that epigenetic marks influence chromatin structure, the potential role of epigenetics in the control of chromosome stability and segregation is still poorly understood. DNA methylation, one of the most important epigenetic marks, influences chromatin dynamics in chromosome areas that are essential for chromosome stability and segregation, such as sub-telomeric and pericentromeric regions, either directly or indirectly through its influence on other epigenetic marks, such as histone acetylation and histone methylation. The mammalian pattern of DNA methylation acquired during embryogenesis is maintained stably but is potentially affected by the environment; these changes can affect chromosome behavior and chromatin organization, resulting in alterations in chromosome segregation. On other hand, it is important to mention that epigenetics, specifically DNA methylation, is not currently considered as a mechanism involved in aneuploidy generation [119]; however, the evidence presented, strongly suggests an important role of epigenetic marks in the control of chromosome segregation and integrity. Also, considering the importance of aneuploidy in the generation of human pathologies, it is important to perform additional investigative efforts to obtain better understanding of the genetic and epigenetic mechanisms that control chromosome integrity and segregation.

ACKNOWLEDGEMENTS

Declared none.

CONFLICT OF INTEREST

The authors confirm that this chapter contents have no conflict of interest.

DISCLOSURE

The chapter submitted for Advances in Genome Science, Vol. 3 entitled **"The Epigenetic Origin of Aneuploidy"** is an update of our article published in Current Genomics, 2008 March; 9(1): 43–50.

REFERENCES

1] Herrera, L. A.; Prada, D.; Andonegui, M. A.; Duenas-Gonzalez, A., The epigenetic origin of aneuploidy. *Curr Genomics* **2008,** *9* (1), 43-50.

[2] Nagaishi, M.; Yamamoto, T.; Iinuma, K.; Shimomura, K.; Berend, S. A.; Knops, J., Chromosome abnormalities identified in 347 spontaneous abortions collected in Japan. *J Obstet Gynaecol Res* **2004,** *30* (3), 237-41.

[3] Torres, E. M.; Williams, B. R.; Amon, A., Aneuploidy: cells losing their balance. *Genetics* **2008,** *179* (2), 737-46.

[4] Gollin, S. M., Mechanisms leading to chromosomal instability. *Semin Cancer Biol* **2005,** *15* (1), 33-42.

[5] Beroukhim, R.; Mermel, C. H.; Porter, D.; Wei, G.; Raychaudhuri, S.; Donovan, J.; Barretina, J.; Boehm, J. S.; Dobson, J.; Urashima, M.; Mc Henry, K. T.; Pinchback, R. M.; Ligon, A. H.; Cho, Y. J.; Haery, L.; Greulich, H.; Reich, M.; Winckler, W.; Lawrence, M. S.; Weir, B. A.; Tanaka, K. E.; Chiang, D. Y.; Bass, A. J.; Loo, A.; Hoffman, C.; Prensner, J.; Liefeld, T.; Gao, Q.; Yecies, D.; Signoretti, S.; Maher, E.; Kaye, F. J.; Sasaki, H.; Tepper, J. E.; Fletcher, J. A.; Tabernero, J.; Baselga, J.; Tsao, M. S.; Demichelis, F.; Rubin, M. A.; Janne, P. A.; Daly, M. J.; Nucera, C.; Levine, R. L.; Ebert, B. L.; Gabriel, S.; Rustgi, A. K.; Antonescu, C. R.; Ladanyi, M.; Letai, A.; Garraway, L. A.; Loda, M.; Beer, D. G.; True, L. D.; Okamoto, A.; Pomeroy, S. L.; Singer, S.; Golub, T. R.; Lander, E. S.; Getz, G.; Sellers, W. R.; Meyerson, M., The landscape of somatic copy-number alteration across human cancers. *Nature* **2010,** *463* (7283), 899-905.

[6] Janssen, A.; van der Burg, M.; Szuhai, K.; Kops, G. J.; Medema, R. H., Chromosome segregation errors as a cause of DNA damage and structural chromosome aberrations. *Science* **2011,** *333* (6051), 1895-8.

[7] Faggioli, F.; Vijg, J.; Montagna, C., Chromosomal aneuploidy in the aging brain. *Mech Ageing Dev* **2011,** *132* (8-9), 429-36.

[8] Duncan, A. W.; Hanlon Newell, A. E.; Smith, L.; Wilson, E. M.; Olson, S. B.; Thayer, M. J.; Strom, S. C.; Grompe, M., Frequent aneuploidy among normal human hepatocytes. *Gastroenterology* **2012,** *142* (1), 25-8.

[9] Weaver, B. A.; Silk, A. D.; Montagna, C.; Verdier-Pinard, P.; Cleveland, D. W., Aneuploidy acts both oncogenically and as a tumor suppressor. *Cancer Cell* **2007,** *11* (1), 25-36.

[10] Solomon, D. A.; Kim, T.; Diaz-Martinez, L. A.; Fair, J.; Elkahloun, A. G.; Harris, B. T.; Toretsky, J. A.; Rosenberg, S. A.; Shukla, N.; Ladanyi, M.; Samuels, Y.; James, C. D.; Yu, H.; Kim, J. S.; Waldman, T., Mutational inactivation of STAG2 causes aneuploidy in human cancer. *Science* **2011,** *333* (6045), 1039-43.

[11] Jin, S. G.; Jiang, Y.; Qiu, R.; Rauch, T. A.; Wang, Y.; Schackert, G.; Krex, D.; Lu, Q.; Pfeifer, G. P., 5-Hydroxymethylcytosine is strongly depleted in human cancers but its levels do not correlate with IDH1 mutations. *Cancer Res* **2011,** *71* (24), 7360-5.

[12] Prada, D.; Gonzalez, R.; Sanchez, L.; Castro, C.; Fabian, E.; Herrera, L. A., Satellite 2 demethylation induced by 5-azacytidine is associated with missegregation of chromosomes 1 and 16 in human somatic cells. *Mutat Res* **2012,** *729* (1-2), 100-5.

[13] Feinberg, A. P., Phenotypic plasticity and the epigenetics of human disease. *Nature* **2007,** *447* (7143), 433-40.

[14] Lengauer, C.; Kinzler, K. W.; Vogelstein, B., Genetic instabilities in human cancers. *Nature* **1998,** *396* (6712), 643-9.

[15] Yuan, B.; Xu, Y.; Woo, J. H.; Wang, Y.; Bae, Y. K.; Yoon, D. S.; Wersto, R. P.; Tully, E.; Wilsbach, K.; Gabrielson, E., Increased expression of mitotic checkpoint genes in breast cancer cells with chromosomal instability. *Clin Cancer Res* **2006,** *12* (2), 405-10.

[16] Thompson, S. L.; Compton, D. A., Examining the link between chromosomal instability and aneuploidy in human cells. *J Cell Biol* **2008,** *180* (4), 665-72.

[17] Thompson, S. L.; Compton, D. A., Chromosome missegregation in human cells arises through specific types of kinetochore-microtubule attachment errors. *Proc Natl Acad Sci U S A* **2011,** *108* (44), 17974-8.

[18] Anand, S.; Penrhyn-Lowe, S.; Venkitaraman, A. R., AURORA-A amplification overrides the mitotic spindle assembly checkpoint, inducing resistance to Taxol. *Cancer Cell* **2003,** *3* (1), 51-62.

[19] Pihan, G. A.; Wallace, J.; Zhou, Y.; Doxsey, S. J., Centrosome abnormalities and chromosome instability occur together in pre-invasive carcinomas. *Cancer Res* **2003,** *63* (6), 1398-404.

[20] D'Assoro, A. B.; Lingle, W. L.; Salisbury, J. L., Centrosome amplification and the development of cancer. *Oncogene* **2002,** *21* (40), 6146-53.

[21] Takai, N.; Hamanaka, R.; Yoshimatsu, J.; Miyakawa, I., Polo-like kinases (Plks) and cancer. *Oncogene* **2005,** *24* (2), 287-91.

[22] Zhou, B. B.; Elledge, S. J., The DNA damage response: putting checkpoints in perspective. *Nature* **2000,** *408* (6811), 433-9.

[23] Smith, L.; Liu, S. J.; Goodrich, L.; Jacobson, D.; Degnin, C.; Bentley, N.; Carr, A.; Flaggs, G.; Keegan, K.; Hoekstra, M.; Thayer, M. J., Duplication of ATR inhibits MyoD, induces aneuploidy and eliminates radiation-induced G1 arrest. *Nat Genet* **1998,** *19* (1), 39-46.

[24] Bassing, C. H.; Suh, H.; Ferguson, D. O.; Chua, K. F.; Manis, J.; Eckersdorff, M.; Gleason, M.; Bronson, R.; Lee, C.; Alt, F. W., Histone H2AX: a dosage-dependent suppressor of oncogenic translocations and tumors. *Cell* **2003,** *114* (3), 359-70.

[25] (a) Weaver, Z.; Montagna, C.; Xu, X.; Howard, T.; Gadina, M.; Brodie, S. G.; Deng, C. X.; Ried, T., Mammary tumors in mice conditionally mutant for Brca1 exhibit gross genomic instability and centrosome amplification yet display a recurring distribution of genomic imbalances that is similar to human breast cancer. *Oncogene* **2002,** *21* (33), 5097-107; (b) Wang, Z.; Cummins, J. M.; Shen, D.; Cahill, D. P.; Jallepalli, P. V.; Wang, T. L.; Parsons, D. W.; Traverso, G.; Awad, M.; Silliman, N.; Ptak, J.; Szabo, S.; Willson, J. K.; Markowitz, S. D.; Goldberg, M. L.; Karess, R.; Kinzler, K. W.; Vogelstein, B.; Velculescu, V. E.; Lengauer, C., Three classes of genes mutated in colorectal cancers with chromosomal instability. *Cancer Res* **2004,** *64* (9), 2998-3001.

[26] der-Sarkissian, H.; Bacchetti, S.; Cazes, L.; Londono-Vallejo, J. A., The shortest telomeres drive karyotype evolution in transformed cells. *Oncogene* **2004,** *23* (6), 1221-8.

[27] Reid, B. J.; Haggitt, R. C.; Rubin, C. E.; Rabinovitch, P. S., Barrett's esophagus. Correlation between flow cytometry and histology in detection of patients at risk for adenocarcinoma. *Gastroenterology* **1987,** *93* (1), 1-11.

[28] Levine, D. S.; Rabinovitch, P. S.; Haggitt, R. C.; Blount, P. L.; Dean, P. J.; Rubin, C. E.; Reid, B. J., Distribution of aneuploid cell populations in ulcerative colitis with dysplasia or cancer. *Gastroenterology* **1991,** *101* (5), 1198-210.

[29] Storchova, Z.; Pellman, D., From polyploidy to aneuploidy, genome instability and cancer. *Nat Rev Mol Cell Biol* **2004,** *5* (1), 45-54.

[30] Minhas, K. M.; Singh, B.; Jiang, W. W.; Sidransky, D.; Califano, J. A., Spindle assembly checkpoint defects and chromosomal instability in head and neck squamous cell carcinoma. *Int J Cancer* **2003,** *107* (1), 46-52.

[31] Cahill, D. P.; Lengauer, C.; Yu, J.; Riggins, G. J.; Willson, J. K.; Markowitz, S. D.; Kinzler, K. W.; Vogelstein, B., Mutations of mitotic checkpoint genes in human cancers. *Nature* **1998,** *392* (6673), 300-3.

[32] (a) Doak, S. H.; Jenkins, G. J.; Parry, E. M.; Griffiths, A. P.; Baxter, J. N.; Parry, J. M., Differential expression of the MAD2, BUB1 and HSP27 genes in Barrett's oesophagus-their association with aneuploidy and neoplastic progression. *Mutat Res* **2004,** *547* (1-2), 133-44; (b) Wu, C. W.; Chi, C. W.; Huang, T. S., Elevated level of spindle checkprotein MAD2 correlates with cellular mitotic arrest, but not with aneuploidy and clinicopathological characteristics in gastric cancer. *World J Gastroenterol* **2004,** *10* (22), 3240-4; (c) Wassmann, K.; Benezra, R., Mitotic checkpoints: from yeast to cancer. *Curr Opin Genet Dev* **2001,** *11* (1), 83-90.

[33] Jurkowska, R. Z.; Jurkowski, T. P.; Jeltsch, A., Structure and function of mammalian DNA methyltransferases. *Chembiochem* **2011,** *12* (2), 206-22.

[34] Schubeler, D.; Lorincz, M. C.; Cimbora, D. M.; Telling, A.; Feng, Y. Q.; Bouhassira, E. E.; Groudine, M., Genomic targeting of methylated DNA: influence of methylation on transcription, replication, chromatin structure, and histone acetylation. *Mol Cell Biol* **2000,** *20* (24), 9103-12.

[35] Costello, J. F.; Plass, C., Methylation matters. *J Med Genet* **2001,** *38* (5), 285-303.

[36] Dhayalan, A.; Rajavelu, A.; Rathert, P.; Tamas, R.; Jurkowska, R. Z.; Ragozin, S.; Jeltsch, A., The Dnmt3a PWWP domain reads histone 3 lysine 36 trimethylation and guides DNA methylation. *J Biol Chem* **2010,** *285* (34), 26114-20.

[37] Jones, P. L.; Veenstra, G. J.; Wade, P. A.; Vermaak, D.; Kass, S. U.; Landsberger, N.; Strouboulis, J.; Wolffe, A. P., Methylated DNA and MeCP2 recruit histone deacetylase to repress transcription. *Nat Genet* **1998,** *19* (2), 187-91.

[38] Fuks, F., DNA methylation and histone modifications: teaming up to silence genes. *Curr Opin Genet Dev* **2005,** *15* (5), 490-5.

[39] Vertino, P. M.; Issa, J. P.; Pereira-Smith, O. M.; Baylin, S. B., Stabilization of DNA methyltransferase levels and CpG island hypermethylation precede SV40-induced immortalization of human fibroblasts. *Cell Growth Differ* **1994,** *5* (12), 1395-402.

[40] Hansen, K. D.; Timp, W.; Bravo, H. C.; Sabunciyan, S.; Langmead, B.; McDonald, O. G.; Wen, B.; Wu, H.; Liu, Y.; Diep, D.; Briem, E.; Zhang, K.; Irizarry, R. A.; Feinberg, A. P., Increased methylation variation in epigenetic domains across cancer types. *Nat Genet* **2011,** *43* (8), 768-75.

[41] Brenner, C.; Fuks, F., DNA methyltransferases: facts, clues, mysteries. *Curr Top Microbiol Immunol* **2006,** *301*, 45-66.

[42] Hermann, A.; Gowher, H.; Jeltsch, A., Biochemistry and biology of mammalian DNA methyltransferases. *Cell Mol Life Sci* **2004,** *61* (19-20), 2571-87.

[43] Rhee, I.; Jair, K. W.; Yen, R. W.; Lengauer, C.; Herman, J. G.; Kinzler, K. W.; Vogelstein, B.; Baylin, S. B.; Schuebel, K. E., CpG methylation is maintained in human cancer cells lacking DNMT1. *Nature* **2000,** *404* (6781), 1003-7.

[44] Sciandrello, G.; Caradonna, F.; Mauro, M.; Barbata, G., Arsenic-induced DNA hypomethylation affects chromosomal instability in mammalian cells. *Carcinogenesis* **2004,** *25* (3), 413-7.

[45] Bird, A. P.; Wolffe, A. P., Methylation-induced repression--belts, braces, and chromatin. *Cell* **1999,** *99* (5), 451-4.

[46] Tuck-Muller, C. M.; Narayan, A.; Tsien, F.; Smeets, D. F.; Sawyer, J.; Fiala, E. S.; Sohn, O. S.; Ehrlich, M., DNA hypomethylation and unusual chromosome instability in cell lines from ICF syndrome patients. *Cytogenet Cell Genet* **2000,** *89* (1-2), 121-8.

[47] Dodge, J. E.; Okano, M.; Dick, F.; Tsujimoto, N.; Chen, T.; Wang, S.; Ueda, Y.; Dyson, N.; Li, E., Inactivation of Dnmt3b in mouse embryonic fibroblasts results in DNA hypomethylation, chromosomal instability, and spontaneous immortalization. *J Biol Chem* **2005,** *280* (18), 17986-91.

[48] (a) Bourc'his, D.; Bestor, T. H., Meiotic catastrophe and retrotransposon reactivation in male germ cells lacking Dnmt3L. *Nature* **2004,** *431* (7004), 96-9; (b) Hata, K.; Kusumi, M.; Yokomine, T.; Li, E.; Sasaki, H., Meiotic and epigenetic aberrations in Dnmt3L-deficient male germ cells. *Mol Reprod Dev* **2006,** *73* (1), 116-22.

[49] Rodriguez, M. J.; Lopez, M. A.; Garcia-Orad, A.; Vig, B. K., Sequence of centromere separation: effect of 5-azacytidine-induced epigenetic alteration. *Mutagenesis* **2001,** *16* (2), 109-14.

[50] Lavia, P.; Ferraro, M.; Micheli, A.; Olivieri, G., Effect of 5-azacytidine (5-azaC) on the induction of chromatid aberrations (CA) and sister-chromatid exchanges (SCE). *Mutat Res* **1985,** *149* (3), 463-7.

[51] Hernandez, R.; Frady, A.; Zhang, X. Y.; Varela, M.; Ehrlich, M., Preferential induction of chromosome 1 multibranched figures and whole-arm deletions in a human pro-B cell line treated with 5-azacytidine or 5-azadeoxycytidine. *Cytogenet Cell Genet* **1997,** *76* (3-4), 196-201.

[52] Fauth, E.; Scherthan, H.; Zankl, H., Frequencies of occurrence of all human chromosomes in micronuclei from normal and 5-azacytidine-treated lymphocytes as revealed by chromosome painting. *Mutagenesis* **1998,** *13* (3), 235-41.

[53] Guttenbach, M.; Schmid, M., Exclusion of specific human chromosomes into micronuclei by 5-azacytidine treatment of lymphocyte cultures. *Exp Cell Res* **1994,** *211* (1), 127-32.

[54] Kokalj-Vokac, N.; Almeida, A.; Viegas-Pequignot, E.; Jeanpierre, M.; Malfoy, B.; Dutrillaux, B., Specific induction of uncoiling and recombination by azacytidine in classical satellite-containing constitutive heterochromatin. *Cytogenet Cell Genet* **1993,** *63* (1), 11-5.

[55] Albanesi, T.; Polani, S.; Cozzi, R.; Perticone, P., DNA strand methylation and sister chromatid exchanges in mammalian cells *in vitro*. *Mutat Res* **1999,** *429* (2), 239-48.

[56] Perticone, P.; Palitti, F.; Cozzi, R.; D'Erme, M.; Bona, R., Persistence of azacytidine-induced SCEs and genomic methylation in CHO cells *in vitro*. *Mutat Res* **1990,** *245* (3), 211-5.

[57] Habib, M.; Fares, F.; Bourgeois, C. A.; Bella, C.; Bernardino, J.; Hernandez-Blazquez, F.; de Capoa, A.; Niveleau, A., DNA global hypomethylation in EBV-transformed interphase nuclei. *Exp Cell Res* **1999,** *249* (1), 46-53.

[58] (a) Blackburn, E. H., Switching and signaling at the telomere. *Cell* **2001,** *106* (6), 661-73; (b) Murnane, J. P., Telomere loss as a mechanism for chromosome instability in human cancer. *Cancer Res* **2010,** *70* (11), 4255-9.

[59] Gonzalo, S.; Jaco, I.; Fraga, M. F.; Chen, T.; Li, E.; Esteller, M.; Blasco, M. A., DNA methyltransferases control telomere length and telomere recombination in mammalian cells. *Nat Cell Biol* **2006,** *8* (4), 416-24.

[60] Bailey, S. M.; Murnane, J. P., Telomeres, chromosome instability and cancer. *Nucleic Acids Res* **2006,** *34* (8), 2408-17.

[61] Garcia-Cao, M.; O'Sullivan, R.; Peters, A. H.; Jenuwein, T.; Blasco, M. A., Epigenetic regulation of telomere length in mammalian cells by the Suv39h1 and Suv39h2 histone methyltransferases. *Nat Genet* **2004,** *36* (1), 94-9.

[62] Garcia-Cao, M.; Gonzalo, S.; Dean, D.; Blasco, M. A., A role for the Rb family of proteins in controlling telomere length. *Nat Genet* **2002,** *32* (3), 415-9.

[63] (a) Brock, G. J.; Charlton, J.; Bird, A., Densely methylated sequences that are preferentially localized at telomere-proximal regions of human chromosomes. *Gene* **1999,** *240* (2), 269-77; (b) Bensaada, M.; Kiefer, H.; Tachdjian, G.; Lapierre, J. M.; Cacheux, V.; Niveleau, A.; Metezeau, P., Altered patterns of DNA methylation on chromosomes from leukemia cell lines: identification of 5-methylcytosines by indirect immunodetection. *Cancer Genet Cytogenet* **1998,** *103* (2), 101-9.

[64] de Lange, T.; Shiue, L.; Myers, R. M.; Cox, D. R.; Naylor, S. L.; Killery, A. M.; Varmus, H. E., Structure and variability of human chromosome ends. *Mol Cell Biol* **1990,** *10* (2), 518-27.

[65] Oh, B. K.; Um, T. H.; Choi, G. H.; Park, Y. N., Frequent changes in subtelomeric DNA methylation patterns and its relevance to telomere regulation during human hepatocarcinogenesis. *Int J Cancer* **2011,** *128* (4), 857-68.

[66] Gisselsson, D.; Shao, C.; Tuck-Muller, C. M.; Sogorovic, S.; Palsson, E.; Smeets, D.; Ehrlich, M., Interphase chromosomal abnormalities and mitotic missegregation of hypomethylated sequences in ICF syndrome cells. *Chromosoma* **2005,** *114* (2), 118-26.

[67] Meeker, A. K.; Hicks, J. L.; Iacobuzio-Donahue, C. A.; Montgomery, E. A.; Westra, W. H.; Chan, T. Y.; Ronnett, B. M.; De Marzo, A. M., Telomere length abnormalities occur early in the initiation of epithelial carcinogenesis. *Clin Cancer Res* **2004,** *10* (10), 3317-26.

[68] Fouladi, B.; Sabatier, L.; Miller, D.; Pottier, G.; Murnane, J. P., The relationship between spontaneous telomere loss and chromosome instability in a human tumor cell line. *Neoplasia* **2000,** *2* (6), 540-54.

[69] Pantic, M.; Zimmermann, S.; El Daly, H.; Opitz, O. G.; Popp, S.; Boukamp, P.; Martens, U. M., Telomere dysfunction and loss of p53 cooperate in defective mitotic segregation of chromosomes in cancer cells. *Oncogene* **2006,** *25* (32), 4413-20.

[70] Martinez, A. C.; van Wely, K. H., Centromere fission, not telomere erosion, triggers chromosomal instability in human carcinomas. *Carcinogenesis* **2011,** *32* (6), 796-803.

[71] Ehrlich, M., DNA hypomethylation, cancer, the immunodeficiency, centromeric region instability, facial anomalies syndrome and chromosomal rearrangements. *J Nutr* **2002,** *132* (8 Suppl), 2424S-2429S.

[72] Ma, Y.; Jacobs, S. B.; Jackson-Grusby, L.; Mastrangelo, M. A.; Torres-Betancourt, J. A.; Jaenisch, R.; Rasmussen, T. P., DNA CpG hypomethylation induces heterochromatin reorganization involving the histone variant macroH2A. *J Cell Sci* **2005,** *118* (Pt 8), 1607-16.

[73] Ehrlich, M., The ICF syndrome, a DNA methyltransferase 3B deficiency and immunodeficiency disease. *Clin Immunol* **2003,** *109* (1), 17-28.

[74] Zhang, R.; Liu, S. T.; Chen, W.; Bonner, M.; Pehrson, J.; Yen, T. J.; Adams, P. D., HP1 proteins are essential for a dynamic nuclear response that rescues the function of perturbed heterochromatin in primary human cells. *Mol Cell Biol* **2007,** *27* (3), 949-62.

[75] Ghosh, S.; Feingold, E.; Dey, S. K., Etiology of Down syndrome: Evidence for consistent association among altered meiotic recombination, nondisjunction, and maternal age across populations. *Am J Med Genet A* **2009,** *149A* (7), 1415-20.

[76] Kamiyama, H.; Suzuki, K.; Maeda, T.; Koizumi, K.; Miyaki, Y.; Okada, S.; Kawamura, Y. J.; Samuelsson, J. K.; Alonso, S.; Konishi, F.; Perucho, M., DNA demethylation in normal colon tissue predicts predisposition to multiple cancers. *Oncogene* **2012**.

[77] Lehnertz, B.; Ueda, Y.; Derijck, A. A.; Braunschweig, U.; Perez-Burgos, L.; Kubicek, S.; Chen, T.; Li, E.; Jenuwein, T.; Peters, A. H., Suv39h-mediated histone H3 lysine 9 methylation directs DNA methylation to major satellite repeats at pericentric heterochromatin. *Curr Biol* **2003,** *13* (14), 1192-200.

[78] Melcher, M.; Schmid, M.; Aagaard, L.; Selenko, P.; Laible, G.; Jenuwein, T., Structure-function analysis of SUV39H1 reveals a dominant role in heterochromatin organization, chromosome segregation, and mitotic progression. *Mol Cell Biol* **2000,** *20* (10), 3728-41.

[79] Peters, A. H.; O'Carroll, D.; Scherthan, H.; Mechtler, K.; Sauer, S.; Schofer, C.; Weipoltshammer, K.; Pagani, M.; Lachner, M.; Kohlmaier, A.; Opravil, S.; Doyle, M.; Sibilia, M.; Jenuwein, T., Loss of the Suv39h histone methyltransferases impairs mammalian heterochromatin and genome stability. *Cell* **2001,** *107* (3), 323-37.

[80] Van Emburgh, B. O.; Robertson, K. D., Modulation of Dnmt3b function *in vitro* by interactions with Dnmt3L, Dnmt3a and Dnmt3b splice variants. *Nucleic Acids Res* **2011,** *39* (12), 4984-5002.

[81] Xie, Z. H.; Huang, Y. N.; Chen, Z. X.; Riggs, A. D.; Ding, J. P.; Gowher, H.; Jeltsch, A.; Sasaki, H.; Hata, K.; Xu, G. L., Mutations in DNA methyltransferase DNMT3B in ICF syndrome affect its regulation by DNMT3L. *Hum Mol Genet* **2006,** *15* (9), 1375-85.

[82] Miniou, P.; Bourc'his, D.; Molina Gomes, D.; Jeanpierre, M.; Viegas-Pequignot, E., Undermethylation of Alu sequences in ICF syndrome: molecular and *in situ* analysis. *Cytogenet Cell Genet* **1997,** *77* (3-4), 308-13.

[83] Hassan, K. M.; Norwood, T.; Gimelli, G.; Gartler, S. M.; Hansen, R. S., Satellite 2 methylation patterns in normal and ICF syndrome cells and association of hypomethylation with advanced replication. *Hum Genet* **2001,** *109* (4), 452-62.

[84] Kondo, T.; Bobek, M. P.; Kuick, R.; Lamb, B.; Zhu, X.; Narayan, A.; Bourc'his, D.; Viegas-Pequignot, E.; Ehrlich, M.; Hanash, S. M., Whole-genome methylation scan in ICF syndrome: hypomethylation of non-satellite DNA repeats D4Z4 and NBL2. *Hum Mol Genet* **2000,** *9* (4), 597-604.

[85] Hagleitner, M. M.; Lankester, A.; Maraschio, P.; Hulten, M.; Fryns, J. P.; Schuetz, C.; Gimelli, G.; Davies, E. G.; Gennery, A.; Belohradsky, B. H.; de Groot, R.; Gerritsen, E. J.; Mattina, T.; Howard, P. J.; Fasth, A.; Reisli, I.; Furthner, D.; Slatter, M. A.; Cant, A. J.; Cazzola, G.; van Dijken, P. J.; van Deuren, M.; de Greef, J. C.; van der Maarel, S. M.; Weemaes, C. M., Clinical spectrum of immunodeficiency, centromeric instability and facial dysmorphism (ICF syndrome). *J Med Genet* **2008,** *45* (2), 93-9.

[86] Andre, N.; Roquelaure, B.; Caillez, M.; Chrestian, M.; Moncla, A.; Blanco-Betancourt, C.; Schiff, C., Macrophage activation syndrome mimicking life-threatening infection in a patient with variable immunodeficiency, centromeric instability, and facial anomalies. *Pediatrics* **2004,** *114* (4), 1127.

[87] van den Brand, M.; Flucke, U. E.; Bult, P.; Weemaes, C. M.; van Deuren, M., Angiosarcoma in a patient with immunodeficiency, centromeric region instability, facial anomalies (ICF) syndrome. *Am J Med Genet A* **2011,** *155A* (3), 622-5.

[88] Schuetz, C.; Barbi, G.; Barth, T. F.; Hoenig, M.; Schulz, A.; Moeller, P.; Smeets, D.; de Greef, J. C.; van der Maarel, S. M.; Vogel, W.; Debatin, K. M.; Friedrich, W., ICF syndrome: high variability of the chromosomal phenotype and association with classical Hodgkin lymphoma. *Am J Med Genet A* **2007,** *143A* (17), 2052-7.

[89] Ley, T. J.; Ding, L.; Walter, M. J.; McLellan, M. D.; Lamprecht, T.; Larson, D. E.; Kandoth, C.; Payton, J. E.; Baty, J.; Welch, J.; Harris, C. C.; Lichti, C. F.; Townsend, R. R.; Fulton, R. S.; Dooling, D. J.; Koboldt, D. C.; Schmidt, H.; Zhang, Q.; Osborne, J. R.; Lin, L.; O'Laughlin, M.; McMichael, J. F.; Delehaunty, K. D.; McGrath, S. D.; Fulton, L. A.; Magrini, V. J.; Vickery, T. L.; Hundal, J.; Cook, L. L.; Conyers, J. J.; Swift, G. W.; Reed, J. P.; Alldredge, P. A.; Wylie, T.; Walker, J.; Kalicki, J.; Watson, M. A.; Heath, S.; Shannon, W. D.; Varghese, N.; Nagarajan, R.; Westervelt, P.; Tomasson, M. H.; Link, D. C.; Graubert, T. A.; DiPersio, J. F.; Mardis, E. R.; Wilson, R. K., DNMT3A mutations in acute myeloid leukemia. *N Engl J Med* **2010,** *363* (25), 2424-33.

[90] Walter, M. J.; Ding, L.; Shen, D.; Shao, J.; Grillot, M.; McLellan, M.; Fulton, R.; Schmidt, H.; Kalicki-Veizer, J.; O'Laughlin, M.; Kandoth, C.; Baty, J.; Westervelt, P.; DiPersio, J. F.; Mardis, E. R.; Wilson, R. K.; Ley, T. J.; Graubert, T. A., Recurrent DNMT3A mutations in patients with myelodysplastic syndromes. *Leukemia* **2011,** *25* (7), 1153-8.

[91] Yan, X. J.; Xu, J.; Gu, Z. H.; Pan, C. M.; Lu, G.; Shen, Y.; Shi, J. Y.; Zhu, Y. M.; Tang, L.; Zhang, X. W.; Liang, W. X.; Mi, J. Q.; Song, H. D.; Li, K. Q.; Chen, Z.; Chen, S. J., Exome sequencing identifies somatic mutations of DNA methyltransferase gene DNMT3A in acute monocytic leukemia. *Nat Genet* **2011,** *43* (4), 309-15.

[92] Kim, S. Y.; Park, S. Y.; Choi, J. W.; Kim do, J.; Lee, S. Y.; Lim, J. H.; Han, J. Y.; Ryu, H. M.; Kim, M. H., Association between MTHFR 1298A>C polymorphism and spontaneous abortion with fetal chromosomal aneuploidy. *Am J Reprod Immunol* **2011,** *66* (4), 252-8.

[93] Valko, M.; Morris, H.; Cronin, M. T., Metals, toxicity and oxidative stress. *Curr Med Chem* **2005,** *12* (10), 1161-208.

[94] Seoane, A. I.; Guerci, A. M.; Dulout, F. N., Malsegregation as a possible mechanism of aneuploidy induction by metal salts in MRC-5 human cells. *Environ Mol Mutagen* **2002,** *40* (3), 200-6.

[95] Guerci, A.; Seoane, A.; Dulout, F. N., Aneugenic effects of some metal compounds assessed by chromosome counting in MRC-5 human cells. *Mutat Res* **2000,** *469* (1), 35-40.

[96] Bouilly, K.; Gagnaire, B.; Bonnard, M.; Thomas-Guyon, H.; Renault, T.; Miramand, P.; Lapegue, S., Effects of cadmium on aneuploidy and hemocyte parameters in the Pacific oyster, Crassostrea gigas. *Aquat Toxicol* **2006,** *78* (2), 149-56.

[97] Loumbourdis, N. S., Hepatotoxic and nephrotoxic effects of cadmium in the frog Rana ridibunda. *Arch Toxicol* **2005,** *79* (8), 434-40.

[98] Poirier, L. A.; Vlasova, T. I., The prospective role of abnormal methyl metabolism in cadmium toxicity. *Environ Health Perspect* **2002,** *110 Suppl 5*, 793-5.

[99] Takiguchi, M.; Achanzar, W. E.; Qu, W.; Li, G.; Waalkes, M. P., Effects of cadmium on DNA-(Cytosine-5) methyltransferase activity and DNA methylation status during cadmium-induced cellular transformation. *Exp Cell Res* **2003,** *286* (2), 355-65.

[100] Adams, R. L., Eukaryotic DNA methyltransferases--structure and function. *Bioessays* **1995,** *17* (2), 139-45.

[101] Reichard, J. F.; Schnekenburger, M.; Puga, A., Long term low-dose arsenic exposure induces loss of DNA methylation. *Biochem Biophys Res Commun* **2007,** *352* (1), 188-92.

[102] Sciandrello, G.; Mauro, M.; Catanzaro, I.; Saverini, M.; Caradonna, F.; Barbata, G., Long-lasting genomic instability following arsenite exposure in mammalian cells: the role of reactive oxygen species. *Environ Mol Mutagen* **2011,** *52* (7), 562-8.

[103] Chen, H.; Ke, Q.; Kluz, T.; Yan, Y.; Costa, M., Nickel ions increase histone H3 lysine 9 dimethylation and induce transgene silencing. *Mol Cell Biol* **2006,** *26* (10), 3728-37.

[104] Chanda, S.; Dasgupta, U. B.; Guhamazumder, D.; Gupta, M.; Chaudhuri, U.; Lahiri, S.; Das, S.; Ghosh, N.; Chatterjee, D., DNA hypermethylation of promoter of gene p53 and p16 in arsenic-exposed people with and without malignancy. *Toxicol Sci* **2006,** *89* (2), 431-7.

[105] Biggart, N. W.; Costa, M., Assessment of the uptake and mutagenicity of nickel chloride in salmonella tester strains. *Mutat Res* **1986,** *175* (4), 209-15.

[106] Ohshima, S., Induction of genetic instability and chromosomal instability by nickel sulfate in V79 Chinese hamster cells. *Mutagenesis* **2003,** *18* (2), 133-7.

[107] Lee, Y. W.; Klein, C. B.; Kargacin, B.; Salnikow, K.; Kitahara, J.; Dowjat, K.; Zhitkovich, A.; Christie, N. T.; Costa, M., Carcinogenic nickel silences gene expression by chromatin condensation and DNA methylation: a new model for epigenetic carcinogens. *Mol Cell Biol* **1995,** *15* (5), 2547-57.

[108] Lee, Y. W.; Broday, L.; Costa, M., Effects of nickel on DNA methyltransferase activity and genomic DNA methylation levels. *Mutat Res* **1998,** *415* (3), 213-8.

[109] Zhang, J.; Li, M.; Wu, Y.; Fan, Y.; Zhou, Y.; Tan, L.; Shao, Z.; Shi, H., Methylation of RAR-beta2, RASSF1A, and CDKN2A genes induced by nickel subsulfide and nickel-carcinogenesis in rats. *Biomed Environ Sci* **2011,** *24* (2), 163-71.

[110] Yan, Y.; Kluz, T.; Zhang, P.; Chen, H. B.; Costa, M., Analysis of specific lysine histone H3 and H4 acetylation and methylation status in clones of cells with a gene silenced by nickel exposure. *Toxicol Appl Pharmacol* **2003,** *190* (3), 272-7.

[111] Costa, M.; Davidson, T. L.; Chen, H.; Ke, Q.; Zhang, P.; Yan, Y.; Huang, C.; Kluz, T., Nickel carcinogenesis: epigenetics and hypoxia signaling. *Mutat Res* **2005,** *592* (1-2), 79-88.

[112] Das, K. C.; Mohanty, D.; Garewal, G., Cytogenetics in nutritional megaloblastic anaemia: prolonged persistence of chromosomal abnormalities in lymphocytes after remission. *Acta Haematol* **1986,** *76* (2-3), 146-54.

[113] Titenko-Holland, N.; Jacob, R. A.; Shang, N.; Balaraman, A.; Smith, M. T., Micronuclei in lymphocytes and exfoliated buccal cells of postmenopausal women with dietary changes in folate. *Mutat Res* **1998,** *417* (2-3), 101-14.

[114] Fenech, M.; Aitken, C.; Rinaldi, J., Folate, vitamin B12, homocysteine status and DNA damage in young Australian adults. *Carcinogenesis* **1998,** *19* (7), 1163-71.

[115] Fenech, M., Cytokinesis-block micronucleus assay evolves into a "cytome" assay of chromosomal instability, mitotic dysfunction and cell death. *Mutat Res* **2006,** *600* (1-2), 58-66.

[116] Fenech, M., Folate (vitamin B9) and vitamin B12 and their function in the maintenance of nuclear and mitochondrial genome integrity. *Mutat Res* **2011**.

[117] Beetstra, S.; Salisbury, C.; Turner, J.; Altree, M.; McKinnon, R.; Suthers, G.; Fenech, M., Lymphocytes of BRCA1 and BRCA2 germ-line mutation carriers, with or without breast cancer, are not abnormally sensitive to the chromosome damaging effect of moderate folate deficiency. *Carcinogenesis* **2006,** *27* (3), 517-24.

[118] Abramsson-Zetterberg, L.; Durling, L. J.; Yang-Wallentin, F.; Rytter, E.; Vessby, B., The impact of folate status and folic acid supplementation on the micronucleus frequency in human erythrocytes. *Mutat Res* **2006,** *603* (1), 33-40.

[119] Gordon, D. J.; Resio, B.; Pellman, D., Causes and consequences of aneuploidy in cancer. *Nat Rev Genet* **2012,** *13* (3), 189-203.

Send Orders for Reprints to reprints@benthamscince.net
Advances in Genome Science, Vol. 3, 2014, 213-232

CHAPTER 8

TWIST1 Gene: First Insights in *Felis Catus*

Cláudia S. Baptista[1,2], Sara Santos[2], Estela Bastos[2], Henrique Guedes-Pinto[2], Ivo G. Gut[3], Fátima Gärtner[4,5] and Raquel Chaves[2,*]

[1]Department of Veterinary Clinics, Institute of Biomedical Sciences Abel Salazar, University of Porto (ICBAS-UP), Porto, Portugal; [2]Institute for Biotechnology and Bioengineering, Centre of Genomics and Biotechnology, University of Trás-os-Montes and Alto Douro (IBB/CGB-UTAD), Vila Real, Portugal; [3]CEA/DSV/IG-Centre National de Génotypage, Evry Cedex, France; [4]Department of Pathology and Molecular Immunology, Institute of Biomedical Sciences Abel Salazar (ICBAS), University of Porto, Porto, Portugal and [5]Institute of Pathology and Immunology (IPATIMUP), University of Porto, Porto, Portugal

Abstract: *TWIST1* is thought to be a novel oncogene. Understanding the molecular mechanisms regulating the *TWIST1* gene expression profiles in tumor cells may give new insights regarding prognostic factors and novel therapeutic targets in veterinary oncology. In the present study we partially isolated the *TWIST1* gene in *Felis catus* and performed comparative studies. Several primer combinations were used based on the alignments of homologous DNA sequences. After PCR amplification, three bands were obtained, purified and sequenced. Several bioinformatic tools were utilized to carry out the comparative studies. Higher similarity was found between the isolated *TWIST1* gene in *Felis catus* and *Homo sapiens* (86%) than between *Homo sapiens* and *Rattus norvegicus* or *Mus musculus* (75%). Partial amino acid sequence showed no change in the four species analyzed. This confirmed that coding sequences presented high similarity (~96%) between man and cat. These results give the first insights regarding the *TWIST1* gene in cat but further studies are required in order to establish, or not, its role in tumor formation and progression in veterinary oncology.

Keywords: Animal model, coding gene, comparative analysis, expression profiles, *Felis catus*, first *TWIST1* isolation in cat, gene localization, *in silico* mapping, metastatic spread, molecular mechanisms, mutations, oncogene, overexpression, phylogenetic analysis, prognostic factors, protein prediction, therapeutic target, tumor formation, *TWIST1* gene, veterinary oncology.

***Address correspondence to Raquel Chaves:** Institute for Biotechnology and Bioengineering, Centre of Genomics and Biotechnology, University of Trás-os-Montes and Alto Douro (IBB/CGB-UTAD), Vila Real, Portugal; Tel: +351 259 350841; E-mail: rchaves@utad.pt

Christian Neri (Ed)

INTRODUCTION

The *TWIST1* gene was originally identified in *Drosophila melanogaster* where its activity was found to be crucial for the establishment of dorsoventral pattern and mesoderm formation of the embryo. This gene encodes a protein called Twist-1, a highly conserved transcription factor that belongs to the family of basic helix–loop–helix (bHLH) proteins [1-3].

In human, this gene (GenBank: NC_000007.13 and Ensembl: ENSG00000122691) has around 2205 bp and is organized in two exons (1002 bp and 664 bp, respectively, exon 1 and exon 2) and one intron (539 bp). The first exon contains an open reading frame encoding a 202 amino acid protein (GenPept: AAC50930.1). Indeed, although ENSEMBL Genome Browser, release 65 (December 2011), reports three transcripts, only one is protein coding (GenBank: NM_000474 and Ensembl: ENST00000242261). In human, the *TWIST1* gene maps to chromosome 7 and is expressed in the head and limb buds in developing embryos [4]. In adult human, its expression is predominantly detected in mesodermally derived tissues [5].

Several studies suggest that the *TWIST1* gene is a novel oncogene that is associated with tumor formation and/or metastatic spread. In fact, besides the regulation of the embryonic morphogenesis, the Twist-1 protein has multiple functions during tumorigenesis, including inhibition of cell safeguard programs, altered cell–cell adhesion, and deregulation of differentiation, contributing to increased cell survival and invasion [5-9].

Mutations on the coding region of the human *TWIST1* gene, leading to haploinsufficiency, have been identified in Saethre–Chotzen syndrome (SCS) [10, 11], in some cases of Baller-Gerold syndrome [12] and in human pediatric osteosarcomas [13]. Although controversial [14], Sahrin and colleagues [15] demonstrated an increased risk of breast cancer in women with Saethre-Chotzen syndrome, suggesting that germline mutations in *TWIST1* may also predispose to breast cancer. *TWIST1* overexpression, both at the protein and mRNA level, was reported in a variety of solid cancers including breast, hepatic, prostate and gastric carcinomas, infiltrative endometrial cancer, gliomas, melanomas, osteosarcomas, rhabdomyosarcomas, nephroblastoma as well as in Sezary syndrome [6, 8, 13, 16-24]. Very recently, it was reported up-regulation of *TWIST1* in chronic myeloid leukemia suggesting that this gene may represent a novel prognostic factor to be considered in this disease [25].

Besides *Drosophila* and human, members of the *TWIST* family have also been identified in different species such as frog [26], mouse [27], leech [28], zebrafish [29], lancelet [30], nematode [31], jellyfish [32] and chicken [33].

In the present work we achieved the partial isolation of *TWIST1* gene in cat. As far as we know, this is the first report regarding the isolation and sequencing of *TWIST1* in cat. We accomplished this purpose in order to carry out comparative studies between *Felis catus* and other species, namely *Homo sapiens*, and to perform *in silico* mapping of this gene in cat. We present a new approach than can be used to *in silico* gene map, and that use several tools that result from the cat and canine genomes sequencing efforts, that is, sequence data and radiation hybrid maps. We believe that this knowledge is a starting point that may permit the research of the molecular mechanisms regulating *TWIST1* gene expression in veterinary oncology.

MATERIALS AND METHODS

Genomic DNA Extraction and PCR Amplification

Blood samples from six different cats were used as the biological material for DNA extraction. After owners consent, blood was collected by venipuncture and stored in tubes with heparin as anticoagulant. DNA extraction was performed from 200 μL of total blood using automatic equipment (Quickgene-810, Fujifilm), and the Quickgene Whole Blood Kit S (Fujifilm) according to instructions of the manufacturer. In order to analyze the integrity of the genomic DNA extracted all samples were subjected to a 1.5% agarose gel electrophoresis. The quantity and quality of the DNA extracted was also determined in a NanoDrop ND-1000 spectrophotometer (NanoDrop Technologies).

In order to isolate the *TWIST1* gene from *Felis catus* genomic DNA, primers were designed based on the alignments of homologous DNA sequences, from species where this gene was already identified, *Homo sapiens* (GenBank NC000007.12) and *Canis familiaris* (GenBank NC 006596.2).

The different primer combinations and corresponding sequences are shown in Table **1**. For combination 1 (Table **1**), DNA was amplified after 95°C denaturation for 5 min with 35 cycles of 95ºC for 30 s, 65ºC for 30 s, 72°C for 30 s and a 72°C final extension step for 10 min. The 25 μL PCR reaction mix contained: 1.5 μL each primer (100 ng/μL), 1.5 μL MgCl2 at 25 mM, 2 μL dNTP mix (2.5 mM), 2μL DNA (78 ng/μL), 0.5 μL (5 U/μL) Taq DNA polymerase (Fermentas), 2.5 μL buffer and 13.5 μL water. For primers combination 2 and 3 (Table **1**), the PCR temperature cycle conditions and concentrations were identical to 1 (Table **1**) for the annealing temperature which was 60ºC instead of 65ºC.

All the amplifications were performed on a T-Personal Thermal Cycler (Biometra) and the PCR products were analyzed by electrophoresis in 1.5% agarose gel, stained with ethidium bromide, visualized under UV light and digitally recorded (UviDoc).

The PCR fragments were excised from the agarose gel and purified, according to the procedure described by Geneclean II Brief Protocol: Purifying DNA (QBiogene).

Table 1: Primer combinations used for the DNA isolation of *Felis catus TWIST1* gene.

Forward Sense Strand	Reverse Sense Strand	Combination
RT-F: 5' - GAGCCCGCAGTCGTACGAG - 3'	**R:** 5' – CTAGTGGGACGCGGACAT - 3'	1
	CAT-R: 5' - GGTCTTCGTGGCTGTTTTCT - 3'	2
	Ex2-R: 5' – CACGCCCTGTTTCTTTGAAT - 3'	3

RNA Extraction and RT-PCR Amplification

Necropsy of a cat with several traumatic lesions due to a car accident was immediately performed after euthanasia for fresh tissue sampling. Other than the skull and cervical vertebral column, no significant internal lesions were observed. Macroscopically normal tissue samples such as skin, liver, spleen, lung, kidney, testicles and bone marrow were stored (-80ºC) in a RNA stabilization solution (RNA Later Tissue Collection, Ambion). Total RNA was extracted with the standard Trizol method (Invitrogen) and quantified by spectrophotometry using the NanoDrop ND-1000 (NanoDrop Technologies). The total reaction volume for cDNA synthesis and amplification (one step reaction) was 20 μL including: 1 μL of RNA extracted from testicles (1 μg/μL), 8 μL of the One-step RT-PCR pre-mix (Intron), 1 μL of each primer (5'-GAGCCCGCAGTCGTACGAG-3' and 5'-CTCTGGAGGACCTG-GTAGAGGA-3'; 100 ng/μL) and 9 μL of water. cDNA was amplified after reverse transcription at 45ºC for 30 min, 94°C denaturation for 5 min with 29 cycles of 94ºC for 30 s, 57ºC for 30 s, 72°C for 30 s, with a 72°C final extension step for 5 min. This amplification was performed on a T-Personal Thermal Cycler (Biometra) and the RT-PCR product was analyzed in a 1.5% agarose gel stained with ethidium bromide and visualized with an ultraviolet transilluminator UviDoc.

The cDNA was excised from the agarose gel and purified with the Geneclean II Brief Protocol: Purifying DNA kit (QBiogene).

Sequencing of the Amplified Products

PCR samples were sequenced in both directions and sequence analysis was performed using the bioinformatic resource at the NCBI: Basic Local Alignment Search Tool (BLAST) and the Vector NTI software (Invitrogen Life Technologies). The feline *TWIST1* DNA sequence and the *TWIST1* cDNA sequence have been submitted to GenBank with the respectively accession number GQ167299 and GQ167300.

Phylogenetic Analysis

For the phylogenetic and evolutionary analysis of the molecular sequence data we used Phylemon [34], an online platform that integrates a suite of more than 20 different tools.

The largest sequenced fragment (960 bp) (fragment 3 from Table **1**) was aligned with *TWIST1* gene from *Homo sapiens* (GenBank NC_000007.12), *Macaca mulatta* (GenBank NC_007860.1), *Pan troglodytes* (Ensembl release 47:

ENSPTRG00000018960, Contig 5.948 and Contig 5.949), *Mus musculus* (GenBank NC_000078.5), *Rattus norvegicus* (GenBank NC_005105.2), *Bos Taurus* (GenBank NC_007302.2) and *Gallus gallus* as the outgroup species (Ensembl release 47: ENSGALT00000010219, Contig 16422).

We utilized the ClustalW v1.83 program [35] for the alignment of the previously mentioned multiple sequences. Basic maximum likelihood (ML) analyses of the DNA sequence data were provided with the DnaML algorithm of the PHYLIP package (version 3.65). Details of the algorithm are published in the paper by Felsenstein and Churchill [36].

In order to confirm the phylogram we used the MEGA4 program [37] that utilizes Maximum Composite Likelihood (MCL) method for estimating evolutionary distances between DNA sequences. The phylogeny reconstruction analysis was performed with the Neighbor-Joining method and the Nucleotide: Maximum Composite Likelihood substitution model.

Protein Prediction

For protein prediction of the 960 bp fragment of *TWIST1* gene, we used the tool TranSeq from EMBOSS (European Molecular Biology Open Software Suite) [38]. This tool translates nucleic acid sequences to the corresponding peptide sequence. We used frame 2 after analysing the six possible reading frames. Simultaneously, in order to confirm the result, we used multiple alignments of the eight sequences performed by the ClustalW program, described on the previous section. We converted the multiple alignment into the multi sequence format (msf) used by GenDoc program, a multiple sequence alignment editor developed by Nicholas *et al.* [39]. The protein prediction was confirmed on the Genedoc program allowing to simultaneously edit the aligned sequences and execute the translation of the coding region into the amino acid sequence.

Gene Localization

The *TWIST1* gene was *in silico* physically mapped in *Felis catus*. In order to perform a comparative analysis we also mapped this gene in the following species: *Homo sapiens*, *Pan troglodytes*, *Macaca mulatta*, *Bos taurus*, *Gallus gallus*, *Mus musculus*, *Rattus norvegicus* and *Canis familiaris*.

The *in silico* analysis of the largest sequenced fragment (960 bp), partial sequence of the *TWIST1* gene, was primarily performed using the "MultiBlast Analysis" (ENSEMBL Genome Browser). All the results obtained were confirmed using the "SyntenyView" (ENSEMBL Genome Browser; ENSEMBL release 66). Since the cat karyotype is not available in the ENSEMBL Genome Browser, the physical map of this gene was established with the Genome Browser for *Felis catus*, the Genome Annotation Resource Fields – GARFIELD [40], the initial sequence and comparative analysis of the cat genome published by Pontius *et al.* [41], and the radiation hybrid maps from Murphy *et al.* [42] and Davis *et al.* [43]. We also used the dog map information in order to improve the localization of the *TWIST1* gene in the cat chromosome. G-banded chromosomal correspondence between cat and dog are in accord with Yang *et al.* comparative chromosomal map [44]. We used the 4249 marker FISH/RH map of the canine genome for chromosome 14 from Breen *et al.* [45] and the FISH/RH maps available from the same author at http://cvm.ncsu.edu/mbs/ breen/dog_map.htm.

The genome versions used in ENSEMBL Genome Browser were: *Homo sapiens* (GRCh37), *Pan troglodytes* (CHIMP2.1.4), *Macaca mulatta* (MMUL1.0), *Bos Taurus* (UMD 3.1), *Gallus gallus* (WASHUC2), *Mus musculus* (NCBI m37), *Rattus norvegicus* (RGSC 3.4), *Canis familiaris* (CanFam2.0) and *Felis catus* (CAT 1.87X assembly).

RESULTS

Genomic DNA Extraction and PCR Amplification

The DNA samples obtained from 200 µl of blood using Quickgene Whole Blood Kit S (Fujifilm) showed a satisfactory concentration (between 67.66 ng/µL and 88.57 ng/µL, with a total yield of 200 µL). The DNA integrity evaluated by agarose gel electrophoresis and by spectrophotometry was satisfactory (OD260/OD280 =1.91).

In order to amplify the *TWIST1* gene from cat genomic DNA extracted from 6 blood samples, we synthesized multiple primers based on the homologous sequences from

Homo sapiens (GenBank NC_000007.12) and *Canis familiaris* (GenBank NC_006596.2). To successfully achieve this goal several primers combinations were used (Table **1**) and, for each sample, the obtained fragments have, approximately, the following lengths: 1=300 bp; 2=800 bp and 3=1000 bp (Fig. **1a**, **1c**).

RNA Extraction and RT-PCR Amplification

To verify if the *TWIST1* gene in *Felis catus* was a coding gene and, if so, in order to localize the coding region, we performed reverse-transcriptase PCR amplification. Total RNA extracted from testicles, liver and bone marrow with the standard Trizol method (Invitrogen) presented a good quantity (281.5 ng/μL, 189 ng/μL and 210.5 ng/μL respectively) and quality evaluated by spectrophotometry (OD260/OD280 = 1.86, 1.87 and 1.83, respectively). Using primers 5'-GAGCCCGCAGTCGTACGAG-3' and 5'-CTCTGGAGGACCTGGTAGAGGA-3', designed based in homologous sequences from *Homo sapiens* (GenBank NC_000007.12) and *Canis familiaris* (GenBank NC_006596.2), in each sample, we amplified a fragment with around 200 bp (Fig. **1b**).

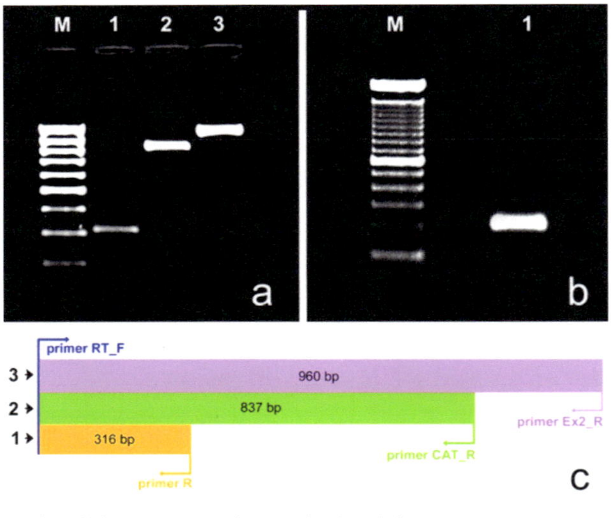

Figure 1: Electrophoresis of the PCR products obtained from cat gDNA extracted from 200 μl of blood in a 1.5% agarose gel (a). Three bands of different sizes were obtained. Lane M: Molecular marker (Gene Ruler 100 bp DNA Ladder, Fermentas). Lanes 1-3: Amplified products with approximately 300, 800 and 1000 bp; (b) Electrophoresis of the RT-PCR product in a 1.5% agarose gel. Lane M: Molecular marker (Mass Ruler DNA Ladder Mix, Fermentas). Lane 1: RT-PCR fragment with approximately 200 bp. (c) Scheme for the sequencing of *Felis catus TWIST1* gene with the identification and localization of primers and the respective product length.

Sequencing of the Amplified Products

After PCR amplification using the multiple primers combinations evidenced in Table **1** and Figs. (**1a** and **1c**), the obtained fragments were excised from the agarose gel, purified and directly sequenced in both directions. After thorough analysis of the 36 sequencing reactions, we isolated 960 bp of the feline *TWIST1* gene DNA. This sequence is shown in Supplementary Fig. (**1**), where it is aligned with the human sequence. Strong similarity (86%) was found between the human sequence and the amplified feline DNA. We can observe major differences in the intron, whereas exon 1 and exon 2 are highly conserved.

To analyze mRNA encoded by the *TWIST1* gene, primers were designed on the feline DNA putative coding region, identified on the basis of similarity between the feline (Supplementary Fig. **1**) and the human DNA sequence. cDNA was obtained from retro-transcription of total RNA isolated from cat testicle. This amplification, allowed us to isolate a fragment with around 200 bp which was also excised from an agarose gel, purified and sequenced in both directions. After sequence evaluation, the resulting 201 bp of feline cDNA was subject to BLAST analysis. The information confirmed that we isolated the cat *TWIST1* gene cDNA. Using the ClustalW program, we aligned this sequence (Supplementary Fig. **1**, underline 201 bp sequence) with the partially isolated predicted coding region (Supplementary Fig. **1**, blue 316 bp sequence) and we concluded that the isolated cDNA is 100% included in the predicted coding region of the *Felis catus TWIST1* gene.

Phylogenetic Analysis

The 960 bp sequence from cat *TWIST1* gene was aligned with homologous sequences from *Homo sapiens*, *Macaca mulatta*, *Pan troglodytes*, *Mus musculus*, *Rattus norvegicus*, *Bos taurus* and *Gallus gallus* in order to obtain the phylogram presented in Fig. (**2**). The results displayed by the phylogenetic analysis allowed us to conclude that there is a higher similarity between the homologous sequences from *Homo sapiens* and *Felis catus* (86%) than similarity between *Homo sapiens* and *Rattus norvegicus* or *Mus musculus* (75%). We also found a high similarity (96%) between *Homo sapiens* and *Felis catus* regarding the 316 bp (partial exon 1) fragment. In fact, we detected a higher similarity between cat and man for the sequence analyzed when comparing with other animal models, as the classical rat and mouse.

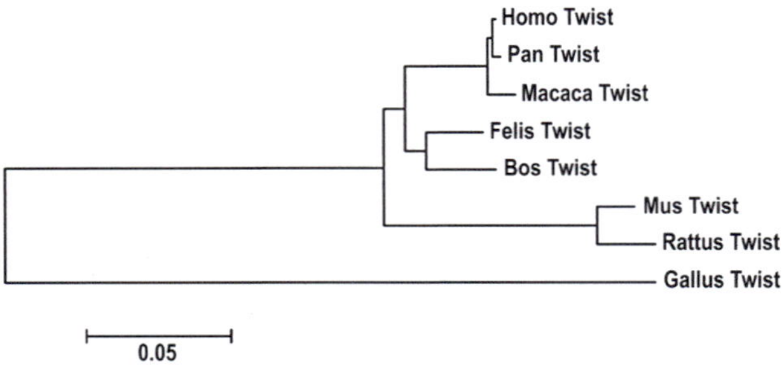

Figure 2: Phylogram constructed using the alignment between the isolated 960 bp sequence from cat *TWIST1* gene and homologous sequences from *Homo sapiens*, *Macaca mulatta*, *Pan troglodytes*, *Mus musculus*, *Rattus norvegicus*, *Bos taurus* and *Gallus gallus*.

Protein Prediction

The deduced partial amino acid sequence of the Twist-1 protein in all species analyzed was generated by the Genedoc program, starting from the Phylemon platform. The maximal length of the partial coding region isolated from cat genomic DNA has 316 bp (Fig. **1c** and Table **1**, primer combination 1) and 104 deduced amino acids from a predicted total of 202 amino acids. The results displayed were confirmed comparing our amino acid sequence with known amino

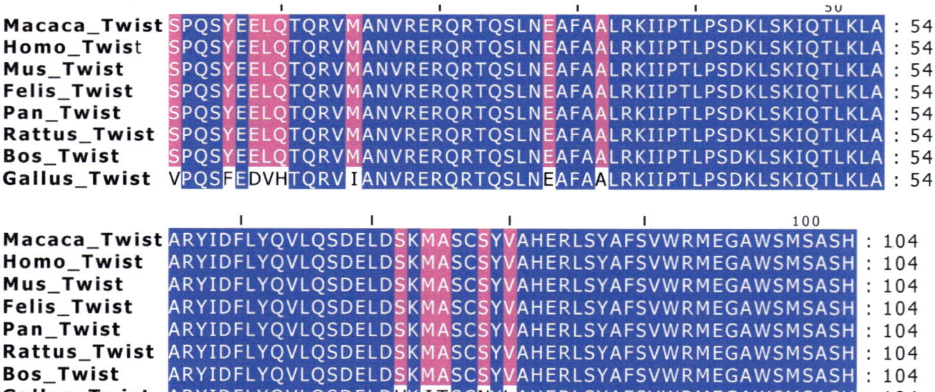

Figure 3: For all species analyzed (*Homo sapiens*, *Rattus norvegicus*, *Mus musculus*, *Macaca mulatta*, *Pan troglodytes*, *Bos Taurus* and *Gallus gallus*) the partial amino acid sequence of the Twist-1 protein was deduced using the Genedoc program, starting from the Phylemon platform, and the multiple alignments were performed using the Clustalw program. We can observe 104 amino acids (316 bp of the codifying region isolated in cat) from a predicted total of 202 amino acids (codifying region with 609 bp). Only the outgroup species (*Gallus gallus*) evidences 13 different amino acids.

acid sequences from GenBank database, from *Homo sapiens* (GenBank NC_000007.12), *Rattus norvegicus* (GenBank NC_005105.2), *Mus musculus* (GenBank NC_000078.5), *Macaca mulatta* (GenBank NC_007860.1), *Pan troglodytes* (Ensembl release 47: ENSPTRG00000018960, Contig 5.948 and Contig 5.949), *Bos Taurus* (GenBank NC_007302.2) and *Gallus gallus* (Ensembl release 47: ENSGALT00000010219, Contig 16422). The multiple alignments were performed by the ClustalW program (Fig. **3**) and they show no change in all the species analyzed, except for *Gallus gallus* (the outgroup species) where 13 amino acids were modified. Between *Homo sapiens* and *Felis catus* we detected 8 transitions and 2 transversions meaning that the conservation at the protein level is different from the scenario observed at the genomic level.

Gene Localization

The results regarding the *in silico* physical mapping of the *TWIST1* gene in *Felis catus, Canis familiaris, Homo sapiens, Pan troglodytes, Macaca mulatta, Mus musculus, Rattus norvegicus, Bos taurus* and *Gallus gallus* species are presented in Fig. (**4**).

The *in silico* analysis of the largest sequenced DNA fragment from the cat *TWIST1* gene (*i.e.* 960 bp DNA fragment) was blasted on the genomes from the species in analysis, using the "MultiBlast Analysis" tool from the ENSEMBL Genome Browser. This analysis allowed the physical mapping of the cat *TWIST1* partial sequence on the following chromosome (chr) species (Fig. **4c**): *Canis familiaris* chr 14, *Homo sapiens* chr 7p21.1, *Pan troglodytes* chr 7, *Macaca mulatta* chr 3, *Mus musculus* chr 12B2, *Rattus norvegicus* chr 6q16 and *Bos taurus* chr 4. Moreover, in all analysis, and for each chromosome, it was possible to determine the localization of the predicted gene in terms of Mb with a higher precision (Fig. **4c**). All the results obtained were confirmed using the "SyntenyView" tool from the ENSEMBL Genome Browser (Ensembl release 66 - February 2012) what allows the search of the syntenic segments between these genomes (Fig. **4c**). The *TWIST1* gene was mapped to chr 2 in *Gallus gallus*; however, this analysis was only possible through the physical mapping of the chicken *TWIST1* gene made in the human genome using the "view synteny regions" tool in the "contigview" menu. Therefore, the localization of the predicted gene in terms of Mb in this chromosome was not as accurate (see Fig. **4c**) as for the other genomes.

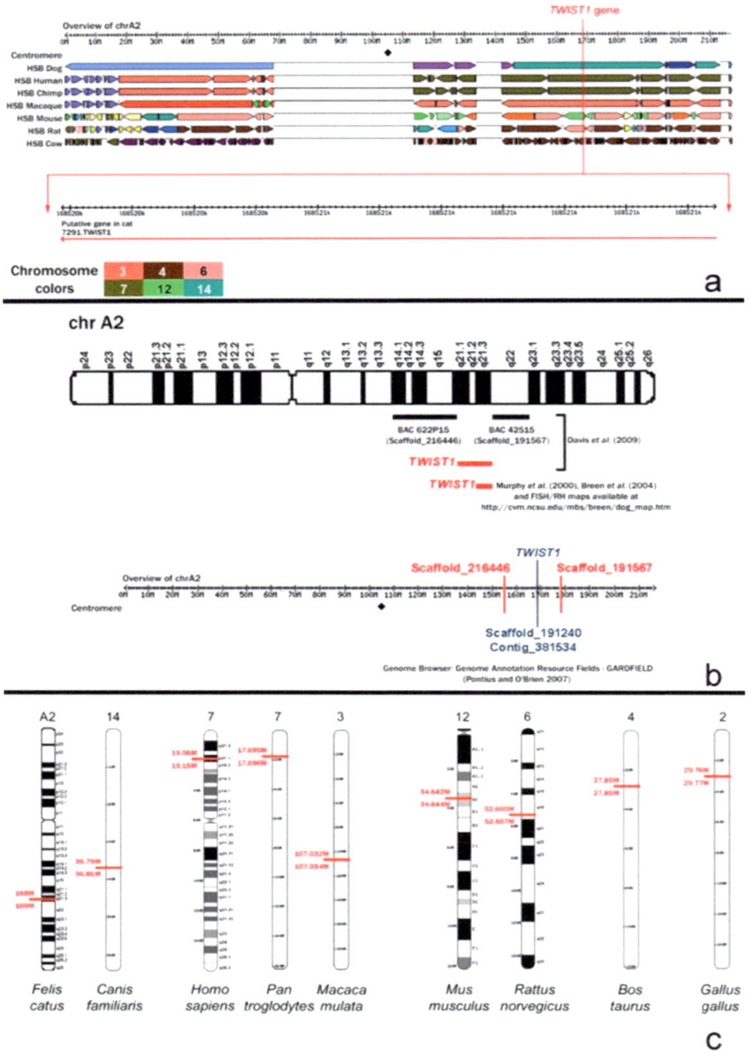

Figure 4: Physical mapping of the *TWIST1* gene in Felis catus, Canis familiaris Homo sapiens, Pan troglodytes, Macaca mulatta, Mus musculus, Rattus norvegicus, Bos taurus and Gallus gallus species. (a) *In silico* comparative analysis of the *TWIST1* gene with the Genome Annotation Resource Fields – GARFIELD [40]. (b) the initial sequence and comparative analysis of the cat genome published by Pontius *et al.* [41], and the radiation hybrid maps from Murphy *et al.* [42] and Davis *et al.* [43]. G-banded chromosomal correspondence between cat and dog are in accord with Yang et al. comparative chromosomal map [44]. We also used the dog map information in order to improve the localization of the *TWIST1* gene in the cat chromosome. We used the 4249 marker FISH/RH map of the canine genome for chromosome 14 from Breen *et al.* [45] and the FISH/RH maps available from the same author at http://cvm.ncsu.edu/mbs/breen/dog_map.htm. (c) Chromosome ideograms showing the *TWIST1* gene localization in several species analysed in the present work.

Finally, the cat *TWIST1* gene was physically mapped to chr A2 (q21.3) in the cat karyotype (Fig. **4b**). In this case we used the Genome Browser for *Felis catus*, the Genome Annotation Resource Fields – GARFIELD [40], the initial sequence and comparative analysis of the cat genome published by Pontius *et al.* [41], and the radiation hybrid maps from Murphy *et al.* [42] and Davis *et al.* [43]. The simple search of the *TWIST1* gene in the cat genome showed the chromosome A2 involved (see Fig. **4b**), with the fine localization in terms of Mb on this chromosome. However, this analysis does not permit to physically map the gene in the A2 ideogram (physical mapping at band level). In order to accomplish this task, we used cat and dog G-banded comparative chromosomal map [44] and the cat radiation hybrid maps [42, 43]; the 4249 marker FISH/RH map of the canine genome for the 14 chromosome [45] and the FISH/RH maps available at <http://cvm.ncsu.edu/mbs/breen/dog_map.htm>. With this comparative analysis it was possible to *in silico* map the *TWIST1* gene to the q21.3 band from cat chromosome A2 (Fig. **4c**).

DISCUSSION

It is commonly accepted that naturally occurring cancers in dogs and cats offer a unique opportunity as models for human cancer biology and translational cancer therapeutics. Several aspects contribute to the advantages of the companion animal model such as the relatively high incidence of some cancers when compared with humans, similar biological behavior with faster rate of progression, large body size that allow imaging studies as well as surgical interventions, comparable responses to cytotoxic agents and radiation therapy, shorter overall lifespan and similar environmental risk factors. It is also important to consider that animal trials are generally much more economical to run than human trials [46-48]. In this work we present evidence that *TWIST1* may be targeted in new studies to understand its role in feline oncology and that, potentially, *Felis catus* could be a very interesting animal model to study human disease, at least regarding this gene.

To the best of our knowledge, this is the first report regarding the isolation and sequencing of *TWIST1* in cat. Therefore, in order to accomplish this purpose and perform comparative studies, several primer combinations were designed to successfully obtain the partial amplification of the *TWIST1* gene in this species. The segments of interest had a high GC content making this task somewhat challenging, but we were able to successfully isolate 960 bp of the cat *TWIST1* gene, namely 358 bp of the exon 1 (including 316 bp of the coding region), 516 bp of the intron and 86 bp of the exon 2 (Fig. **5**). When compared to the homologous sequence in *Homo sapiens*, this isolated sequence presented major genomic differences in the nonconserved intronic region, as expected (80% similarity). Evolutionarily conserved regions, meaning exon 1 and exon 2, evidenced 96% similarity between both these species.

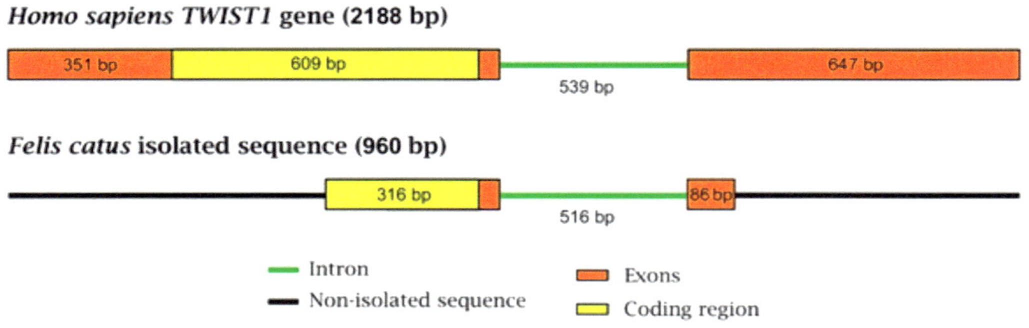

Figure 5: Schematic representation of the *Homo sapiens TWIST1* gene and the orthologous gene sequence isolated in the present work from *Felis catus* (960 bp in total). We can observe that we have isolated 358 bp of exon 1 (including 316 bp of the coding region), 516 bp of the intron and 86 bp of the exon 2.

According to the ENSEMBL Genome Browser release 65 (December 2011), in cat, the homologous region with respect to the human *TWIST1* gene predicts a novel pseudogene (ENSFCAG00000006288). These genes are defined as genomic DNA sequences that bear significant homology to functional genes but have lost their potential as DNA templates for functional products [49, 50]. In this research work, the RT-PCR results demonstrated that the feline *TWIST1* gene is a transcribed gene, evidencing, at least, one transcript. However, additional studies are required in order to study its expression profile in normal and tumor cells, in this species.

The phylogenetic analysis performed with the alignment of largest cat DNA fragment from *TWIST1* gene and homologous sequences from several species (Fig. **2**), detected a higher similarity between the sequences of cat and man. This information is extremely important, suggesting the cat as an attractive model, at least for *TWIST1* gene studies, and that should be used instead of the classical animal models (*e.g.* the rat and mice). This data is supporting a growing body of evidence that propose alternative animal models, as the cat, relative to the classical ones. In fact, the evolutionary history of the genomes indicates great conservation between cat and human, with around 30 homologous segments separating both genomes. In contrast, genomes as the ones of rat, mouse and dog are four times more reshuffled than those of species retaining the more slowly evolving genomes (*e.g.* cat and human) [51, 52].

The partial amino acid sequence (104 aa from 202 aa, in total) inferred from the coding region within fragment 3 (Fig. **1c**), showed no change in all the species analyzed, except for *Gallus gallus* (the outgroup species) where 13 aminoacids were modified. This conservation at the protein level is different from the scenario observed at the genomic level. We detected 8 transitions and 2 transversions when comparing the coding fragment between *Homo sapiens* and *Felis catus*. All these sequence variations correspond to synonymous or silent mutations. Considering these evidences, it is encouraging to think that the ongoing studies regarding the *TWIST1* gene in *Homo sapiens* can be extrapolated to *Felis catus*. The *in silico* analysis presented allowed the physical map of *TWIST1* gene to several species: *Felis catus* chr A2q21.3, *Canis familiaris* chr 14, *Homo sapiens* chr 7p21.1, *Pan troglodytes* chr 7, *Macaca mulatta* chr 3, *Mus musculus* chr 12B2, *Rattus norvegicus* chr 6q16 and *Bos taurus* chr 4. The importance of gene mapping in genetic clinical studies is well known for diagnosis purposes. In this paper we present a new approach that can be used to *in silico* gene map, and can use several tools that result from the cat and canine genomes sequencing efforts, that is, sequence data and radiation hybrid maps.

CONCLUDING REMARKS

We believe that this investigation is the first to isolate the partial sequence of the *TWIST1* gene in the *Felis catus* and to perform comparative analysis between this oncogene in cat with other species, namely *Homo sapiens*. We observed that, regarding this gene, there is a higher similarity between cat and man than between man and other widely used animal models such as rat or mouse, and that *TWIST1* maps to the q21.3 band from cat chromosome A2. This work gives the first insights regarding the *TWIST1* gene in *Felis catus,* so we may contribute to the study of the molecular mechanisms affecting its expression profile in cat tumor cells [53].

ACKNOWLEDGEMENTS

We thank to Portuguese Foundation for Science and Technology (FCT) for funding our work, namely project POCI/CVT/62940/2004 and PhD grants SFRH/BD/31754/2006 and BD/SFRH/23406/2005.

CONFLICT OF INTEREST

The authors confirm that this chapter contents have no conflict of interest.

DISCLOSURE

The chapter submitted for Advances in Genome Science, Vol. 3 entitled "***TWIST1* gene: first insights in *Felis catus***" is an update of our article published in Current Genomics, 2010, 11: 212-220.

REFERENCES

[1] Nusslein-Volhard, C.; Wieschaus, E.; Kluding, H. Mutations affecting the pattern of the larval cuticle in *Drosophila melanogaster*. 1. Zygotic loci on the 2nd chromosome. *Roux's Arch. Dev. Biol.*, **1984**, *193*, 267-282.

[2] Simpson, P. Maternal-zygotic gene interactions of the dorsoventral pattern in *Drosophila* embryos. *Genetics*, **1983**, *105*, 615-632.

[3] Thisse, B.; Messal, M.; Perrin-Schmitt, F. The twist gene: isolation of a Drosophila zygotic gene necessary for the establishment of dorsoventral pattern. *Nucleic Acids Res.*, **1987**, *15*, 3439-3453.

[4] Rose, C.S.; Malcolm, S. A TWIST in development. *Trends Genet.*, **1997**, *13*, 384-387.

[5] Wang, S.M.; Coljee, V.W.; Pignolo, R.J.; Rotenberg, M.O.; Cristofalo, V.J.; Sierra, F. Cloning of the human *twist* gene: its expression is retained in adult mesodermally-derived tissues. *Gene*, **1997**, *187*, 83-92.

[6] Maestro, R.; Dei Tos, A.P.; Hamamori, Y.; Krasnokutsky, S.; Sartorelli, V.; Kedes, L.; Doglioni, C.; Beach, D.H.; Hannon, G.J. *Twist* is a potential oncogene that inhibits apoptosis. *Genes Dev.*, **1999**, *13*, 2207-2217.

[7] Puisieux, A.; Valsesia-Wittmann, S.; Ansieau, S. A twist for survival and cancer progression. *Br. J. Cancer*, **2006**, *94*, 13-17.

[8] van Doorn, R.; Dijkman, R.; Vermeer, M.H.; Out-Luiting, J.J.; van der Raaij-Helmer, E.M.; Willemze, R.; Tensen, C.P. Aberrant expression of the tyrosine kinase receptor EphA4 and

the transcription factor twist in Sezary syndrome identified by gene expression analysis. *Cancer Res.*, **2004**, *64*, 5578- 5586.

[9] Yang, J., Mani, S.A.; Donaher, J.L.; Ramaswamy, S.; Itzykson, R.A.; Come, C.; Savagner, P.; Gitelman, I.; Richardson, A.; Weinberg, R.A. Twist, a master regulator of morphogenesis, plays an essential role in tumor metastasis. *Cell*, **2004**, *117*, 927-939.

[10] Krebs, I.; Weis, I.; Hudler, M.; Rommens, J.M.; Roth, H.; Scherer, S.W.; Tsui, L-C.; Füchtbauer, E-M.; Grzeschik, K-H.; Tsuji, K.; Kunz, J. Translocation breakpoint maps 5 kb 3′ from TWIST in a patient affected with Saethre-Chotzen syndrome. *Hum. Mol. Genet.*, **1997**, *6*, 1079-1086.

[11] Yousfi, M.; Lasmoles, F.; Ghouzzi, V.E.; Marie, P.J. Twist haploinsufficiency in Saethre-Chotzen syndrome induces calvarial osteoblast apoptosis due to increased TNFα expression and caspase-2 activation. *Hum. Mol. Genet.*, **2002**, *1*, 359-369.

[12] Gripp, K.W.; Stolle, C.A.; Celle, L.; McDonald-McGinn, D.M.; Whitaker, L.A.; Zackai, E.H. TWIST gene mutation in a patient with radial aplasia and craniosynostosis: further evidence for heterogeneity of Baller-Gerold syndrome. *Am. J. Med. Genet.*, **1999**, *82*, 170-176.

[13] Entz-Werle, N.; Stoetzel, C.; Berard-Marec, P.; Kalifa, C.; Brugiere, L.; Pacquement, H.; Schmitt, C.; Tabone, M.D.; Gentet,J.C.; Quillet, R.; Oudet, P.; Lutz, P.; Babin-Boilletot, A.; Gaub, M.P.; Perrin-Schmitt, F. Frequent genomic abnormalities at TWIST in human pediatric osteosarcomas. *Int. J. Cancer*, **2005**, *117*, 349-355.

[14] James, P.A.; Culling, B.; Mullan, G.; Jenkins, M.; Elakis, G.; Turner, A.M.; Mowat, D.M.; Wilson, M.; Anderson, P.; Savarirayan, R.; Cliffe, S.T.; Caramins, M.; Buckley, M.F.; Tucker, K.; Roscioli, T. Breast cancer risk is not increased in individuals with *TWIST1* mutation confirmed Saethre-Chotzen syndrome: an Australian multicenter study. *Genes Chromosomes Cancer*, **2009**; *48*, 533-538.

[15] Sahlin, P ; Windh, P.; Lauritzen, C.; Emanuelsson, M.; Grönberg, H.; Stenman, G. Women with Saethre-Chotzen syndrome are at increased risk of breast cancer. *Genes Chromosomes Cancer,* **2007**, *46*, 656-660.

[16] Hoek, K.; Rimm, D.L.; Williams, K.R.; Zhao, H.; Ariyan, S.; Lin, A.; Kluger, H.M.; Berger, A.J.; Cheng, E.; Trombetta, E.S.; Wu, T.; Niinobe, M.; Yoshikawa, K.; Hannigan, G.E.; Halaban, R. Expression profiling reveals novel pathways in the transformation of melanocytes to melanomas. *Cancer Res.*, **2004**, *64*, 5270-5282.

[17] Kwok, W.K.; Ling, M.T.; Lee, T.W.; Lau, T.C.; Zhou, C.; Zhang, X.; Chua, C.W.; Chan, K.W.; Chan, F.L.; Glackin, C.; Wong, Y.C.; Wang, X. Up-regulation of TWIST in prostate cancer and its implication as a therapeutic target. *Cancer Res.*, **2005**, *65*, 5153-5162.

[18] Martin, T.A.; Goyal, A.; Watkins, G.; Jiang, W.G. Expression of the transcription factors snail, slug, and twist and their clinical significance in human breast cancer. *Ann. Surg. Oncol.*, **2005**, *12*, 488-496.

[19] Rosivatz, E.; Becker, I.; Specht, K.; Fricke, E.; Luber, B.; Busch, R.; Hofler, H.; Becker, K.F. Differential expression of the epithelial-mesenchymal transition regulators snail, SIP1, and twist in gastric cancer. *Am. J. Pathol.*, **2002**, *161*, 1881-1891.

[20] Vernon, A.E.; LaBonne, C. Tumor metastasis: a new twist on epithelial-mesenchymal transitions. *Curr. Biol.*, **2004**, *14*, 719-721.

[21] Watanabe, O.; Imamura, H.; Shimizu, T.; Kinoshita, J.; Okabe, T.; Hirano, A.; Yoshimatsu, K.; Konno, S.; Aiba, M.; Ogawa, K. Expression of twist and wnt in human breast cancer. *Anticancer Res.*, **2004**, *24*, 3851-3856.

[22] Kyo, S.; Sakaguchi, J.; Ohno, S.; Mizumoto, Y.; Maida, Y.; Hashimoto, M.; Nakamura, M.; Takakura, M.; Nakajima, M.; Masutomi, K.; Inoue, M. High Twist expression is involved in infiltrative endometrial cancer and affects patient survival. *Hum. Pathol.,* **2006**; *37*, 431–438.

[23] Man, T.K.; Chintagumpala, M.; Visvanathan, J.; Shen, J.; Perlaky, L.; Hicks, J.; Johnson, M.; Davino, N.; Murray, J.; Helman, L.; Meyer, W.; Triche, T.; Wong, K.K.; Lau, C.C. Expression profiles of osteosarcoma that can predict response to chemotherapy. *Cancer Res.,* **2005**, *65*, 8142–8150.

[24] Pajer, P.; Pecenka, V.; Karafiat, V.; Kralova, J.; Horejsi, Z.; Dvorak, M. The twist gene is a common target of retroviral integration and transcriptional deregulation in experimental nephroblastoma. *Oncogene,* **2003**, *22*, 665–673.

[25] Cosset, E.; Hamdan, G.; Jeanpierre, S.; Voeltzel, T.; Sagorny, K.; Hayette, S.; Mahon, F.X.; Dumontet, C.; Puisieux, A.; Nicolini, F.E.; Maguer-Satta, V. Deregulation of TWIST-1 in the CD34+ compartment represents a novel prognostic factor in chronic myeloid leukemia. *Blood,* **2011**, *117*, 1673-1676.

[26] Hopwood, N.D.; Pluck, A.; Gurdon, J.B. Xenopus twist is expressed in response to induction in the mesoderm and the neural crest. *Cell*, **1989**, *59*, 893-903.

[27] Wolf, C.; Thisse, C.; Stoetzel, C.; Thisse, B.; Gerlinger, P.; Perrin-Schmitt, F. The M-twist gene of Mus is expressed in subsets of mesodermal cells and is closely related to the Xenopus X-twi and the Drosophila twist genes. *Dev. Biol.*, **1991**, *143*, 363-373.

[28] Soto, J.G.; Nelson, B.H.; Weisblat, D.A. A leech homolog of twist: evidence for its inheritance as a maternal mRNA. *Gene*, **1997**, *199*, 31-37.

[29] Morin-Kensicki, E.M.; Eisen, J.S. Sclerotome development and peripheral nervous system segmentation in embryonic zebrafish. *Development*, **1997**, *124*, 159-167.

[30] Yasui, K.; Zhang, S.; Uemura, M.; Aizawa, S.; Ueki, T. Expression of a twist-related gene, Bbtwist, during the development of a lancelet species and its relation to cephalochordate anterior structures. *Dev. Biol.*, **1998**, *195*, 49-59.

[31] Harfe, B.D.; Fire, A. Muscle and nerve-specific regulation of a novel NK-2 class homeodomain factor in Caenorhabditis elegans. *Development*, **1998**, *125*, 421-429.

[32] Spring, J.; Yanze, N.; Middel, A.M.; Stierwald, M.; Gröger, H.; Schmid, V. The mesoderm specification factor Twist in the life cycle of Jellyfish. *Dev. Biol.*, **2000**, *28*, 363-375.

[33] Tavares, A.T.; Izpisua-Belmonte, J.C.; Rodriguez-Leon, J. Developmental expression of Chick twist and its regulation during limb patterning. *Int. J. Dev. Biol.*, **2001**, *45*, 707-713.

[34] Tárraga, J.; Medina, I.; Arbiza, L.; Huerta-Cepas, J.; Gabaldón, T.; Dopazo, J.; Dopazo, H. Phylemon: a suite of web tools for molecular evolution, phylogenetics and phylogenomics. *Nucleic Acids Res.*, **2007**, *35*, W38 - W42.

[35] Thompson, J.D.; Higgins, D.G.; Gibson, T.J. CLUSTAL W: Improving the sensitivity of progressive multiple sequence alignment through sequence weighting, position-specific gap penalties and weight matrix choice. *Nucleic Acids Res.*, **1994**, *22*, 4673-4680.

[36] Felsenstein, J.; Churchill, G.A. A Hidden Markov Model approach to variation among sites in rate of evolution. *Mol. Biol. Evol.*, **1996**, *13*, 93-104.

[37] Tamura, K.; Dudley, J.; Nei, M.; Kumar, S. MEGA4: Molecular Evolutionary Genetics Analysis (MEGA) software version 4.0. *Mol. Biol. Evol.*, **2007**, *24*, 1596-1599.

[38] Rice, P.; Longden, I.; Bleasby, A. EMBOSS: the European Molecular Biology Open Software Suite. *Trends Genet.*, **2000**, *16*, 276-277.

[39] Nicholas, K.B.; Nicholas, H.B. Gendoc: a tool for editing and annotating multiple sequence alignments. Multiple Sequence Alignment Editor and Shading Utility. Version 2.0.001. Distributed by the authors. **1997**.

[40] Pontius, J.U.; O'Brien, S.J. Genome Annotation Resource Fields- GARFIELD: a genome browser for *Felis catus*. *J. Hered.*, **2007**, *98*, 386-389.

[41] Pontius, J.U.; Mullikin, J.C.; Smith, D.R.; Agencourt Sequencing Team; Lindblad-Toh, K.; Gnerre, S.; Clamp, M.; Chang, J.; Stephens, R.; Neelam, B.; Volfovsky, N.; Schäffer, A.A.; Agarwala, R.; Narfström, K.; Murphy, W.J.; Giger, U.; Roca, A.L.; Antunes, A.; Menotti-Raymond, M.; Yuhki, N.; Pecon-Slattery, J.; Johnson, W.E.; Bourque, G.; Tesler, G.; NISC Comparative Sequencing Program; O'Brien, S.J. Initial sequence and comparative analysis of the cat genome. *Genome Res.*, **2007**, *17*, 1675-1689.

[42] Murphy, W.J.; Davis, B.; David, V.A.; Agarwala, R.; Schäffer, A.A.; Wilkerson, A.J.P.; Neelam, B.; O'Brien, S.J.; Menotti- Raymond, M. A 1.5-Mb-resolution radiation hybrid map of the cat genome and comparative analysis with the canine and human genomes. *Genomics*, **2007**, *89*, 189-196.

[43] Davis, B.W.; Raudsepp, T.; Pearks Wilkerson, A.J.; Agarwala, R.; Schäffer, A.A.; Houck, M.; Chowdhary, B.P.; Murphy, W.J. A high-resolution cat radiation hybrid and integrated FISH mapping resource for phylogenomic studies across Felidae. *Genomics*, **2009**, *93*, 299-304.

[44] Yang, F.; O'Brien, P.C.; Milne, B.S.; Graphodatsky, A.S.; Solanky, N.; Trifonov, V.; Rens, W.; Sargaqn, D., Fergunson-Smith, M.A. A complete comparative chromosome map for the dog, red fox, and human and its integration with canine genetic maps. *Genomics*, **1999**, 62: 189–202.

[45] Breen, M.; Hitte, C.; Lorentzen, T.D.; Thomas, R.; Cadieu, E.; Sabacan, L.; Scott, A.; Evanno, G.; Parker, H.G.; Kirkness, E.F.; Hudson, R.; Guyon, R.; Mahairas, G.G.; Gelfenbeyn, B.; Fraser, C.M.; André, C.; Galibert, F.; Ostrander, E.A. An integrated 4249 marker FISH/RH map of the canine genome. *BMC Genomics*, **2004**, *5*, 65 doi:10.1186/1471-2164-5-65.

[46] Baptista, C.S.; Villagrasa, M.H.; Marinho, A.A. Standardised B-scan and A-scan echographic evaluation of spontaneous anterior uveal melanomas in the dog. *Vet. J.*, **2006**, *171*, 322-330.

[47] Paoloni, M.; Khanna, C. Translation of new cancer treatments from pet dogs to humans. *Nat. Rev. Cancer*, **2008**, *8*, 147-156.

[48] Vail, D.M.; MacEwen, E.G. Spontaneously occurring tumors of companion animals as models for human cancer. *Cancer Invest.*, **2000**, *18*, 781-792.

[49] Vanin, E.F. Processed pseudogenes: characteristics and evolution. *Ann. Rev. Genet.*, **1985**, *19*, 253-272.

[50] Zheng, D.; Zhang, Z.; Harrison, P.M.; Karro, J.; Carriero, N.; Gerstein, M. Integrated pseudogene annotation for human chromosome 22: Evidence for transcription. *J. Mol. Biol.*, **2005**, *349*, 27-45.

[51] Murphy, W.J.; Stanyon, R.; O'Brien, S.J. Evolution of mammalian genome organization inferred from comparative gene mapping. *Genome Biol.*, **2001**, *2*(6), reviews0005.1-0005.8.

[52] O´Brien, S.J.; Wienberg, J.; Lyons, L.A. Comparative genomics: lessons from cats. *Trends Genet.*, **1997**, *13*, 393-399

[53] Baptista, C.S.; Santos, S.; Laso, A.; Bastos, E.; Ávila, S.; Guedes-Pinto, H.; Gärtner, F.; Gut, I.G.; Castrillo, J.L.; Chaves, R. Sequence variation and mRNA expression of the *TWIST1* gene in cats with mammary hyperplasia and neoplasia. *Vet J.,* **2012**, *191,* 203-207.

SUPPLEMENTARY MATERIAL

Supplementary Fig. 1. Using the ClustalW program we aligned the 960 bp sequence isolated in *Felis catus* with the similar region of the *TWIST1* gene in *Homo sapiens*. Exon 1, codifying region, intron and exon 2 are evidenced. Underline is the sequence amplified by RT-PCR.

Send Orders for Reprints to reprints@benthamscince.net
Advances in Genome Science, Vol. 3, **2014, 233-270**

CHAPTER 9

Advances in the Study of Brain Aging and Alzheimer's Disease Using Microarray and Next-Generation Sequencing: Focus on Selective Neuronal Vulnerability

Xinkun Wang*, Mary L. Michaelis and Elias K. Michaelis

Higuchi Biosciences Center and Department of Pharmacology and Toxicology, The University of Kansas, Lawrence, KS 66047, USA

Abstract: Pivotal brain functions, such as neurotransmission, cognition, and memory, decline with advancing age and, especially, in neurodegenerative conditions associated with aging, such as Alzheimer's disease (AD). Yet, deterioration in structure and function of the nervous system during aging or in AD is not uniform throughout the brain. Selective neuronal vulnerability (SNV) is a general but sometimes overlooked characteristic of brain aging and AD. There is little known at the molecular level to account for the phenomenon of SNV. Functional genomic analyses, through unbiased whole genome expression studies, could lead to new insights into a complex process such as SNV. Microarray and next-generation sequencing (RNA-Seq) data generated thus far (as of March 2012) using both human brain tissue and brains from animal models of aging and AD were analyzed in this chapter. Convergent trends that have emerged from these data sets were considered in identifying possible molecular and cellular pathways involved in SNV. It appears that during normal brain aging and in AD, neurons vulnerable to injury or cell death are characterized by significant decreases in the expression of genes related to mitochondrial metabolism and energy production. In AD, vulnerable neurons also exhibit down-regulation of genes related to synaptic neurotransmission and vesicular transport, cytoskeletal structure and function, and neurotrophic factor activity. A prominent category of genes that are up-regulated in AD are those related to inflammatory response and some components of calcium signaling. These genomic differences between sensitive and resistant neurons can now be used to explore the molecular underpinnings of previously suggested mechanisms of cell injury in aging and AD.

Keywords: Alzheimer's disease, bioinformatics, biological pathway, brain aging, calcium signaling, cytoskeleton, energy metabolism, gene expression, gene ontology, genome, genomics, microarray, mitochondrion, neurobiology,

***Address corresponding to Xinkun Wang:** Higuchi Biosciences Center, The University of Kansas, 2099 Constant Avenue, Lawrence, KS 66047, USA; Tel: (785) 864-4589; Fax: (785) 864-5738; Email: xwang@ku.edu

Christian Neri (Ed)

neurodegeneration, neurodegenerative disease, neuroinflammation, neurotrophic factor, next-generation sequencing, oxidative stress, selective neuronal vulnerability, synaptic neurotransmission, systems biology, transcriptome.

1. INTRODUCTION

Brain aging and associated neurodegenerative diseases such as Alzheimer's disease (AD), do not affect all neurons equally. For example, in the hippocampus, neurons in the CA1 region are vulnerable to brain aging and AD, but those in the nearby CA3 region are not nearly as heavily damaged as the CA1 neurons [1-3]. A pattern of selective loss of synapses and neurons in certain brain regions has been described for both the aging process [4-6] and AD [7-10]. These studies have been performed, for the most part, at the microanatomical level and have identified relatively few neurochemical changes that correlate with either neuronal vulnerability or resistance to age- or AD-associated injury or death. The selective vulnerability of certain brain neurons appears to be an intrinsic characteristic of these neurons. Besides aging and AD, this phenomenon, of selective neuronal vulnerability (SNV), is also a characteristic of many other neural insults, such as Parkinson's disease [11], amyotrophic lateral sclerosis (ALS) [12], ischemia [13], epileptic seizures [14], and oxidative stress (OS) [15, 16]. Yet, SNV is often overlooked in the study of brain aging and neurodegenerative diseases. By definition, SNV refers to the fact that, only select populations of neurons are uniquely vulnerable to injury or death under adverse conditions, whereas other neurons are relatively resistant to such stresses in their environment. The selective vulnerability of some neurons is often manifested in structural and functional changes that may or may not lead to the death of the cells. For example, vulnerable neurons often suffer loss of dendrites that leads to a significant impairment of synaptic transmission, but the cells may still survive for a time in this altered state. Understanding the mechanisms underlying SNV is an essential step in efforts to develop strategies to moderate the deleterious impact of aging and neurodegenerative diseases on the overall quality of life.

The aging process and age-associated disease conditions, including AD, are marked by genomic instability and consequential or compensatory changes in gene expression patterns [17-19]. Since the functional status of cells is determined to a large extent by their genomic activity, genomic studies of neurons that are selectively vulnerable to brain aging and AD, are expected to yield new insights into the intrinsic biochemical and cell biological processes that make some neurons susceptible to a wide variety of stresses. In this chapter, we first describe the brain regions and/or neuronal populations that are currently known to be most affected by aging and AD. Secondly, we collect and carry out an analysis of the published functional genomic studies on the phenomenon of SNV in brain aging and AD. Finally, we attempt to integrate and discuss the findings from multiple studies into some common patterns or characteristics that seem to offer the most likely explanations for the differential vulnerability of neuronal populations to stresses due to both aging and to disease. The functional genomics studies and data presented here are those available as of March 2012.

2. SELECTIVE NEURONAL VULNERABILITY IN BRAIN AGING

2.1. Brain Regions that are Vulnerable to Normal Aging

As indicated above, several brain regions uniquely susceptible to age-dependent cell damage ultimately disrupt normal function and compromise behavioral performance. The frontal cortex is one such region that plays a pivotal role in cognition and memory, and even subtle changes in the neuronal environment of selectively vulnerable neurons appear to lead to the cognitive impairment characteristic of normal brain aging [20, 21]. In efforts to identify the most age-sensitive regions in the brain, non-invasive techniques such as structural brain imaging and functional magnetic resonance imaging are being widely used and combined with knowledge derived from post-mortem analyses of aging human brain. These studies, based mostly on volumetric measurements in human brain, show that the association cortex, the neostriatum, and the cerebellum are the most vulnerable regions to age-dependent loss of volume, whereas the primary sensory cortices (such as the visual cortex), the entorhinal cortex, the paleostriatum, and the pons show much less shrinkage [22, 23]. (See Fig. (**1**) for location of several of these brain regions).

It is important to note that there are variable patterns in different brain regions with regard to volumetric changes during aging, or even during development. For example, the cerebellum does not change in volume from middle to old age.

However, significant cerebellar volume decreases occur between young adulthood and middle age, with much less change taking place between middle and old age. Thus the volume shrinkage of the cerebellum occurs early in life and then slows down in the mid-50's in humans [24]. In addition, it has been observed that during development, some cortical regions exhibit a continuous increase in volume, whereas in the frontal and parietal cortices the increase is followed by a decrease in volume during the transition from adolescence to young adulthood [25]. These changes in brain structure are not accompanied, of course, by cognitive decline in young individuals thus revealing the complexity faced with trying to link brain imaging changes and the cause of alterations in neurological function. Other indices of brain activity besides volume changes, such as positron emission tomography (PET) of cerebral metabolic rate of glucose (CMR_{Gl}), might provide additional correlative measures of structure to function in the human brain.

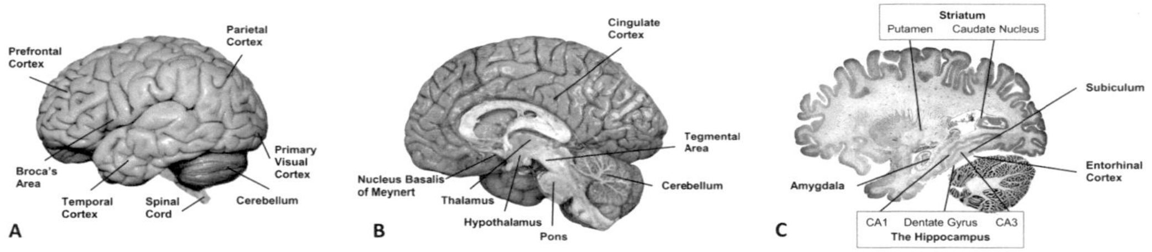

Figure 1: Human brain structures examined in this review. Different views are shown in three panels: (A) lateral view of the left hemisphere, (B) medial view of the right hemisphere, and (C) sagittal view of one hemisphere. (The image in Panel C is adapted with permission from http://www.brains.rad.msu.edu, supported by the US National Science Foundation).

The brain volumetric measurements as well as the PET measures of CMR_{Gl} reinforce the notion that aging affects some regions of the brain more than others. In addition to the volume changes in select brain regions of the brain during aging, it has been repeatedly shown that the frontal cortex shows the greatest and most consistent decrements in CMR_{Gl} as compared with all other regions in the cortex or subcortical components of the aging brain [26, 27]. These changes in metabolic activity in select cortical areas during aging are either related to altered neuronal expression of some key enzymes controlling the overall metabolic state of neurons and associated glial cells, or they are the result of altered activation of synapses or of the disruptive effects of abnormal neuronal excitability during aging. Gene expression analyses, in combination with neuro-imaging studies (including PET scan), as well as detailed microanatomical investigations, are providing new windows onto the molecular and cellular changes that might account for such differential patterns of neuronal susceptibility to the aging process and to age-related diseases. It should be emphasized, however, that neuronal losses during aging even in select, sensitive regions are relatively modest, whereas decreases in the number of synapses in the same regions appear to be a more prominent characteristic of brain aging. These observations have led to the assertion that most of the functional decline associated with normal aging is caused by relatively subtle changes, such as loss of dendrites, reductions in spine densities, altered spine morphologies, or changes in the molecular profile of synapses [21, 28-31].

2.2. Functional Genomic Studies on Brain Regions Most Vulnerable to Aging

2.2.1. Human Studies

In order to study how brain regions differentially respond to the stresses associated with increasing age, some investigators have used functional genomics approaches, though the number of these studies is still rather limited. Nevertheless, such studies have provided further support for the concept of regional heterogeneity with respect to the differential rates of aging-associated changes. An initial report of differential gene expression patterns in three human brain regions was that by Evans *et al.* [32] who examined gene expression in cerebellar cortex and two cerebral cortical regions (anterior cingulate cortex and dorsolateral prefrontal cortex). The microarray gene expression patterns show that the two regions of the cerebral cortex had similar levels of expression of most genes and that these two cortical regions differed significantly in terms of gene expression patterns from those of the cerebellar cortex [32]. More than one thousand transcripts were differentially expressed in the cortical *vs.* cerebellar regions and the most prominent ontological categories among the differentially expressed genes were those of signal transduction, neurogenesis, synaptic transmission, and transcription factor regulation. In another study of regional differences in gene expression in human brain, Khaitovich *et al.* [33] analyzed patterns of expression in six areas of the brain: cerebellum, caudate nucleus, dorsolateral prefrontal cortex, anterior cingulate cortex, primary visual cortex, and Broca's area. Consistent with the study by Evans *et al.* [32], Khaitovich *et al.* also found that while the four regions of the cerebral cortex are similar to each other, the overall transcriptomic profiles of the cerebral cortex, the caudate nucleus, and the cerebellum differ significantly from each other [33]. The gene ontology categories that differed most in terms of expression in the various brain regions were those of synaptic transmission, signal transduction, neurogenesis, neuronal development, and calcium ion [Ca^{2+}] regulation. Because the analyses of gene expression patterns in both the Evans *et al.* and Khaitovich *et al.* studies contained too few brain samples across the aging spectrum (45 to 88 years of age), Fraser *et al.* [34] conducted a meta-analysis on the results of these two studies using the aging-related pattern of changes in gene expression in the frontal pole of the human brain identified in the study by Lu *et al.* [18]. Using the 841 genes identified as showing a pattern of either increasing or decreasing expression with advancing age in the Lu *et al.* study, they calculated the correlation coefficients resulting from comparisons of aging-related profiles between two tissues (Spearman rank correlation, r, and significance *P* values), the frontal pole of the Lu *et al.* study and one of the tissues studied by Khaitovich *et al.* The investigators described a highly significant correlation between the frontal pole in the Lu *et al.* study and each of the four regions of the cerebral cortex in the Khaitovich *et al.* study (anterior cingulate cortex, Broca's area, prefrontal cortex, and primary visual cortex, $r>0.8$ and $p<0.02$), but a lack of correlation between the frontal pole and the two non-cerebral-cortical regions (caudate nucleus and cerebellum, $r<0.1$ and $p>0.4$) in the Khaitovich *et al.* study. The meta-analysis confirmed the heterogeneity in age-dependent genome-wide gene expression profiles across different brain regions and demonstrated that the expression patterns of the five cerebral cortical regions are quite similar to each other in terms of the rates of appearance of age-related changes, whereas the cerebellum and caudate nucleus respond to aging very differently. Further analyses of the difference between the cerebral cortical regions and cerebellum revealed that the number of genes whose expression pattern changes with advancing age is lowest in the cerebellum.

A more recent study traced gene expression changes in four regions of postmortem brains in 55 normal individuals aged 20-99 years [35]. Similar to the studies described above, GeneChip data from this study again showed the regional heterogeneity in gene expression patterns in human brain during the aging process. However, unlike previous studies that reported significantly higher numbers of genes whose expression was changing with advancing age in vulnerable brain regions as compared with less vulnerable regions, this study indicated that two regions that are vulnerable to aging and AD, the hippocampus and the entorhinal cortex, displayed fewer aging-responsive genes than the superior frontal gyrus and post-central gyrus, regions that are less vulnerable to age-associated neuronal loss. It is important to note, however, that the relative differences in age-related neuronal injury or loss in these neocortical and paleocortical regions are less pronounced than the differences between, for example, cerebral, cortical and cerebellar regions. Therefore, the results of the study by Berchtold *et al.* [35] may have identified more subtle differences among generally vulnerable brain regions.

Principal component analysis of gene expression in single neurons captured by laser capture microdissection (LCM) from the two vulnerable regions of hippocampus and entorhinal cortex revealed that these neurons expressed 102 genes at lower levels than did neurons in regions that are less susceptible to the aging process, such as the superior frontal gyrus and posterior cingulated cortex [36]. Beyond the fact that the results from this study confirm the differential pattern of gene expression across several brain regions, they also point to some specific genes whose relatively low level of expression in susceptible regions in the aged brain (mean age: 79.8 ± 9.1 years) may predispose these neurons to differential vulnerability to age and AD-associated injury and death. Among the top categories of genes that were under-expressed in neuronal populations in hippocampus and entorhinal cortex were those related to Ca^{2+} homeostasis, such as *Slc8A1* (solute carrier family 8—sodium/calcium exchanger—member 1), *Ppp2ca* (protein phosphatase 2, catalytic subunit, α isoform) and *Calm2* (calmodulin 2), and those associated with intracellular signaling.

2.2.2. Animal Studies

As described above, the aging process is associated with specific gene expression patterns in the human brain, but the genes expressed differentially in other species during the aging process differ from those detected in human brain even among close relatives of humans, such as the chimpanzee [34]. Despite the differences in specific gene expression patterns among various species, neurons in certain brain regions, such as the hippocampus CA1 field and the basal forebrain cholinergic neuronal region, are selectively vulnerable in all species [16, 37-41]. Using a rodent model of brain aging, Xu *et al.* [42] studied five central nervous system regions (cerebral cortex, hippocampus, striatum, cerebellum, and spinal cord) across three age groups, 6 (young), 16 (middle age), and 24 (old) month old animals. Each of these regions was shown to have a unique transcriptomic signature. Spinal cord has the most genes whose expression changes as a result of aging (600+), the cerebral cortex and hippocampus are next (each with 400+ genes), while the striatum and cerebellum have the fewest age-related changes in gene expression (with only 100+ differential genes). Although these gene expression patterns differ somewhat from those detected in the human brain, they nevertheless were consistent with the idea of heterogeneity in brain aging across different regions, and with the notion that certain regions, such as the cerebellum, exhibit the lowest levels of change in gene expression related to the aging process. Therefore, patterns of gene expression in other species besides humans may provide additional information about specific molecular changes that are related to the vulnerability of some cells to the aging process, even though the specific genes affected by aging may differ among animal species.

Functional categorization of aging-associated gene expression changes in rodent brain indicated that all regions show decreases in the expression of genes related to mitochondrial function with the decreases being more significant in the cerebral cortex, hippocampus, and spinal cord [42]. The transcriptomic data also match fairly well with the observed pattern of decreases with age in mitochondrial metabolic capacity of small to medium-sized mitochondria in human cerebellum [43]. The mitochondrial metabolic capacity decreases from young to middle age but stays stable from middle to old age. The transcriptomic studies in mice and the mitochondrial metabolism studies in human cerebellum might indicate that this region of the brain is relatively plastic with respect to regulation of gene transcription and metabolism before middle age but becomes relatively insensitive, transcriptionally and metabolically, to the aging process after middle age. The differential response of cerebellar neurons at middle age and old age might be due to the lower metabolic rate of cerebellar neurons as has been observed for both human and primate brains [44-46]. An associated effect derived from the low levels of mitochondrial metabolic activity in aging cerebellum would be a lowering of the levels of reactive oxygen species (ROS) and OS, as well as of DNA oxidative damage that results from elevations in ROS levels in neurons [47, 48]. Decreased levels of oxidatively modified DNA would protect cerebellar neurons from damage that might otherwise occur in this brain region.

Increased inflammation is another characteristic of brain aging [49]. This is especially evident in some brain regions that are prone to aging, such as the medial prefrontal cortex, in which immune response genes are significantly up-regulated [50]. To probe for the possible cellular source of this up-regulation, cells that are positive for a microglial marker CD11b (or ITGAM) were isolated from this region and subjected to further transcriptomic analysis by next-generation sequencing. The RNA-Seq approach showed that many of the up-regulated genes detected in the selected region were also enriched in this cell population. This suggests that much of the up-regulation in the expression of immune response genes occurs in microglial cells. Thus, the study of gene expression patterns in animal and human brain during aging has enhanced our understanding of the changes that occur at the genomic level, which might identify the molecular causes for the age-associated structural, biochemical, and physiological changes.

3. SELECTIVE NEURONAL VULNERABILITY IN ALZHEIMER'S DISEASE

3.1. Brain Regions that are Vulnerable to Alzheimer's Disease

AD is characterized by the accumulation of oligomeric or aggregated amyloid β peptides (Aβ) in senile plaques (SPs), fibrils of hyperphosphorylated tau protein in neurofibrillary tangles (NFTs), and more importantly, by the loss or injury of neurons in select brain regions [51]. The cerebral cortex, especially the association areas, and the hippocampus, are the predominant regions of AD pathology [4, 52, 53]. Among these regions, the most vulnerable neurons are those in layer II of the entorhinal cortex, the subiculum, and the CA1 region of the hippocampus; these neuronal populations consistently display high levels of NFTs and are first lost in the early phases of the disease [54-56]. With regard to the distribution of NFTs across the different regions of the cerebral cortex, neurons of the temporal cortex contain more NFTs than those (in decreasing order) of the frontal, parietal, occipital, and posterior cingulate cortex [57, 58]. Besides the neurons in the cerebral cortex and the hippocampus, many subcortical structures that have connections to the cerebral cortex also contain neurons with large amounts of NFTs, including the amygdala, nucleus basalis of Meynert, ventral tegmental area, dorsal raphe, locus coeruleus, olfactory bulb, and some midline thalamic and hypothalamic nuclei [59, 60]. In contrast to neurons in these regions that contain high levels of NFT's, primary sensory and motor neurons are largely spared in the disease [61, 62].

Among the most vulnerable neurons in AD are the large pyramidal neurons in the areas mentioned above, particularly association neurons with long projections [63, 64]. Examples of such neurons are those in the entorhinal cortex layer II which provide the major cortical input to the hippocampus through the perforant pathway, and those in the subiculum which represent the major output of the hippocampus to the prefrontal cortex and other brain regions [65]. The underlying factors for the vulnerability of these large-sized neurons are thought to be their high demand for energy that leads to high levels of oxidative phosphorylation. Oxidative phosphorylation depends on the activity of the electron transport chain (ETC) enzymes, and hyperactivation of these enzymes is known to lead to OS and may thus bring about protein, DNA, and lipid modifications that are injurious to these neurons. In addition, the relatively large surface area of these neurons may increase exposure to extracellular toxic agents while their long processes would put increased demands on axoplasmic transport over long distances. Finally, neurons with long processes also have high levels of neurofilaments whose protein subunits tend to form aggregates during cellular stress.

3.2. Functional Genomic Studies on Selective Neuronal Vulnerability in Alzheimer's Disease

Several investigators have explored the mechanisms of region-specific neurodegeneration in AD by performing functional genomic studies on the most vulnerable brain regions affected by the disease. A list of these studies (up to

March 2012) is presented in Table **1** for human subjects and Table **2** for animal models of AD. The studies summarized in these two tables vary in their experimental approaches and include transcriptomic studies of individual regions of the brain as well as cross-sectional studies comparing regions that exhibit differential neuronal vulnerabilities. As will be noted, some of the studies are based on an analysis of gene expression patterns in a whole brain region, whereas others represent high-resolution analyses of gene expression patterns in individual neurons within a given region. Despite these different experimental approaches, several common patterns are apparent, and these commonalities provide some mechanistic insights into the question why only certain regions of the brain or certain neuronal populations are most vulnerable in AD.

Table 1: Transcriptomic studies and gene categories that characterize selectively vulnerable regions in human AD brains.

Sample Source	Technology	Gene Categories with Altered Expression in AD	Refs.
I. Region-to-Region Comparative Studies			
Amygdala (Vulnerable) *vs.* Cerebellum and Parietal-Occipital Neocortex (Resistant)			
19 AD and 10 control brains (Demographic and clinical info not available)	GeneChip Human Exon 1.0 ST Arrays	Down-regulation in AD-vulnerable region of SorCS1, a SORL1 homolog involved in γ-secretase processing and trafficking APP	[141]
Amygdala and Cingulate Cortex (Vulnerable) *vs.* Striatum and Cerebellum (Resistant)			
6 AD (83.0±2.6 [mean±SD] yrs; 3Female+3Male; CERAD* diagn.), 9 controls (65.9±13.5yrs; 2F+7M)	InCyte Human Unigene Lifearray	Up-regulation of chronic inflammation and down-regulation of signal transduction and energy metabolism in vulnerable regions	[90]
Entorhinal Cortex (Vulnerable) *vs.* Dentate Gyrus (Resistant)			
6 AD and 6 controls (Demographic and clinical info not available)	GeneChip Human Genome HG-U133A Arrays	VPS35 and retromer complex down-regulation in brain regions selectively vulnerable to AD	[142]
Entorhinal Cortex (EC), Hippocampus (HIP), Middle Temporal Gyrus (MTG), and Posterior Cingulate Cortex (PCC) (Vulnerable) *vs.* Superior Frontal Gyrus (SFG) and Primary Visual Cortex (VCX) (Resistant)			
33 AD (79.9±6.9 yrs; 18F+15M; Braak III-VI), 14 controls (79.8±9.1 yrs; 4F+10M) (neurons obtained by LCM)	GeneChip Human Genome U133 Plus 2.0 Arrays	Synaptic transmission and synaptic vesicle transport down-regulated in EC and HIP. Microtubule-based movement and axon cargo transport decreased in MTG and PCC. Energy metabolism genes significantly reduced in all vulnerable regions compared to resistant regions	[67, 69]
Hippocampus (Vulnerable) *vs.* Parietal Cortex (Resistant)			
4 AD (81.5±7.1 yrs; 1F+3M; Braak III), 4 controls (87.8±4.7 yrs; 3F+1M; Braak I-II)	Incyte Human UniGem V Array	Calcineurin Aβ up-regulated in vulnerable region	[130]

Table 1: contd…

II. Single Vulnerable Region				
Cerebral Cortex				
Frontal Lobe				
	21 AD (77.2±9.6 yrs; 11F+10M; Braak IV-VI), 28 controls (78.1±8.7 yrs; 15F+13M; Braak 0-III)	Agilent 44K Whole Human Genome Arrays	Synaptic activity up-regulated at early pre-symptomatic stages, and down at late stages. Similar pattern for APP processing and Aβ clearance	[143]
	61 AD (78.0yrs [mean]; 31F+30M), 23 controls (73.0 yrs, no gender information)	Two Custom cDNA Arrays	Calcium dysregulation in AD	[76]
	6 AD (85±7 yrs), 6 controls (74±10 yrs)	Clontech Atlas Human 12K Arrays	Increased production of interferon-γ and up-regulation of interferon-induced genes in AD brain	[99]
	13 AD (82.8±6.2 yrs; 10F+3M), 11 controls (75.8±10.5 yrs; 5F+6M)	Affymetrix HuFL GeneChips	Down-regulation of synaptic vesicle trafficking in AD brains	[119]
	10 AD (85.3±5.1 yrs; 4F+6M), 14 non-demented controls with varied levels of NFT and/or SP (83.7±8.1 yrs; 7F+7M)	GeneChip Human Genome U95Av2 Arrays	Up-regulation of inflammation markers, including MHC I/II genes, CD14 and TLR2, in cases with mild/moderate dementia compared to controls	[105]
	1 AD (87 yrs; M), 5 controls (26.4 yrs [mean; range: 22-29 yrs]; all M)	RNA-Seq (Nexg-Gen Sequencing)	Synaptic transmission, neurotransmitter transport, response to metal ion, metal ion transport, regulation of synaptic plasticity, induction of apoptosis and axon transport significantly changed	[79]
Parietal Lobe				
	10 AD (82.3±6.7 yrs; 4F+6M; Braak V-VI), 6 demented but non-AD (70.7±9.0 yrs; 2F+4M), 10 non-demented controls (84.9±5.4 yrs; 7F+3M; Braak≤III)	Agilent 44K Whole Human Genome Arrays	Immune response, CNS development, and Aβ processing and disposition among the most changed in AD brains	[101]
Temporal Lobe				
	8 AD (72.6±9.1 yrs; 3F+5M; Braak V/VI (except 2 III/IV)), 8 controls (80.1±7.1 yrs; 3F+5M; Braak 0-II (except 1 III/IV))	GeneChip Human Exon 1.0 ST Arrays	Synaptic dysfunction, perturbed neurotransmission, and activation of neuroinflammation in AD brains	[103]
	5 AD (moderate dementia; CDR** 2), 6 controls (CDR 0)	Incyte Human UniGem V1 Arrays	Synaptic vesicle protein synapsin a-type isoform selectively decreased in vulnerable regions of AD brains	[118]
	22 AD (82.6±5.1 yrs; 15F+7M; Braak IV-VI), 9 controls (82.0±7.9 yrs; 5F+4M)	Illumina Sentrix HumanRef-8 BeadChips	Up-regulation of inflammation and transcription regulation related genes, down-regulation of genes involved in synaptic transmission and ion channels	[104]
	1 AD (87 yrs; M), 5 controls (26.4 yrs [mean; range: 22-29 yrs]; all M)	RNA-Seq (Nexg-Gen Sequencing)	Axon transport, neurotransmitter transport and induction of apoptosis significantly changed. Alternative splicing and differential promoter use of APOE detected	[79]
Entorhinal Cortex (Comparison of NFT-Containing *vs*. Adjacent Non-NFT-Containing Neurons)				

Table 1: contd…

	19 AD (84.7±7.5 yrs; 10F+9M; Braak III-IV), 14 non-demented controls (80.1±7.9 yrs; 6F±9M[sic]; Braak 0-II) (neurons obtained by LCM; paired analysis for NFT- and non-NFT neurons in AD cases)	GeneChip Human Genome HG-U133A and U133 Plus 2.0 Arrays	Multiple bio-processes dysfunctional in the process of NFT formation. Key up-regulated genes included APOJ (altering mitochondrial function and inducing apoptosis), TIMP3 (pro-apoptotic), and IRAK1 (inflammatory response)	[66]
Hippocampus				
Whole Hippocampus				
	2 AD (80,82 yrs; both M), 3 controls (58,68,78 yrs; all M)	GeneChip Human Genome HG-U133A	Down-regulation of genes related to energy metabolism in the AD-affected region	[93]
	AD (83.2±8.1 yrs; 10F+8M; CDR 3-5) and controls (78.5±14.7 yrs; 6F+5M; CDR 0)	Genome-wide cDNA Array	Impaired glucose/energy metabolism and attenuation of PGC-1α expression; hypothesized to lead to Aβ generation and amyloidogenesis	[94]
	10 AD (85.3±5.1 yrs; 4F+6M), 14 non-demented controls with varied levels of NFT and/or SP (83.7±8.1 yrs; 7F+7M)	GeneChip Human Genome U95Av2 Arrays	Up-regulation of inflammatory molecules, mostly MHC II, in AD patients with early stage dementia	[105]
Hippocampal CA1				
	22 AD (7 Incipient: 90.0±5.6 yrs, MMSE*** 20-26; 8 moderate: 83.4±3.1 yrs, MMSE 14-19; 7 severe: 84.0±10.6 yrs, MMSE<14), 9 controls (85.3±8.1 yrs; MMSE>25)	GeneChip Human Genome HG-U133A Arrays	Up-regulation of tumor suppressors, oligodendrocyte growth factors, protein kinase A modulators, cell adhesion, apoptosis, lipid metabolism, and inflammation; down-regulation of protein folding/metabolism/transport, energy metabolism and signaling in AD	[95]
	6 AD (70.3±3.3 yrs; 3F+3M; CDR 2-3), 6 controls (69.0±1.8 yrs; 3F+3M; CDR 0)	GeneChip Human Genome U95Av2 Arrays	Dysregulation of metal ion homeostasis; down-regulation of transcription factor signaling and neurotrophic support; upregulation of apoptotic and neuroinflammatory signaling in AD	[102]
	5 AD (77.2±7.4 yrs; 2F+3M), 5 controls (73.2±9.8 yrs; 2F+3M) (CA1 neurons obtained by LCM)	Incyte GDA and custom cDNA Arrays	Down-regulation of phosphatases/kinases, cytoskeletal proteins, synaptic proteins, glutamate receptors, and dopamine receptors in AD	[68]
Nucleus Basalis of Meynert				
	10 AD (84.5±4.0 yrs; 7F+3M; MMSE 14.8±8.6), 6 controls (81.5±7.2 yrs; 4F+2M; MMSE 27.4±1.5) (cholinergic neurons collected by single-cell aspiration)	Custom cDNA Arrays	Neurotrophin receptors, synaptic proteins and protein phosphatases down-regulated in AD brain	[120]

*CERAD: Consortium to Establish a Registry for Alzheimer's Disease; **CDR: clinical demention rating; ***MMSE: Mini-Mental State Examination.

Table 2: Transcriptomic studies that characterize selective vulnerability in brains from mouse models of AD.

Sample Source	Technology	Gene Categories with Altered Expression in AD	Refs
I. Region-to-Region Comparisons			
Hippocampus and Cortex (Vulnerable) *vs.* Cerebellum, Striatum and Brainstem (Resistant)			
4 APP+PS1 transgenic, 4 non-transgenic controls (17-18 mos littermates)	InCyte Rat LifeArray 1 & 2	Down-regulated memory consolidation and up-regulated inflammatory genes in transgenic mice	[98]
II. Single Vulnerable Region			
Cerebral Cortex			
Whole cerebral cortex			
10 Balb/c with icv inj. of aggregated Aβ25-35, 10 with saline (weight 22-24 g)	Atlas Mouse 1.2 Expression Arrays	Apoptosis, energy metabolism, calcium ion homeostasis dysfunction, cell adhesion and neuronal dystrophy in Aβ-treated mice	[91]
3 AD model genotypes: APP/PS-1$^{P264L/P264L}$(4 & 18 mos, n=3 each age), Tg2576/PS-1$^{P264L/P264L}$ & Tg2576/PS-1$^{P264L/+}$(2 & 12 mos, n=3 each age/genotype) and wild-type mice (n=3 corresponding to each genotype)	GeneChip Murine Genome U74A Arrays	Up-regulated in the 3 AD models: immune response, carbohydrate metabolism, and proteolysis. Down-regulated: pituitary adenylate cyclase-activating peptide (PACAP), brain-derived neurotrophic factor (BDNF), and insulin-like growth factor I receptor (IGF-IR)	[92]
Frontal Cortex			
3 PS1/PS2 double knockout mice and 3 wild-type controls (2, 4, 6, 7 & 8 mos; 3 animals pooled at each age)	GeneChip Mouse Genome 430 Arrays	Progressive, age-dependent up-regulation of neuroinflammation and cathepsin in transgenic mice	[100]
Hippocampus			
Whole Hippocampus			
5 APPsw transgenic mouse brains, 5 non-transgenic controls (12 mos littermates)	ABI Mouse Genome Survey Arrays	Up-regulated: signal transduction and protein binding; down-regulated: extracellular space, protein binding, cell communication, and transporter activity in transgenic mice	[132]
3 PS1/PS2 double knockout mice and 3 wild-type controls (2, 4, 6, 7 & 8 mos; 3 animals pooled at each age)	GeneChip Mouse Genome 430 Arrays	Progressive, age-dependent up-regulation of neuroinflammation and cathepsin in transgenic mice	[100]

3.2.1. Gene Expression Patterns in Different Brain Regions in AD

The region-specific studies cover most brain areas affected by AD, including the entorhinal cortex, hippocampus, temporal cortex, frontal cortex, parietal cortex, cingulate cortex, amygdala, and nucleus basalis of Meynert. Because of their significance in cognition and memory formation, gene expression patterns in the hippocampus and frontal cortex have been examined in several of these studies. An effective approach to study SNV in AD is to compare vulnerable regions with resistant regions. As listed on the top of Table **1**, a number of vulnerable *vs.* resistant regional comparisons were conducted, covering the majority of the vulnerable regions listed above. Regions of the brain populated with neurons that are resistant to AD-related damage included the striatum, visual cortex, cerebellum, and brainstem. However, many of the existing genomic expression studies shown in Table **1** were not direct, side-by-side comparisons of the patterns of gene expression in vulnerable *vs.* resistant regions. For many of these studies, only vulnerable regions were used and the corresponding analyses were focused on how the patterns observed in these vulnerable regions differed between brains from AD subjects and those from age-matched controls. Although the majority of the gene analyses were carried out with RNA isolated from dissected brain regions, some were done with RNA isolated from cells selected by LCM [66-69].

3.2.2. Methodological Considerations with Regard to Gene Expression Studies in AD

It is well known that within a brain region characterized by overall sensitivity to stress there are cells that are resistant to the stress stimuli. A good example of this is the high level of sensitivity to OS, ischemia-induced injury, or glutamate toxicity of pyramidal neurons in the CA1 subfield of the hippocampus whereas the proximal neurons in the CA3 subfield are resistant to the same adverse conditions [70-73]. With regard to AD-associated neuronal injury or loss, the hippocampus as a whole is usually considered as a vulnerable region but the CA1 pyramidal neurons are much more susceptible to the pathogenic condition in AD than are neurons in the CA3 subfield or in the dentate gyrus [8]. Therefore, studies of brain metabolism and or gene expression that examine regional brain differences are actually reporting on the average metabolic or transcriptomic activity of a region that is composed of both sensitive and resistant neurons.

Although it is ideal to conduct studies on SNV using only vulnerable neurons rather than all of the cells in a given region, the technical difficulties involved in collecting individual neurons still poses a considerable challenge to investigators. For this reason, most published genomic expression studies of SNV in AD are studies of whole brain regions. In such studies, unique markers of vulnerable neurons might be masked by signals from stress-resistant neurons, and differences between vulnerable and resistant neurons can be diluted by background noise from other cells in a given brain region, such as glial cells. But, at the single cell level, the ratio of genuine signals to false signals or background noise is much greater making the results more dependable. Recent developments in selective cell procurement technologies, such as LCM, have made high-resolution analyses a reality [74]. The low mRNA yield that accompanies the usually small number of cells captured by LCM can be remedied by linear amplification of the messenger molecules [75]. As shown in Table **1**, among existing studies, several groups employed the target cell capture approach and demonstrated the feasibility and power of gene expression analysis performed on a uniform population of cells [66-69].

The vast majority of functional genomic studies related to regional vulnerability in AD have used postmortem brain tissue from human AD subjects and the respective age-matched controls (Table **1**). In most of these studies, fewer than 10 human brains per group were analyzed. The largest population of AD cases reported thus far involved analysis of samples from 61 AD patients [76]. In this chapter, we have also examined the patterns of gene expression in experimental animals that are used as surrogates of human AD so that we may determine whether animals thought to represent models of AD exhibit similar patterns in terms of selectively vulnerable or resistant neuronal populations as do humans. The animal models used are, for the most part, transgenic mice that over-express one to three genes associated with familial cases of AD. These are the transgenic mice for amyloid precursor protein with Swedish mutation (APPsw) and those with APP plus presenilin 1 (PS1), as well as the double gene knockout mice for PS1 plus PS2. In addition, brain tissue from wild type mice injected with β-amyloid has been analyzed in order to determine the gene expression patterns in vulnerable *vs.* resistant neuronal populations. The findings from these animal studies are summarized in Table **2** and the observed gene expression patterns integrated into the overall pattern of SNV in AD.

Early transcriptomic studies of AD-vulnerable regions focused mainly on the identification of differentially expressed genes that investigators believed to be important for the disease process. Although attention to single gene changes may provide valuable information about potential targets for therapeutic intervention in AD, such an approach does not allow for the visualization of possible gene interactions and the effect that such interactions may have on disease-relevant molecular pathways in cells. To achieve an understanding of the disease process, a "systems approach" to analyzing transcriptomic data should provide a more comprehensive view of the disease process [77] and would, therefore, be the preferable approach for data analysis in such studies. Bioinformatic analyses of transcriptomic data now encompass the identification of genes, gene ontologies, bio-functions, biological pathways, and gene networks. More recent studies of SNV in AD brains have made use of gene ontological annotations, expert annotated and compiled bio-functions (such as those offered by Ingenuity Systems), and the currently known biological pathways, in order to identify key gene ontologies, bio-functions and pathways that can differentiate vulnerable from resistant neurons in AD. These higher-level analyses have provided more insight into the underlying mechanisms of SNV in AD than the early studies did.

With regard to genomic technologies that have been employed in the study of the transcriptomes of AD-vulnerable regions, the vast majority of the studies utilized the microarray approach. Various forms of microarrays were used in these studies, including oligomeric and cDNA arrays, one- and two-color arrays, customized and commercially available standard arrays (depending on probe type, detection mode, and the manner they are designed and produced, respectively). Despite the fact that these different forms of microarrays have their own unique characteristics and that different studies have their own experimental specificities, concordant results and common mechanistic themes have been reported from the studies presented here. This is especially true when the lists of identified genes are analyzed as a group in terms of gene ontologies, pathways, and gene networks (see next section for common gene functional categories or mechanistic factors).

As the cost of next-generation sequencing is decreasing rapidly and dramatically, the use of new sequencing-based approaches, such as RNA-Seq, has begun to emerge as a common approach for transcriptomic studies within the last few years [78]. And, very recently, we have seen the first report on the use of RNA-Seq in the study of two brain regions that are selectively vulnerable to AD, *i.e.*, frontal and temporal lobes [79]. Although the number of AD cases in this study was rather limited (with only one AD brain), the RNA-Seq approach uncovered many of the same gene expression patterns as those that had been reported in previous studies using the microarray approach.

3.2.3. Possible Common Mechanistic Factors Determining Neuronal Vulnerability in AD

It is usually not easy to determine whether the findings from a single transcriptomic study can be used in deriving general principles about the state of neuronal injury in brain. Added to this uncertainty are possibly confounding experimental parameters such as brain region, cellular resolution, brain tissue origin, and others already mentioned above. In an effort to reach the goal of understanding the key mechanistic factors underlying SNV in AD, we have collected data from existing studies and have attempted to determine if there are common mechanistic factors that have been identified by many of the studies we have examined. Shared key bio-functions and pathways identified in the studies we have summarized were used to arrive at possible mechanisms for differential vulnerability of neurons in AD. The factors we have identified represent the most prominent characteristics of AD-affected regions and are factors that were reproducibly observed in many studies, despite the use of different regions or diverse neuronal populations from the brains of either human subjects or animal models. The key biological functions and molecular pathways that seem to distinguish vulnerable from resistant neurons in AD are outlined below, and some of the key genes in these functions are presented in Table **3**.

Table 3: Functionally categorized key genes identified in transcriptomic studies as being characteristic of vulnerable regions in human AD or animal models of AD (Green: down-regulated genes; Red: up-regulated genes).

Genes	Selection Criteria	Protein Confirmation	Species	Study
Mitochondrial Metabolism and Energy Production				
COX4I1, TPI1, TPI1P1, LMO4, NDUFS5, LOC100130794, LOC652797, PKM2, ENO3, UQCRC2, OXCT1, SLC25A4	FC* > 2.0	nc**	Human	[90]
OXPHOS: ATP5A1, ATP5B, ATP5C1, ATP5G3, ATP5O, COX5B, NDUFA1, NDUFA10, NDUFA11, NDUFA13, NDUFB8, SDHB, UQCRC1, UQCRC2, *etc.* Mito membrane translocases: TIMM17A, TIMM23, TIMM50, TOMM22, TOMM34	P < 0.01 (FDR***)	Subunits of Complex I-V	Human	[67]
Glycolysis: GAPD, GPI, LDHA; TCA: ACLY, FH, MDH1, MDH2, SUCLA2; OXPHOS: ATP5B,COX4I1, COX6A1, NDUFS4, UQCRC2, SLC25A14	P < 0.05	nc	Human	[93]
Glycolysis: GAPD, GPI, PFKM, PGK1; TCA: ACLY, CS; OXPHOS: COX6C, MTATP6; Pyruvate metabolism: PC, PCK2, PDHA1, PDK3; Others: PPARGC1A (PGC-1α), UCP2	P ≤ 0.01	PPARGC1A (PGC-1α)	Human	[94]
ACAD8, ALDH2, COX7A2L, COX7C, HADH, NDUFB3, PDHA1, PDHX, SDHA,SLC25A6	P < 0.05 (FDR)	nc	Human	[95]
Carbohydrate metabolism: FUCA1, GUSB, MAN2B1		nc	Mouse	[92]
Neuroinflammatory Response				
C3, HLA-DRB5, CD74, CD99	FC > 2.0	nc	Human	[90]
IFITM3	FC > 5.0	nc	Human	[99]
CCL5, C4A, IL1F7, IL1R1, NFKB2, TGFB1, TGFB3, TNFSF7	FDR ≤ 10%		Human	[103]
CCL27, CCR3, CCR5, CXCR2 (IL-8Rβ), CXCR4, IL-28A	P < 0.02 (Permutation)	CCR3, CXCR4	Human	[101]
IFNA5, IFNG, IL10RA, IL10RB, IL13RA1, IL17R, IL18, IL2RG, IL6R	P < 0.05 (FDR)	nc	Human	[95]
APP, B94, CEX1, DPP1, HB15, HUMJE, IFNIND, IL1A (IL-1α), IL1B (IL-1β), NF-IL6, NFKBIA, PLA2G4A, PTGS2, RELB	FC ≥ 3.0& P ≤ 0.05	nc	Human	[102]
C1QA, C1QB, C4A, CBS, THRA, GFAP	FC > 1.4& P < 0.05	nc	Mouse	[98]
AKT1, CSF2RB2, GPI1	FC ≥ 2.0	nc	Mouse	[91]
C1QA, C4A, CD14, FCGR1, GFAP, IGH-6	FC ≥ 1.2& P ≤ 0.05	nc	Mouse	[92]
C1QA, C1QB, C1QC, C3AR1, C4B	FC > 2.0& P < 0.05	C1QA	Mouse	[100]
BLNK, CD14, CD53, CD37, CD74, CX3CR1, DRB1, FCGBP, FCGR1A, FCGR3A, FCGR3B, HLA-A, HLA-DPA1, HLA-DRA, HLA-DRB1, HLA-DRB3, ICAM2, IFI16, IFITM1, IL15RA, LCP1, LY96, OLR1, PPBP, TGFBR2, TLR2	DChip: P<0.05, FC >1.2 & CyberT: P<0.01, FC >1.2	HLA-DR+DP+DQ products	Human	[105]
IRAK1	Paired permutation P < 0.05	IRAK1	Human	[66]

Table 3: contd…

Synaptic Transmission and Synaptic Vesicle Transport				
CHRNB2, DRD1IP, GABRA1, GABRA4, GABRG2, GAD1, GAD2, GRIA2, GRIA3, GRIA4, GRIK1, GRIK4, GRINA, GRIN2A, GRIN2D, GRM4, HTR1B, HTR2A, HTR3B, NRN1, SNAP25, STX12, SV2A, SV2B, SYNJ1, SYT8, SYT11, SYT12, SYT13, VAMP1	FDR ≤ 10%	GAD65 (GAD2)	Human	[103]
DNM1, STX1A, SYT1	P < 0.05	DNM1	Human	[119]
SYN2(a)	FC ≥ 1.8	Synapsin Ia, IIa, IIIa	Human	[118]
CHAT, CSPG5, NFL, SLIT2, SYN1, SYP	FC ≥ 3.0& P ≤ 0.05	nc	Human	[102]
ARC, SLC6A3 (DAT), DRD1, DRD2, DRD3, DRD4, DRD5, GRIA1, GRIA2, GRIN2B, SYN1, SYP, SNCA, SNCB, SYTx	FC ≥ 2.0	nc	Human	[68]
SYP, SYTx	P ≤ 0.05	nc	Human	[120]
ARC, ATP1A3, EGR1 (Zif268), NR4A1 (Nur77/TR3)	FC > 1.4& P < 0.05	nc	Mouse	[98]
Calcium Regulation				
CAMK2A, CAMK2G, ITPKB, PPP2CA, PPP3CA, PVALB	FDR ≤ 10%	CAMKII, PP2AC (PPP2CA)	Human	[103]
PPP3CB (Calcineurin Aβ)	FC>2.0	nc	Human	[130]
ITPKB, RGS4	P < 0.05	nc	Human	[76]
CAST, DAPK2, EP300, S100A4	P < 0.05 (FDR)	nc	Human	[95]
Cytoskeletal Function				
VPS35	P < 0.01	VPS35	Human	[142]
NF-L, NF-M, NF-H, TUBB	FC ≥ 2.0	nc	Human	[68]
Signal Transduction				
AHNAK, CHL1, COPS8, ENSA, FKBP4, GSK3A, ITPR1, MAPK10, MAP2K1, MAP3K6, NOMO3,NOMO1,NOMO2, NPTN, NR4A1, PGRMC1, PIP4K2B, PPAP2C, PPP2CA, PRKACB, PRKCZ, PTK2, RAN, RCAN2, RGS5, RUNDC3A, SIRPA,YWHAB	FC > 2.0	nc	Human	[90]
AIM1, CDK2AP1, FZR1, GSK3B, SFRP1, TGFBR3	P < 0.05 (FDR)	nc	Human	[95]
AKT3, CDK5, GNA12, GNB5, GNG5, GSK3B, MAPK4, MAPK6, MAPK8, MAPK9, PLCB2, PPP3CB, PPP2R1B, PRKCB1, PRKCD, PRKCE, PRKCZ, RGS4	FDR ≤ 10%	AKT3, GSK3β	Human	[103]
ADAM9, CDP5, DAXX, FAS	FC ≥ 3.0& P ≤ 0.05	nc	Human	[102]
ERK1, ERK2, FAK, GSK3B, NOS2, PPP1CA, PPP1CC, PPP2R1A, PPP2CA	FC ≥ 2.0	nc	Human	[68]
PPP1CA, PPP1CC	P ≤ 0.05	nc	Human	[120]
AKT1, CDH2, ERBB2, FGFR1, FZD3, FZD7, ITGAL, ITGAM, JUN, LIMK1	FC ≥ 2.0	ITGAM	Mouse	[91]

Table 3: contd…

ADORA2A, ATP2A1, BMP2, CACNG8, CAMTA2, CAP1, CIDEA, DUSP9, GAL, GNG8, GPR27, GPR77, GPR151, HCN2, KCNA5, MAP3K8, OXT, PDE6G, PLA1A, PRKAG3, RBL1, S100A8, SHC1, SMAD4	FC ≥ 2.0& P < 0.05	nc	Mouse	[132]
Protein Folding and Proteolytic Activity				
AHSA1, CCT2, CCT3,CCT4, CCT5, DNAJA1, HSP90AB1, HSP90AA2, HSP90AA1, PDIA6	P < 0.05 (FDR)	nc	Human	[95]
CTSD	FC ≥ 2.0	nc	Human	[68]
CTSD	P ≤ 0.05	nc	Human	[120]
CTSD, CTSH, CTSS	FC ≥ 1.2& P ≤ 0.05	nc	Mouse	[92]
CTSC, CTSD, CTSH, CTSS, CTSZ	FC > 2.0& P < 0.05	CTSD	Mouse	[100]
CLU (APOJ), TIMP3	Paired permutation P < 0.05	CLU, TIMP3	Human	[66]
Neurotrophic Support				
BDNF, HER2B, P2XR	FC ≥ 3.0& P ≤ 0.05	nc	Human	[102]
NTRK2 (trkB), NTRK3 (trkC)	P ≤ 0.05	nc	Human	[120]
ADCYAP1 (PACAP), BDNF, IGF1R	FC ≥ 1.2& P ≤ 0.05	nc	Mouse	[92]

*FC: Fold Change; **nc: no confirmation; ***FDR: Benjamini and Hochberg False Discovery Rate.

a. *Decreases in Mitochondrial Metabolism and Energy Production.* Since the metabolic activity of the brain is very high, accounting for an estimated 20-25% of glucose and oxygen utilization in the body, it is not surprising that reduced metabolism and energy production might be significantly correlated with cognitive decline in AD [80, 81]. The CMR_{Gl} is significantly lower in AD-affected brain regions when compared with the same brain regions of age-matched healthy controls [82-85]. On the other hand, the same measurements of the metabolic rate in regions spared by the disease process are equivalent to those in the brains of control subjects. Decreases in CMR_{Gl} in AD have been reported for regions that are most vulnerable as defined by neuropathological changes, including those of the frontal, parietal, temporal, and posterior cingulate cortex, as well as the hippocampus and entorhinal cortex [85-87]. It is also highly significant that the regions that tend to accumulate the highest amounts of Aβ in AD, or in cognitively normal individuals with amyloid plaques in their brain, differ from other brain regions in terms of the overall levels of glycolytic activity [88, 89]. Glucose metabolism in brain is generally through the pathway that includes oxidative phosphorylation in mitochondria but, apparently, in the most vulnerable regions of the brain, glucose metabolism is through the glycolytic pathway, a pathway that leads to more rapid utilization of glucose but less ATP generation. Thus, differences in energy metabolism and glucose utilization in brain are emerging as prominent features of an AD brain as compared with non-AD brains. As shown in Tables **1** and **2**, the down-regulation of genes related to metabolic homeostasis and energy production is a recurring theme in AD brains [67, 69, 90-95]. For example, Liang *et al.* [67] compared expression of 80 metabolically relevant genes in brain regions differentially affected by AD. Compared with controls, vulnerable regions in brains of AD cases exhibited significant decreases in the expression of many genes related to cellular metabolism. In the posterior cingulate cortex, 70% of the nuclear genes that encode mitochondrial ETC subunits were expressed at significantly lower levels in the AD cases compared with controls. In middle temporal gyrus and hippocampal CA1 subfield, expression levels of 61% to 65% of these genes were significantly lower in AD brains *vs*. those in age-matched controls. In comparison, in a region that is relatively spared by AD, *i.e.*, the visual cortex, only 16% of the genes were expressed at lower levels than in controls. This detailed study, along with many others, demonstrates that there is a decline in mitochondrial energy metabolism in the AD brain, and that the regions that suffer the greatest decline in genes related to cell metabolism and energy production are those that are most vulnerable to the disease process in AD. Among the mitochondrial ETC proteins whose genes are down-regulated in AD, the genes *Cox4* (Cytochrome c oxidase, subunit IV) and *Uqcrc2* (Cytochrome bc-1 complex core protein) are the most consistently down-regulated genes. It is important to point out that in addition to the lower expression levels of genes related to ETC enzymes in vulnerable neurons in the AD brain, those for subunits of the translocases of the outer and inner mitochondrial membranes (*Tomm* and *Timm*) are also down-regulated in these regions [96]. These translocases play an important role in the import of mitochondria-targeted proteins from the cytoplasm through the outer and inner mitochondrial membrane. Included in the group of proteins transported into mitochondria are many of the ETC enzymes. To add to the importance of TOMM proteins in AD, new polymorphisms associated with AD (multiple T bases) have now been identified in an intron of the gene for one of the subunits of the TOMM complex, *Tomm40* [97]. This genetic variation leads to the development of late onset AD at an earlier age than most such cases of late onset AD.

b. *Increases in Gene Expression of Inflammatory Response Genes.* In many of the studies listed in Table **1** [90, 92, 95, 98-105], inflammatory responses are among the most pronounced bio-functions or pathways associated with up-regulated genes in affected AD neurons. Aβ aggregation and deposition in AD brains is known to be neurotoxic and to elicit the production of cytokines, chemokines, and other neuroinflammatory mediators from microglia, the immune system cells of the brain [106, 107]. Up-regulated genes involved in this response, as uncovered by functional genomic studies, include: 1) cytokines (such as interleukin[IL]-1α, IL-1β, IL1F7, interferon α5, interferon γ, interferon λ2, transforming growth factors β1 and β3, and tumor necrosis factor α); 2) chemokines (such as CCL5 and CCL27); 3) cytokine/chemokine receptors (such as chemokine C-X-C motif receptor 2 [CXCR2], CXCR4, CCR3, IL-1 receptor type I, IL-2 receptor γ, IL-10 receptor α, IL-10 receptor β, IL-13 receptor α1, and receptors of IL-6, IL-17, and IL-18); 4) proinflammatory transcription factors (such as NF-IL6 and NF-κB 2) and other pro-inflammatory proteins (such as interferon-induced protein 3 or IFITM3). Not all of these up-regulated genes (or their products) are pro-inflammatory, of course, since some of the gene products, such as those of the IL-10 receptor, are anti-inflammatory [108]. Overall, these findings are consistent with other experimental evidence showing that in AD brain, inflammation is particularly localized in regions most affected by the disease [109-116]. The up-regulation of some anti-inflammatory genes may be indicative of a compensatory response to inflammation in AD brains. Based on the fact that increased expression of inflammatory response genes is primarily focused on the AD-vulnerable regions, it is highly likely that neuroinflammation is a significant contributor to the pathogenic cascade that is activated in AD [117]. Although the activation of inflammatory responses in AD-affected regions might be neuroprotective at the early stage of the disease by controlling and minimizing the damaging effects of Aβ, chronic inflammation and long exposure to inflammatory response effectors might well be detrimental to neurons and glial cells in later stages of the disease.

c. *Dysfunction of Synaptic Transmission and Synaptic Vesicle Transport.* The third common transcriptional profile identified in many of the studies is the alteration in levels of gene expression related to synaptic activity in vulnerable neurons [67-69, 103, 104, 118-120]. Since synaptic activity is fundamental to memory formation and cognitive function, the region-specific decreases in gene expression related to synaptic neurotransmission uncovered in these studies can be viewed as a neurobiological correlate of AD-associated cognitive decline. Specifically, these studies demonstrate that the expression of many important genes for synaptic transmission or vesicular transport are significantly down-regulated, including the genes for neuritin, synapsin I, synapsin II, synaptobrevin, synaptojanin, synaptophysin, synaptopodin, synaptotagmin, syntaxin, α-synuclein, and β-synuclein. The down-regulation in sensitive regions of genes related to synaptic vesicle docking and recycling at the synapse, such as dynamin I, is also likely to be linked to the decline in cognitive function in AD, because of the ensuing decreases in neurotransmitter release at synapses. This observation suggests a selective weakening or loss of neuronal connections in AD-affected brain regions, and may help explain the decreases in synaptic density observed by quantitative morphometric analyses performed on brain regions vulnerable to AD-induced damage [121-123]. Biochemical analyses of these protein products are in agreement with the transcriptional profile described above [124-126]. Region-specific dysfunction in synaptic transmission may result not only from the down-regulation of the genes described above, but also from direct effects of Aβ on synaptic macromolecules [127, 128], or from the effects of products of chronic inflammatory response, such as tumor necrosis factor [129], or from other factors some of which are discussed below.

d. *Calcium Regulation, Cytoskeletal Function, Signal Transduction, and Proteolytic Activity.* Additional transcriptomic alterations that might also play a role in AD-associated SNV were identified in the functional genomic studies. These include: 1) Up-regulation in AD-vulnerable regions of genes related to Ca^{2+} homeostasis and signaling pathways, such as the inositol trisphosphate (IP_3)-receptor kinase and the Ca^{2+}-regulated phosphatase calcineurin (*Ppp3c*), as well as the down-regulation of genes encoding Ca^{2+} binding proteins [76, 90, 91, 100, 130]. The decrease in Ca^{2+} binding proteins results in gradual loss of Ca^{2+} buffering capacity, which leads to pathological increases in intracellular Ca^{2+} in vulnerable neurons, thereby enhancing NFT formation and neurodegeneration [131]; 2) Down-regulation of the expression of genes for cytoskeletal proteins, *e.g.*, β-tubulin and neurofilament subunits *Nf-L, Nf-M, Nf-H* and *Nf-66*, as well as those for some axonal cargo transporters in AD brains [68, 90, 95]; 3) Down-regulation in AD brains of genes involved in signal transduction pathways, such as G-protein-coupled receptor (GPCR) signaling [90, 95, 132]; 4) Down-regulation of genes for neurotrophic factors and their receptors, such as *Bdnf* (brain-derived neurotrophic factor) and the receptors *TrkB* and *TrkC* [92, 102, 120]; and 5) Up-regulation of the genes for cathepsin D, H, S and Y, a family of endolysosomal proteases involved in the proteolytic cleavage of APP and the formation of Aβ [68, 92, 98, 100, 120].

4. CONCLUDING REMARKS

In the preceding sections we have presented information from microanatomical, brain imaging, functional imaging, and functional genomic studies in order to begin to define the intrinsic characteristics of selectively vulnerable neurons in the aging brain and in AD. Potential candidate neurochemical processes have been suggested in the past as possible determinants of differential vulnerability, especially the expression of high levels of neurofilament proteins in vulnerable neurons and of calcium-binding proteins in resistant neurons [6]. However, the information gleaned from those studies, although accurate, was relatively limited when compared with the large data sets available to us today as a result of whole genome expression studies. The key functional gene categories that have emerged from our analysis of studies that distinguish vulnerable from resistant neurons are summarized in Fig. (**2**). The summary shown in Fig. (**2**) highlights what appear to be intrinsic differences between vulnerable and resistant neurons in terms of gene expression, regardless of the site of their location in brain.

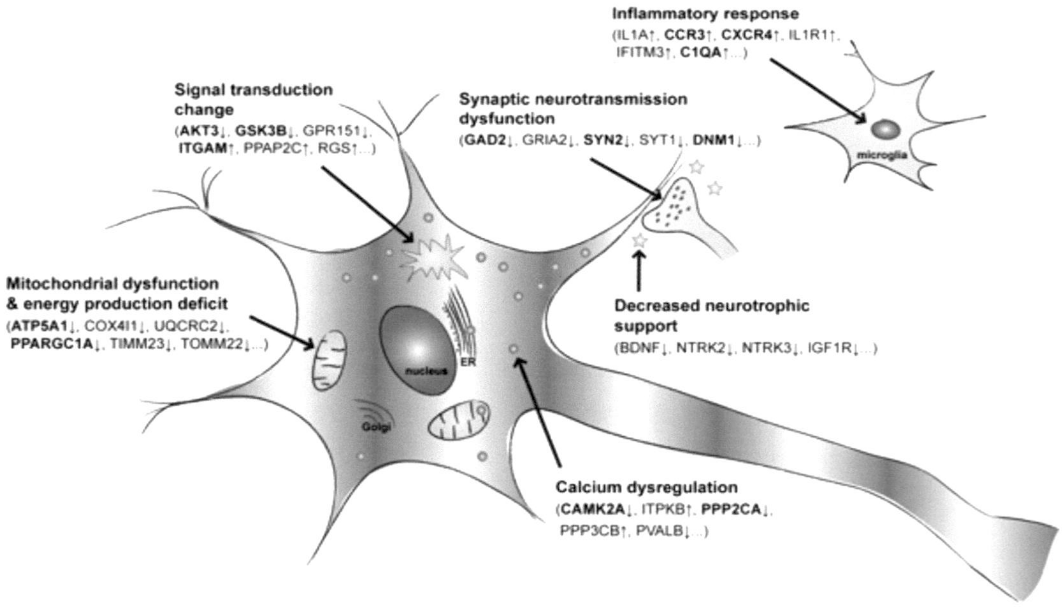

Figure 2: Major functional categories of genes whose expression is altered in neurons frequently damaged in AD. Genes in parentheses are some of the most representative and those that are most frequently detected as being over- or under-expressed in AD brains in more than one study (see Table **3**). The genes in boldface type are those for which the respective protein levels have been

confirmed as ones that change in AD in the same direction as the gene expression levels. The arrows after each gene represent the direction of their respective change in vulnerable regions during AD pathogenesis. Some of the major gene expression patterns characteristic of vulnerable neurons in AD are also shared by neurons that are susceptible to the effects of the aging process, as well as neurons sensitive to the damaging effects of oxidative stress (see text).

The summary of studies at the genomic level presented in this chapter, points to some commonalities between aging and AD, the primary ones being the decline in energy metabolism, the altered Ca^{2+} homeostasis and signaling, the decreases in synaptic neurotransmission, and the enhanced inflammatory responses. The common genomic differences between vulnerable and resistant neurons in both aging and AD brains fit quite well within the hypothetical schemes already developed in the literature to account for selective neuronal losses in both processes, *i.e.*, the mitochondrial cascade hypothesis [133], the Ca^{2+} hypothesis of aging and AD [134, 135], the synaptic disruption hypothesis of AD and aging [136, 137], and the neuro-inflammation hypothesis of aging and AD [138]. The genomic commonalities discussed here between normal aging and AD are consistent with findings from other studies that examined transcriptional overlap between these two states [104, 139], and more importantly, with the fact that advanced age is the major risk factor for AD.

Several of the functional categories of genes shown in Fig. (**2**), are also differentially expressed in neurons that are vulnerable to OS, regardless of their location in the brain [71, 72]. Included among these categories are those of enhanced immune/inflammatory response, decreased energy generation, reduced signal transduction, and dysregulation of Ca^{2+} signaling. The oxidative stress hypothesis of aging is one of the predominant theories of age-associated cell damage and death in many organs, including the brain [140]. Thus, the functional genomics approach to defining some of the molecular characteristics of neuronal vulnerability to aging and AD has confirmed most of the major hypotheses of molecular and cellular changes proposed as key processes leading to neurodegeneration in aging and AD. What the summary of the functional genomic studies of brain in aging and AD have offered beyond the identification of potential pathways to neuronal vulnerability are key genes and gene products related to these pathways and whose expression is intrinsically different in vulnerable *vs*. resistant neurons. The identification of specific genes in these pathways that may determine vulnerability of neurons can be used future for drug development to prevent the loss of neurons in neurodegenerative diseases such as AD. For example, the development of drugs against specific targets that control calcium signaling or against some of the inflammatory response factors may provide the means to protect selectively vulnerable neurons from further damage in AD. Mitochondrial metabolic dysfunction, either because of down-regulation of the expression of ETC enzymes or mitochondrial membrane translocases, such as TOMM40, may also become an attractive target for the development of agent that delay the stress of the aging process or the initiation and continuation of the damage induced in AD-sensitive neuronal populations.

ACKNOWLEDGEMENTS

The work was supported by grants from the National Institute on Aging (NIA), AG12993, AG035982 and AG025350, the Eunice Kennedy Shriver National Institute of Child Health and Human Development (NICHD), HD02528, the National Center for Research Resources (NCRR), RR-P20 17708, the Kansas Technology Enterprise Corporation, and the Miller-Hedwig-Wilbur Fund.

CONFLICT OF INTEREST

The authors confirm that this chapter contents have no conflict of interest.

DISCLOSURE

The chapter submitted for Advances in Genome Science, Vol. 3 entitled **"Functional genomics of brain aging and Alzheimer's disease: focus on selective neuronal vulnerability by X. Wang"** is an update of our article published in Current Genomics, 2010, 11: 618-633.

ABBREVIATIONS

Aβ = Amyloid β peptide

AD = Alzheimer's disease

APP = amyloid precursor protein

CDR = clinical demention rating

CERAD = Consortium to Establish a Registry for Alzheimer's Disease

CMR_{Gl} = cerebral metabolic rate of glucose

CNS = central nervous system

ETC = electron transport chain

FC = fold change

FDR = Benjamini and Hochberg false discovery rate

LCM = laser capture microdissection

MMSE = Mini-Mental State Examination

nc = no confirmation

NFT = neurofibrillary tangle

OS = oxidative stress

PET = positron emission tomography

PS1 = presenilin 1

ROS = reactive oxygen species

SNV = selective neuronal vulnerability.

REFERENCES

[1] Mueller, S.G.; Stables, L.; Du, A.T.; Schuff, N.; Truran, D.; Cashdollar, N., *et al.* Measurement of hippocampal subfields and age-related changes with high resolution MRI at 4T. *Neurobiol. Aging,* **2007**, *28*(5), 719-726.

[2] O'Banion, M.K.; Coleman, P.D.; Callahan, L.M. Regional neuronal loss in aging and Alzheimer's disease: a brief review. *Semin. Neurosci.,* **1994**, *6*, 307-314.

[3] Giannakopoulos, P.; Gold, G.; Michel, J.P.; Bouras, C. Cellular vulnerability in brain aging and Alzheimer's disease. Clinical correlates and molecular background. *Ann. Med. Interne (Paris).* **1998**, *149*(4), 187-191.

[4] Morrison, J.H.; Hof, P.R. Life and death of neurons in the aging brain. *Science,* **1997**, *278*(5337), 412-419.

[5] Terry, R.D.; DeTeresa, R.; Hansen, L.A. Neocortical cell counts in normal human adult aging. *Ann Neurol,* **1987**, *21*(6), 530-539.

[6] Morrison, B.M.; Hof, P.R.; Morrison, J.H. Determinants of neuronal vulnerability in neurodegenerative diseases. *Ann Neurol,* **1998**, *44*(3 Suppl 1), S32-44.

[7] Morrison, J.H.; Hof, P.R. Life and death of neurons in the aging cerebral cortex. *Int Rev Neurobiol,* **2007**, *81*, 41-57.

[8] Braak, H.; Braak, E. Neuropathological stageing of Alzheimer-related changes. *Acta Neuropathol,* **1991**, *82*(4), 239-259.

[9] Braak, H.; Braak, E. The human entorhinal cortex: normal morphology and lamina-specific pathology in various diseases. *Neurosci Res,* **1992**, *15*(1-2), 6-31.

[10] Price, D.L.; Cork, L.C.; Struble, R.G.; Kitt, C.A.; Walker, L.C.; Powers, R.E., *et al.* Dysfunction and death of neurons in human degenerative neurological diseases and in animal models. *Ciba Found Symp,* **1987**, *126*, 30-48.

[11] Hirsch, E.; Graybiel, A.M.; Agid, Y.A. Melanized dopaminergic neurons are differentially susceptible to degeneration in Parkinson's disease. *Nature,* **1988**, *334*(6180), 345-348.

[12] Rowland, L.P.; Shneider, N.A. Amyotrophic lateral sclerosis. *N. Engl. J. Med.,* **2001**, *344*(22), 1688-1700.

[13] Schmidt-Kastner, R.; Freund, T.F. Selective vulnerability of the hippocampus in brain ischemia. *Neuroscience,* **1991**, *40*(3), 599-636.

[14] Mathern, G.W.; Kuhlman, P.A.; Mendoza, D.; Pretorius, J.K. Human fascia dentata anatomy and hippocampal neuron densities differ depending on the epileptic syndrome and age at first seizure. *J. Neuropathol. Exp. Neurol.,* **1997**, *56*(2), 199-212.

[15] Wang, X.; Pal, R.; Chen, X.W.; Limpeanchob, N.; Kumar, K.N.; Michaelis, E.K. High intrinsic oxidative stress may underlie selective vulnerability of the hippocampal CA1 region. *Brain Res.,* **2005**, *140*(1-2), 120-126.

[16] Wilde, G.J.; Pringle, A.K.; Wright, P.; Iannotti, F. Differential vulnerability of the CA1 and CA3 subfields of the hippocampus to superoxide and hydroxyl radicals *in vitro*. *J. Neurochem.,* **1997**, *69*(2), 883-886.

[17] Bohr, V.A.; Ottersen, O.P.; Tonjum, T. Genome instability and DNA repair in brain, ageing and neurological disease. *Neuroscience,* **2007**, *145*(4), 1183-1186.

[18] Lu, T.; Pan, Y.; Kao, S.Y.; Li, C.; Kohane, I.; Chan, J., *et al.* Gene regulation and DNA damage in the ageing human brain. *Nature,* **2004**, *429*(6994), 883-891.

[19] Busuttil, R.; Bahar, R.; Vijg, J. Genome dynamics and transcriptional deregulation in aging. *Neuroscience,* **2007**, *145*(4), 1341-1347.

[20] Crivello, N.A.; Rosenberg, I.H.; Shukitt-Hale, B.; Bielinski, D.; Dallal, G.E.; Joseph, J.A. Aging modifies brain region-specific vulnerability to experimental oxidative stress induced by low dose hydrogen peroxide. *Age (Dordr),* **2007**, *29*(4), 191-203.

[21] Morrison, J.H.; Hof, P.R. Selective vulnerability of corticocortical and hippocampal circuits in aging and Alzheimer's disease. *Prog. Brain Res.,* **2002**, *136*, 467-486.

[22] Raz, N.; Rodrigue, K.M. Differential aging of the brain: patterns, cognitive correlates and modifiers. *Neurosci. Biobehav. Rev.,* **2006**, *30*(6), 730-748.

[23] Raz, N.; Rodrigue, K.M.; Haacke, E.M. Brain aging and its modifiers: insights from *in vivo* neuromorphometry and susceptibility weighted imaging. *Ann. N. Y. Acad. Sci.,* **2007**, *1097*, 84-93.

[24] Raz, N.; Lindenberger, U.; Rodrigue, K.M.; Kennedy, K.M.; Head, D.; Williamson, A., *et al.* Regional brain changes in aging healthy adults: general trends, individual differences and modifiers. *Cereb. Cortex,* **2005**, *15*(11), 1676-1689.

[25] Giedd, J.N.; Blumenthal, J.; Jeffries, N.O.; Castellanos, F.X.; Liu, H.; Zijdenbos, A., *et al.* Brain development during childhood and adolescence: a longitudinal MRI study. *Nat. Neurosci.,* **1999**, *2*(10), 861-863.

[26] Loessner, A.; Alavi, A.; Lewandrowski, K.U.; Mozley, D.; Souder, E.; Gur, R.E. Regional cerebral function determined by FDG-PET in healthy volunteers: normal patterns and changes with age. *J Nucl Med,* **1995**, *36*(7), 1141-1149.

[27] Reiman, E.M.; Chen, K.; Langbaum, J.B.S.; Lee, W.; Reschke, C.; Bandy, D., *et al.* Higher serum total cholesterol levels in late middle age are associated with glucose hypometabolism in brain regions affected by Alzheimer's disease and normal aging. *Neuroimage,* **2010**, *49*(1), 169-176.

[28] Burke, S.N.; Barnes, C.A. Neural plasticity in the ageing brain. *Nat Rev Neurosci,* **2006**, *7*(1), 30-40.

[29] Rapp, P.R.; Gallagher, M. Preserved neuron number in the hippocampus of aged rats with spatial learning deficits. *Proc. Natl. Acad. Sci. U. S. A.,* **1996**, *93*(18), 9926-9930.

[30] West, M.J.; Coleman, P.D.; Flood, D.G.; Troncoso, J.C. Differences in the pattern of hippocampal neuronal loss in normal ageing and Alzheimer's disease. *Lancet,* **1994**, *344*(8925), 769-772.

[31] Mattson, M.P.; Magnus, T. Ageing and neuronal vulnerability. *Nat Rev Neurosci,* **2006**, *7*(4), 278-294.

[32] Evans, S.J.; Choudary, P.V.; Vawter, M.P.; Li, J.; Meador-Woodruff, J.H.; Lopez, J.F., *et al.* DNA microarray analysis of functionally discrete human brain regions reveals divergent transcriptional profiles. *Neurobiol. Dis.,* **2003**, *14*(2), 240-250.

[33] Khaitovich, P.; Muetzel, B.; She, X.; Lachmann, M.; Hellmann, I.; Dietzsch, J., *et al.* Regional patterns of gene expression in human and chimpanzee brains. *Genome Res.,* **2004**, *14*(8), 1462-1473.

[34] Fraser, H.B.; Khaitovich, P.; Plotkin, J.B.; Paabo, S.; Eisen, M.B. Aging and gene expression in the primate brain. *PLoS Biol,* **2005**, *3*(9), e274.

[35] Berchtold, N.C.; Cribbs, D.H.; Coleman, P.D.; Rogers, J.; Head, E.; Kim, R., *et al.* Gene expression changes in the course of normal brain aging are sexually dimorphic. *Proc. Natl. Acad. Sci. U. S. A.,* **2008**, *105*(40), 15605-15610.

[36] Liang, W.S.; Dunckley, T.; Beach, T.G.; Grover, A.; Mastroeni, D.; Walker, D.G., *et al.* Gene expression profiles in anatomically and functionally distinct regions of the normal aged human brain. *Physiol Genomics,* **2007**, *28*(3), 311-322.

[37] Datson, N.A.; Morsink, M.C.; Steenbergen, P.J.; Aubert, Y.; Schlumbohm, C.; Fuchs, E., *et al.* A molecular blueprint of gene expression in hippocampal subregions CA1, CA3, and DG is conserved in the brain of the common marmoset. *Hippocampus,* **2009**, *19*(8), 739-752.

[38] Cowen, T. Selective vulnerability in adult and ageing mammalian neurons. *Autonomic Neuroscience,* **2002**, *96*(1), 20.

[39] Haces, M.L.; Montiel, T.; Massieu, L. Selective vulnerability of brain regions to oxidative stress in a non-coma model of insulin-induced hypoglycemia. *Neuroscience,* **2010**, *165*(1), 28-38.

[40] Kadar, T.; Dachir, S.; Shukitt-Hale, B.; Levy, A. Sub-regional hippocampal vulnerability in various animal models leading to cognitive dysfunction. *J. Neural Transm.,* **1998**, *105*(8-9), 987-1004.

[41] Wu, C.K.; Nagykery, N.; Hersh, L.B.; Scinto, L.F.; Geula, C. Selective age-related loss of calbindin-D28k from basal forebrain cholinergic neurons in the common marmoset (Callithrix jacchus). *Neuroscience,* **2003**, *120*(1), 249-259.

[42] Xu, X.; Zhan, M.; Duan, W.; Prabhu, V.; Brenneman, R.; Wood, W., *et al.* Gene expression atlas of the mouse central nervous system: impact and interactions of age, energy intake and gender. *Genome Biol,* **2007**, *8*(11), R234.

[43] Bertoni-Freddari, C.; Fattoretti, P.; Giorgetti, B.; Spazzafumo, L.; Solazzi, M.; Balietti, M. Age-related decline in metabolic competence of small and medium-sized synaptic mitochondria. *Naturwissenschaften,* **2005**, *92*(2), 82-85.

[44] Bentourkia, M.; Bol, A.; Ivanoiu, A.; Labar, D.; Sibomana, M.; Coppens, A., *et al.* Comparison of regional cerebral blood flow and glucose metabolism in the normal brain: effect of aging. *J. Neurol. Sci.,* **2000**, *181*(1-2), 19-28.

[45] Noda, A.; Ohba, H.; Kakiuchi, T.; Futatsubashi, M.; Tsukada, H.; Nishimura, S. Age-related changes in cerebral blood flow and glucose metabolism in conscious rhesus monkeys. *Brain Res.,* **2002**, *936*(1-2), 76-81.

[46] Sakamoto, S.; Ishii, K. Low cerebral glucose extraction rates in the human medial temporal cortex and cerebellum. *J. Neurol. Sci.,* **2000**, *172*(1), 41-48.

[47] Corral-Debrinski, M.; Horton, T.; Lott, M.T.; Shoffner, J.M.; Beal, M.F.; Wallace, D.C. Mitochondrial DNA deletions in human brain: regional variability and increase with advanced age. *Nat. Genet.,* **1992**, *2*(4), 324-329.

[48] Mecocci, P.; MacGarvey, U.; Kaufman, A.E.; Koontz, D.; Shoffner, J.M.; Wallace, D.C., *et al.* Oxidative damage to mitochondrial DNA shows marked age-dependent increases in human brain. *Ann. Neurol.,* **1993**, *34*(4), 609-616.

[49] Chung, H.Y.; Cesari, M.; Anton, S.; Marzetti, E.; Giovannini, S.; Seo, A.Y., *et al.* Molecular inflammation: underpinnings of aging and age-related diseases. *Ageing research reviews,* **2009**, *8*(1), 18-30.

[50] Bordner, K.A.; Kitchen, R.R.; Carlyle, B.; George, E.D.; Mahajan, M.C.; Mane, S.M., *et al.* Parallel declines in cognition, motivation, and locomotion in aging mice: association with immune gene upregulation in the medial prefrontal cortex. *Experimental gerontology,* **2011**, *46*(8), 643-659.

[51] Minati, L.; Edginton, T.; Bruzzone, M.G.; Giaccone, G. Current concepts in Alzheimer's disease: a multidisciplinary review. *Am J Alzheimers Dis Other Demen,* **2009**, *24*(2), 95-121.

[52] Gomez-Isla, T.; Price, J.L.; McKeel, D.W., Jr.; Morris, J.C.; Growdon, J.H.; Hyman, B.T. Profound loss of layer II entorhinal cortex neurons occurs in very mild Alzheimer's disease. *J. Neurosci.,* **1996**, *16*(14), 4491-4500.

[53] Gomez-Isla, T.; Hollister, R.; West, H.; Mui, S.; Growdon, J.H.; Petersen, R.C., *et al.* Neuronal loss correlates with but exceeds neurofibrillary tangles in Alzheimer's disease. *Ann. Neurol.,* **1997**, *41*(1), 17-24.

[54] Hyman, B.T.; Van Hoesen, G.W.; Damasio, A.R.; Barnes, C.L. Alzheimer's disease: cell-specific pathology isolates the hippocampal formation. *Science,* **1984**, *225*(4667), 1168-1170.

[55] Price, J.L.; Ko, A.I.; Wade, M.J.; Tsou, S.K.; McKeel, D.W.; Morris, J.C. Neuron number in the entorhinal cortex and CA1 in preclinical Alzheimer disease. *Arch. Neurol.,* **2001**, *58*(9), 1395-1402.

[56] Kordower, J.H.; Chu, Y.; Stebbins, G.T.; DeKosky, S.T.; Cochran, E.J.; Bennett, D., *et al.* Loss and atrophy of layer II entorhinal cortex neurons in elderly people with mild cognitive impairment. *Ann. Neurol.,* **2001**, *49*(2), 202-213.

[57] Arnold, S.E.; Hyman, B.T.; Flory, J.; Damasio, A.R.; Van Hoesen, G.W. The topographical and neuroanatomical distribution of neurofibrillary tangles and neuritic plaques in the cerebral cortex of patients with Alzheimer's disease. *Cereb. Cortex,* **1991**, *1*(1), 103-116.

[58] Vogt, B.A.; Van Hoesen, G.W.; Vogt, L.J. Laminar distribution of neuron degeneration in posterior cingulate cortex in Alzheimer's disease. *Acta Neuropathol,* **1990**, *80*(6), 581-589.

[59] Braak, H.; Braak, E. Alzheimer's disease affects limbic nuclei of the thalamus. *Acta Neuropathol,* **1991**, *81*(3), 261-268.

[60] Kovacs, T.; Cairns, N.J.; Lantos, P.L. beta-amyloid deposition and neurofibrillary tangle formation in the olfactory bulb in ageing and Alzheimer's disease. *Neuropathol. Appl. Neurobiol.,* **1999**, *25*(6), 481-491.

[61] Lewis, D.A.; Campbell, M.J.; Terry, R.D.; Morrison, J.H. Laminar and regional distributions of neurofibrillary tangles and neuritic plaques in Alzheimer's disease: a quantitative study of visual and auditory cortices. *J. Neurosci.,* **1987**, *7*(6), 1799-1808.

[62] Pearson, R.C.; Esiri, M.M.; Hiorns, R.W.; Wilcock, G.K.; Powell, T.P. Anatomical correlates of the distribution of the pathological changes in the neocortex in Alzheimer disease. *Proc. Natl. Acad. Sci. U. S. A.,* **1985**, *82*(13), 4531-4534.

[63] De Lacoste, M.C.; White, C.L., 3rd The role of cortical connectivity in Alzheimer's disease pathogenesis: a review and model system. *Neurobiol. Aging,* **1993**, *14*(1), 1-16.

[64] Hof, P.R.; Cox, K.; Morrison, J.H. Quantitative analysis of a vulnerable subset of pyramidal neurons in Alzheimer's disease: I. Superior frontal and inferior temporal cortex. *J. Comp. Neurol.,* **1990**, *301*(1), 44-54.

[65] Witter, M.P.; Groenewegen, H.J.; Lopes da Silva, F.H.; Lohman, A.H. Functional organization of the extrinsic and intrinsic circuitry of the parahippocampal region. *Prog. Neurobiol.,* **1989**, *33*(3), 161-253.

[66] Dunckley, T.; Beach, T.G.; Ramsey, K.E.; Grover, A.; Mastroeni, D.; Walker, D.G., *et al.* Gene expression correlates of neurofibrillary tangles in Alzheimer's disease. *Neurobiol. Aging,* **2006**, *27*(10), 1359-1371.

[67] Liang, W.S.; Reiman, E.M.; Valla, J.; Dunckley, T.; Beach, T.G.; Grover, A., *et al.* Alzheimer's disease is associated with reduced expression of energy metabolism genes in posterior cingulate neurons. *Proc. Natl. Acad. Sci. U. S. A.,* **2008**, *105*(11), 4441-4446.

[68] Ginsberg, S.D.; Hemby, S.E.; Lee, V.M.; Eberwine, J.H.; Trojanowski, J.Q. Expression profile of transcripts in Alzheimer's disease tangle-bearing CA1 neurons. *Ann. Neurol.,* **2000**, *48*(1), 77-87.

[69] Liang, W.S.; Dunckley, T.; Beach, T.G.; Grover, A.; Mastroeni, D.; Ramsey, K., *et al.* Altered neuronal gene expression in brain regions differentially affected by Alzheimer's disease: a reference data set. *Physiological genomics,* **2008**, *33*(2), 240-256.

[70] Wang, X.; Pal, R.; Chen, X.W.; Limpeanchob, N.; Kumar, K.N.; Michaelis, E.K. High intrinsic oxidative stress may underlie selective vulnerability of the hippocampal CA1 region. *Brain Res Mol Brain Res,* **2005**, *140*(1-2), 120-126.

[71] Wang, X.; Zaidi, A.; Pal, R.; Garrett, A.S.; Braceras, R.; Chen, X.W., *et al.* Genomic and biochemical approaches in the discovery of mechanisms for selective neuronal vulnerability to oxidative stress. *BMC Neurosci,* **2009**, *10*, 12.

[72] Wang, X.; Michaelis, E.K. Selective neuronal vulnerability to oxidative stress in the brain. *Front Aging Neurosci,* **2010**, *2*, 12.

[73] Wang, X.; Bao, X.; Pal, R.; Agbas, A.; Michaelis, E.K. Transcriptomic responses in mouse brain exposed to chronic excess of the neurotransmitter glutamate. *BMC Genomics,* **2010**, *11*, 360.

[74] Bohm, C.; Newrzella, D.; Sorgenfrei, O. Laser microdissection in CNS research. *Drug discovery today,* **2005**, *10*(17), 1167-1174.

[75] King, C.; Guo, N.; Frampton, G.M.; Gerry, N.P.; Lenburg, M.E.; Rosenberg, C.L. Reliability and reproducibility of gene expression measurements using amplified RNA from laser-microdissected primary breast tissue with oligonucleotide arrays. *The Journal of molecular diagnostics : JMD,* **2005**, *7*(1), 57-64.

[76] Emilsson, L.; Saetre, P.; Jazin, E. Alzheimer's disease: mRNA expression profiles of multiple patients show alterations of genes involved with calcium signaling. *Neurobiol. Dis.,* **2006**, *21*(3), 618-625.

[77] Noorbakhsh, F.; Overall, C.M.; Power, C. Deciphering complex mechanisms in neurodegenerative diseases: the advent of systems biology. *Trends Neurosci.,* **2009**, *32*(2), 88-100.

[78] Wang, Z.; Gerstein, M.; Snyder, M. RNA-Seq: a revolutionary tool for transcriptomics. *Nat Rev Genet,* **2009**, *10*(1), 57-63.

[79] Twine, N.A.; Janitz, K.; Wilkins, M.R.; Janitz, M. Whole transcriptome sequencing reveals gene expression and splicing differences in brain regions affected by Alzheimer's disease. *PLoS One,* **2011**, *6*(1), e16266.

[80] Small, G.W.; Ercoli, L.M.; Silverman, D.H.; Huang, S.C.; Komo, S.; Bookheimer, S.Y., *et al.* Cerebral metabolic and cognitive decline in persons at genetic risk for Alzheimer's disease. *Proc. Natl. Acad. Sci. U. S. A.,* **2000**, *97*(11), 6037-6042.

[81] Perneczky, R.; Hartmann, J.; Grimmer, T.; Drzezga, A.; Kurz, A. Cerebral metabolic correlates of the clinical dementia rating scale in mild cognitive impairment. *J. Geriatr. Psychiatry Neurol.,* **2007**, *20*(2), 84-88.

[82] Fukuyama, H.; Ogawa, M.; Yamauchi, H.; Yamaguchi, S.; Kimura, J.; Yonekura, Y., *et al.* Altered cerebral energy metabolism in Alzheimer's disease: a PET study. *J. Nucl. Med.,* **1994**, *35*(1), 1-6.

[83] Desgranges, B.; Baron, J.C.; de la Sayette, V.; Petit-Taboue, M.C.; Benali, K.; Landeau, B., *et al.* The neural substrates of memory systems impairment in Alzheimer's disease. A PET study of resting brain glucose utilization. *Brain,* **1998**, *121 (Pt 4)*, 611-631.

[84] Stein, D.J.; Buchsbaum, M.S.; Hof, P.R.; Siegel, B.V., Jr.; Shihabuddin, L. Greater metabolic rate decreases in hippocampal formation and proisocortex than in neocortex in Alzheimer's disease. *Neuropsychobiology,* **1998**, *37*(1), 10-19.

[85] Alexander, G.E.; Chen, K.; Pietrini, P.; Rapoport, S.I.; Reiman, E.M. Longitudinal PET Evaluation of Cerebral Metabolic Decline in Dementia: A Potential Outcome Measure in Alzheimer's Disease Treatment Studies. *A. J. Psychiatry,* **2002**, *159*(5), 738-745.

[86] de Leon, M.J.; Convit, A.; Wolf, O.T.; Tarshish, C.Y.; DeSanti, S.; Rusinek, H., *et al.* Prediction of cognitive decline in normal elderly subjects with 2-[(18)F]fluoro-2-deoxy-D-glucose/poitron-emission tomography (FDG/PET). *Proc Natl Acad Sci U S A,* **2001**, *98*(19), 10966-10971.

[87] De Santi, S.; de Leon, M.J.; Rusinek, H.; Convit, A.; Tarshish, C.Y.; Roche, A., *et al.* Hippocampal formation glucose metabolism and volume losses in MCI and AD. *Neurobiol Aging,* **2001**, *22*(4), 529-539.

[88] Vlassenko, A.G.; Vaishnavi, S.N.; Couture, L.; Sacco, D.; Shannon, B.J.; Mach, R.H., *et al.* Spatial correlation between brain aerobic glycolysis and amyloid-β (Aβ) deposition. *Proceedings of the National Academy of Sciences,* **2010**, *107*(41), 17763-17767.

[89] Vaishnavi, S.N.; Vlassenko, A.G.; Rundle, M.M.; Snyder, A.Z.; Mintun, M.A.; Raichle, M.E. Regional aerobic glycolysis in the human brain. *Proceedings of the National Academy of Sciences,* **2010**, *107*(41), 17757-17762.

[90] Loring, J.F.; Wen, X.; Lee, J.M.; Seilhamer, J.; Somogyi, R. A gene expression profile of Alzheimer's disease. *DNA Cell Biol.,* **2001**, *20*(11), 683-695.

[91] Kong, L.N.; Zuo, P.P.; Mu, L.; Liu, Y.Y.; Yang, N. Gene expression profile of amyloid beta protein-injected mouse model for Alzheimer disease. *Acta Pharmacol Sin,* **2005**, *26*(6), 666-672.

[92] Wu, Z.-L.; Ciallella, J.R.; Flood, D.G.; O'Kane, T.M.; Bozyczko-Coyne, D.; Savage, M.J. Comparative analysis of cortical gene expression in mouse models of Alzheimer's disease. *Neurobiol. Aging,* **2006**, *27*(3), 377-386.

[93] Brooks, W.M.; Lynch, P.J.; Ingle, C.C.; Hatton, A.; Emson, P.C.; Faull, R.L.M., *et al.* Gene expression profiles of metabolic enzyme transcripts in Alzheimer's disease. *Brain Res.,* **2007**, *1127*(1), 127-135.

[94] Qin, W.; Haroutunian, V.; Katsel, P.; Cardozo, C.P.; Ho, L.; Buxbaum, J.D., *et al.* PGC-1{alpha} expression decreases in the Alzheimer disease brain as a function of dementia. *Arch. Neurol.,* **2009**, *66*(3), 352-361.

[95] Blalock, E.M.; Geddes, J.W.; Chen, K.C.; Porter, N.M.; Markesbery, W.R.; Landfield, P.W. Incipient Alzheimer's disease: microarray correlation analyses reveal major transcriptional and tumor suppressor responses. *Proc. Natl. Acad. Sci. U. S. A.,* **2004**, *101*(7), 2173-2178.

[96] Liang, W.S.; Reiman, E.M.; Valla, J.; Dunckley, T.; Beach, T.G.; Grover, A., *et al.* Alzheimer's disease is associated with reduced expression of energy metabolism genes in posterior cingulate neurons. *Proc Natl Acad Sci U S A,* **2008**, *105*(11), 4441-4446.

[97] Roses, A.D.; Lutz, M.W.; Amrine-Madsen, H.; Saunders, A.M.; Crenshaw, D.G.; Sundseth, S.S., *et al.* A TOMM40 variable-length polymorphism predicts the age of late-onset Alzheimer's disease. *Pharmacogenomics J,* **2009**.

[98] Dickey, C.A.; Loring, J.F.; Montgomery, J.; Gordon, M.N.; Eastman, P.S.; Morgan, D. Selectively Reduced Expression of Synaptic Plasticity-Related Genes in Amyloid Precursor Protein + Presenilin-1 Transgenic Mice. *J. Neurosci.,* **2003**, *23*(12), 5219-5226.

[99] Ricciarelli, R.; d'Abramo, C.; Massone, S.; Marinari, U.; Pronzato, M.; Tabaton, M. Microarray analysis in Alzheimer's disease and normal aging. *IUBMB life,* **2004**, *56*(6), 349-354.

[100] Mirnics, K.; Norstrom, E.M.; Garbett, K.; Choi, S.H.; Zhang, X.; Ebert, P., *et al.* Molecular signatures of neurodegeneration in the cortex of PS1/PS2 double knockout mice. *Mol Neurodegener,* **2008**, *3*, 14.

[101] Weeraratna, A.T.; Kalehua, A.; Deleon, I.; Bertak, D.; Maher, G.; Wade, M.S., *et al.* Alterations in immunological and neurological gene expression patterns in Alzheimer's disease tissues. *Exp. Cell Res.,* **2007**, *313*(3), 450-461.
[102] Colangelo, V.; Schurr, J.; Ball, M.J.; Pelaez, R.P.; Bazan, N.G.; Lukiw, W.J. Gene expression profiling of 12633 genes in Alzheimer hippocampal CA1: transcription and neurotrophic factor down-regulation and up-regulation of apoptotic and pro-inflammatory signaling. *J. Neurosci. Res.,* **2002**, *70*(3), 462-473.
[103] Tan, M.G.; Chua, W.T.; Esiri, M.M.; Smith, A.D.; Vinters, H.V.; Lai, M.K. Genome wide profiling of altered gene expression in the neocortex of Alzheimer's disease. *J. Neurosci. Res.,* **2010**, *88*(6), 1157-1169.
[104] Avramopoulos, D.; Szymanski, M.; Wang, R.; Bassett, S. Gene expression reveals overlap between normal aging and Alzheimer's disease genes. *Neurobiol. Aging,* **2011**, *32*(12), 2319 e2327-2334.
[105] Parachikova, A.; Agadjanyan, M.G.; Cribbs, D.H.; Blurton-Jones, M.; Perreau, V.; Rogers, J., *et al.* Inflammatory changes parallel the early stages of Alzheimer disease. *Neurobiol. Aging,* **2007**, *28*(12), 1821-1833.
[106] Kalaria, R.N. Microglia and Alzheimer's disease. *Curr Opin Hematol,* **1999**, *6*(1), 15-24.
[107] Akiyama, H.; Barger, S.; Barnum, S.; Bradt, B.; Bauer, J.; Cole, G.M., *et al.* Inflammation and Alzheimer's disease. *Neurobiol. Aging,* **2000**, *21*(3), 383-421.
[108] Moore, K.W.; de Waal Malefyt, R.; Coffman, R.L.; O'Garra, A. Interleukin-10 and the interleukin-10 receptor. *Annu. Rev. Immunol.,* **2001**, *19*, 683-765.
[109] Rozemuller, J.M.; Eikelenboom, P.; Stam, F.C. Role of microglia in plaque formation in senile dementia of the Alzheimer type. An immunohistochemical study. *Virchows Arch. B. Cell Pathol. Incl. Mol. Pathol.,* **1986**, *51*(3), 247-254.
[110] Rogers, J.; Luber-Narod, J.; Styren, S.D.; Civin, W.H. Expression of immune system-associated antigens by cells of the human central nervous system: relationship to the pathology of Alzheimer's disease. *Neurobiol. Aging,* **1988**, *9*(4), 339-349.
[111] Eikelenboom, P.; Hack, C.E.; Rozemuller, J.M.; Stam, F.C. Complement activation in amyloid plaques in Alzheimer's dementia. *Virchows Arch. B. Cell Pathol. Incl. Mol. Pathol.,* **1989**, *56*(4), 259-262.
[112] Haga, S.; Akai, K.; Ishii, T. Demonstration of microglial cells in and around senile (neuritic) plaques in the Alzheimer brain. An immunohistochemical study using a novel monoclonal antibody. *Acta Neuropathol,* **1989**, *77*(6), 569-575.
[113] McGeer, P.L.; Akiyama, H.; Itagaki, S.; McGeer, E.G. Immune system response in Alzheimer's disease. *Can. J. Neurol. Sci.,* **1989**, *16*(4 Suppl), 516-527.
[114] Perlmutter, L.S.; Scott, S.A.; Barron, E.; Chui, H.C. MHC class II-positive microglia in human brain: association with Alzheimer lesions. *J. Neurosci. Res.,* **1992**, *33*(4), 549-558.
[115] Strauss, S.; Bauer, J.; Ganter, U.; Jonas, U.; Berger, M.; Volk, B. Detection of interleukin-6 and alpha 2-macroglobulin immunoreactivity in cortex and hippocampus of Alzheimer's disease patients. *Lab. Invest.,* **1992**, *66*(2), 223-230.
[116] Overmyer, M.; Helisalmi, S.; Soininen, H.; Laakso, M.; Riekkinen, P., Sr.; Alafuzoff, I. Reactive microglia in aging and dementia: an immunohistochemical study of postmortem human brain tissue. *Acta Neuropathol,* **1999**, *97*(4), 383-392.
[117] Eikelenboom, P.; Veerhuis, R.; Scheper, W.; Rozemuller, A.J.; van Gool, W.A.; Hoozemans, J.J. The significance of neuroinflammation in understanding Alzheimer's disease. *J Neural Transm,* **2006**, *113*(11), 1685-1695.

[118] Gluck, M.R.; Moy, L.Y.; Jayatilleke, E.; Hogan, K.A.; Manzino, L.; Sonsalla, P.K. Parallel increases in lipid and protein oxidative markers in several mouse brain regions after methamphetamine treatment. *J. Neurochem.,* **2001**, *79*(1), 152.

[119] Yao, P.J.; Zhu, M.; Pyun, E.I.; Brooks, A.I.; Therianos, S.; Meyers, V.E., *et al.* Defects in expression of genes related to synaptic vesicle trafficking in frontal cortex of Alzheimer's disease. *Neurobiol. Dis.,* **2003**, *12*(2), 97-109.

[120] Mufson, E.J.; Counts, S.E.; Ginsberg, S.D. Gene expression profiles of cholinergic nucleus basalis neurons in Alzheimer's disease. *Neurochem. Res.,* **2002**, *27*(10), 1035-1048.

[121] Davies, C.A.; Mann, D.M.; Sumpter, P.Q.; Yates, P.O. A quantitative morphometric analysis of the neuronal and synaptic content of the frontal and temporal cortex in patients with Alzheimer's disease. *J. Neurol. Sci.,* **1987**, *78*(2), 151-164.

[122] Scheff, S.W.; Price, D.A.; Schmitt, F.A.; DeKosky, S.T.; Mufson, E.J. Synaptic alterations in CA1 in mild Alzheimer disease and mild cognitive impairment. *Neurology,* **2007**, *68*(18), 1501-1508.

[123] Clinton, J.; Blackman, S.E.; Royston, M.C.; Roberts, G.W. Differential synaptic loss in the cortex in Alzheimer's disease: a study using archival material. *Neuroreport,* **1994**, *5*(4), 497-500.

[124] Sze, C.I.; Bi, H.; Kleinschmidt-DeMasters, B.K.; Filley, C.M.; Martin, L.J. Selective regional loss of exocytotic presynaptic vesicle proteins in Alzheimer's disease brains. *J. Neurol. Sci.,* **2000**, *175*(2), 81-90.

[125] Perdahl, E.; Adolfsson, R.; Alafuzoff, I.; Albert, K.A.; Nestler, E.J.; Greengard, P., *et al.* Synapsin I (protein I) in different brain regions in senile dementia of Alzheimer type and in multi-infarct dementia. *J. Neural Transm.,* **1984**, *60*(2), 133-141.

[126] Honer, W.G.; Dickson, D.W.; Gleeson, J.; Davies, P. Regional synaptic pathology in Alzheimer's disease. *Neurobiol. Aging,* **1992**, *13*(3), 375-382.

[127] Haass, C.; Selkoe, D.J. Soluble protein oligomers in neurodegeneration: lessons from the Alzheimer's amyloid beta-peptide. *Nat Rev Mol Cell Biol,* **2007**, *8*(2), 101-112.

[128] Rowan, M.J.; Klyubin, I.; Wang, Q.; Hu, N.W.; Anwyl, R. Synaptic memory mechanisms: Alzheimer's disease amyloid beta-peptide-induced dysfunction. *Biochem. Soc. Trans.,* **2007**, *35*(Pt 5), 1219-1223.

[129] Lue, L.F.; Brachova, L.; Civin, W.H.; Rogers, J. Inflammation, A beta deposition, and neurofibrillary tangle formation as correlates of Alzheimer's disease neurodegeneration. *J. Neuropathol. Exp. Neurol.,* **1996**, *55*(10), 1083-1088.

[130] Hata, R.; Masumura, M.; Akatsu, H.; Li, F.; Fujita, H.; Nagai, Y., *et al.* Up-regulation of calcineurin Abeta mRNA in the Alzheimer's disease brain: assessment by cDNA microarray. *Biochem. Biophys. Res. Commun.,* **2001**, *284*(2), 310-316.

[131] Riascos, D.; de Leon, D.; Baker-Nigh, A.; Nicholas, A.; Yukhananov, R.; Bu, J., *et al.* Age-related loss of calcium buffering and selective neuronal vulnerability in Alzheimer's disease. *Acta neuropathologica,* **2011**, *122*(5), 565-576.

[132] Jee, S.W.; Cho, J.S.; Kim, C.K.; Hwang, D.Y.; Shim, S.B.; Lee, S.H., *et al.* Oligonucleotide-based analysis of differentially expressed genes in hippocampus of transgenic mice expressing NSE-controlled APPsw. *Neurochem. Res.,* **2006**, *31*(8), 1035-1044.

[133] Swerdlow, R.H.; Burns, J.M.; Khan, S.M. The Alzheimer's disease mitochondrial cascade hypothesis. *J Alzheimers Dis,* **2010**, *20 Suppl 2*, S265-279.

[134] Khachaturian, Z.S. Hypothesis on the Regulation of Cytosol Calcium Concentration and the Aging Brain. *Neurobiology of Aging*, *8*(4), 345-346.

[135] Khachaturian, Z.S. Calcium Hypothesis of Alzheimer's Disease and Brain Aginga. *Annals of the New York Academy of Sciences,* **1994**, *747*(1), 1-11.

[136] Selkoe, D.J. Soluble oligomers of the amyloid [beta]-protein impair synaptic plasticity and behavior. *Behavioural Brain Research,* **2008**, *192*(1), 106-113.

[137] Terry, R.D.; Masliah, E.; Salmon, D.P.; Butters, N.; DeTeresa, R.; Hill, R., *et al.* Physical basis of cognitive alterations in Alzheimer's disease: synapse loss is the major correlate of cognitive impairment. *Ann. Neurol.,* **1991**, *30*(4), 572-580.

[138] McGeer, P.L.; McGeer, E.G. Inflammation, autotoxicity and Alzheimer disease. *Neurobiol Aging,* **2001**, *22*(6), 799-809.

[139] Miller, J.A.; Oldham, M.C.; Geschwind, D.H. A systems level analysis of transcriptional changes in Alzheimer's disease and normal aging. *The Journal of neuroscience : the official journal of the Society for Neuroscience,* **2008**, *28*(6), 1410-1420.

[140] Harman, D. Aging: a theory based on free radical and radiation chemistry. *J. Gerontol.,* **1956**, *11*(3), 298-300.

[141] Reitz, C.; Tokuhiro, S.; Clark, L.N.; Conrad, C.; Vonsattel, J.P.; Hazrati, L.N., *et al.* SORCS1 alters amyloid precursor protein processing and variants may increase Alzheimer's disease risk. *Annals of neurology,* **2011**, *69*(1), 47-64.

[142] Small, S.A.; Kent, K.; Pierce, A.; Leung, C.; Kang, M.S.; Okada, H., *et al.* Model-guided microarray implicates the retromer complex in Alzheimer's disease. *Ann. Neurol.,* **2005**, *58*(6), 909-919.

[143] Bossers, K.; Wirz, K.T.; Meerhoff, G.F.; Essing, A.H.; van Dongen, J.W.; Houba, P., *et al.* Concerted changes in transcripts in the prefrontal cortex precede neuropathology in Alzheimer's disease. *Brain : a journal of neurology,* **2010**, *133*(Pt 12), 3699-3723.

Send Orders for Reprints to reprints@benthamscince.net
Advances in Genome Science, Vol. 3, 2014, 271-287

CHAPTER 10

Investigating Molecular Mechanisms of Chronic Pain in the Anterior Cingulate Cortex Through Genetically Engineered Mice

Susan S. Kim[1], Giannina Descalzi[1] and Min Zhuo[1,2,*]

[1]Department of Physiology, Faculty of Medicine, University of Toronto Centre for the Study of Pain, 1 King's College Circle, Toronto, Ontario, Canada and [2]Department of Brain and Cognitive Sciences, Seoul National University, Seoul 151-746, Korea

Abstract: Recent advances into the understanding of molecular mechanism of chronic pain have been largely developed through the use of genetic manipulations. This is in part due to the scarcity of selective pharmacological tools, which can be readily solved by creating knockout or transgenic mice. By identifying new genes that are of import, our efforts can then be aimed at studying relevant signaling pathways, and combination of pharmacological manipulations with genetic models can be used to further examine the specific mechanisms involved in chronic pain. In this review, we will examine the genetic models that are currently in use to study chronic pain in the anterior cingulate cortex: knockout mice; transgenic mice; and the strength of combining pharmacology with these genetic models.

Keywords: Adenylyl cyclase, AMPARs, anterior cingulate cortex, behavior, calcium calmodulin, chronic pain, electroporation, gene deletion, genetic manipulation, *in vitro*, *in vivo*, inflammatory pain, knockout models, molecular mechanisms, mouse models, neuropathic pain, NMDARs, regional inhibition, siRNA.

INTRODUCTION

Pain is an essential sensory mechanism that warns us of imminent tissue damage. However, clinical problems associated with pain occur when pain thresholds are dramatically altered. Pain is divided into two major categories: Acute or physiological pain, and chronic or pathological pain [1, 2]. Unlike physiological pain, which helps us determine potentially dangerous stimuli and avoid it, making

***Address correspondence to Min Zhuo:** Department of Physiology, University of Toronto, 1 King's College Circle, Toronto, Ontario M5S 1A8, Canada; Tel: 416-978-4018; E-mail: min.zhuo@utoronto.ca

Christian Neri (Ed)

it essential for survival, pathological pain has no benefit and only occurs after insult (*e.g.*, tissue or nerve injury or disease) [3, 4]. In addition to spontaneous occurrences of pain, there are two common pathological conditions that develop after tissue or nerve injury: Allodynia, where the pain threshold has been reduced, allowing non-noxious stimuli that normally do not cause pain to induce pain; and hyperalgesia, in which there is an enhanced response to noxious stimuli [2].

Recent studies using animal models as well as human brain imaging techniques have consistently revealed that chronic pain is likely due to sensitization along the somatosensory pathways, including long-lasting changes in peripheral tissues, the spinal cord, and the cortex [2]. Furthermore, descending biphasic modulatory systems that control the amount of pain information reaching the brain also undergo plastic changes [5]. Still, much about the molecular mechanisms of chronic pain remain to be explored. Integrative experimental approaches including electrophysiological, pharmacological, biochemical, anatomical, and behavioral methods have proven to be useful for investigating the molecular mechanisms of chronic pain. In this review, we will focus on the use of genetically manipulated mice in studying synaptic and molecular mechanisms of chronic pain in the anterior cingulate cortex (ACC), an area known to mediate the emotional component of pain. We believe that the development of novel and effective therapeutic agents to combat chronic pain will be greatly enhanced in the near future as researchers increasingly use genetic manipulations in combination with more traditional methods.

MOUSE MODELS FOR INVESTIGATING CHRONIC PAIN

In the laboratory, studies investigating the molecular mechanisms involved in the development and maintenance of chronic pain have employed several well established animal models of neuropathic and inflammatory pain. As mice and humans share approximately 99% of their genes [6], genetically manipulated mice offer a unique opportunity to investigate the roles of specific genes in models of human physiology. Various stimuli and well established behavioral paradigms are used that mimic clinical pain in humans [7]. Inflammatory pain, which results from tissue damage and is reversible when the underlying cause has been rectified, is most commonly studied using the formalin test and injections of complete Freund's adjuvant (CFA). The formalin test, a common test of tissue injury and inflammation on a timescale of hours, is used to model "acute" inflammation. Formalin is injected into the dorsum of the hindpaw, then spontaneous responses to the injection is measured [8, 9]. Formalin-induced behavioral responses are typically categorized into two distinct phases: Phase one (0-10 min) and phase two (10-55 min) [9], although we have previously shown that mice continue to exhibit behavioral responses during the 55-120 min period, "phase three" [10]. Formalin-induced behavioral responses also depend on NMDA receptors at different levels of the brain [11]. CFA, on the other hand, is a water-in-oil emulsion that contains heat killed *Mycobacterium tuberculosis* cell wall components. Much like formalin, CFA is injected into the dorsum of the hindpaw, although its effects can last for days. Following CFA injection, allodynia is typically observed on day 1 and 3 after injection by applying von Frey filaments to the dorsum of the hindpaw to measure paw withdrawal responses.

Neuropathic pain results from damage to components of the nervous system, such as primary afferent nerves, spinal cord, and cortical regions [12]. As the onset of neuropathic pain may be delayed after nerve injury, pain may still be present even after healing is complete [4, 12]. Moreover, neuropathic pain commonly occurs as a secondary symptom in diseases like diabetes and cancer, and may also occur with treatments such as chemotherapeutics or cytotoxic drugs [12]. Furthermore, due to the dynamic nature of the pain system, the signs and symptoms of neuropathic pain change over time, and injury to a peripheral nerve causes functional and biochemical changes not only at the site of injury, but also to other parts of the affected nerve, and eventually to higher order neurons in the spinal cord and brain [4]. A number of animal models have been reported to simulate human peripheral neuropathic pain conditions. The peripheral nerve injury models have been the most extensively studied, with the exact methods differing in the location and form of injury, including transaction and ligation [13]. Partial injury to the sciatic nerve is most commonly used to induce neuropathic pain behavior, including loose ligation of the sciatic nerve, developed by Bennett and Xie [14], as well as spinal L5/L6 nerve ligation, developed by Kim and Chung [15]. Recently, we have discovered a novel mouse model of neuropathic pain, a simple and less invasive procedure where the common peroneal nerve is ligated, leaving motor function intact [16]. The presence of neuropathic pain is mainly assessed by hyperalgesia to thermal and mechanical stimuli and allodynia to cold and tactile stimuli [13, 15, 17-19].

In addition to inflammatory pain and neuropathic pain, a wide variety of other pain assays have been developed to assess physiological and pathological pain in mice, such as thermal nociception and visceral pain. A list of the most used models has been provided below (see Table **1**).

Table 1: Tests for studying physiological and pathological pain

Physiological (Acute) Pain	**Pathological (Chronic) Pain**
Thermal nociception	*Inflammatory pain model*
Hot plate test	CFA
Tail flick test	Formalin test
Tail immersion test	Carrageenan
Thermal test (Hargreaves' test)	
Cold plate test	*Neuropathic pain model*
	Chronic constriction injury of sciatic nerve
Mechanical nociception	Partial sciatic nerve ligation
Tail pinch	Spinal nerve ligation
Paw pressure test	Common peroneal nerve ligation
	Spared nerve injury model
Visceral nociception	Infraorbital nerve ligation
Colorectal distension	Loose ligation of sciatic nerve
Writhing test	
	Diabetic model
	Streptozotocin injection
	Alloxan injection
	Chemotherapeutic agent-induced model
	Paclitaxel
	Vincristine
	Suramin
	Cisplatin
	Oxaliplatin
	Carboplatin

	Muscle pain model
	Carrageenan
	Capsaicin
	Acidic saline

DELETION OF SIGNALING GENES

The availability of genetic mouse models has revolutionized biomedical research, and alterations to classical gene modification techniques have made possible the generation of conditional, inducible, and multi-gene knockout mouse mutants. These techniques deliver a more detailed and informative look into the *in vivo* functions and endogenous expression patterns of individual genes.

The basic premise behind the development of a knockout mouse is to replace the normal functioning gene with one that is nonfunctional (*i.e.*, a null mutation) using homologous recombination. By comparing the phenotypes of these knockout mice to those of wild-type mice, the function of the targeted gene could then be deduced. If phenotypic differences exist between the knockout mice and their wild-type counterparts, it strongly suggests the gene is responsible for or regulates that particular function. However, knockout technology has its limitations. Many knockout mice are embryonically lethal due to developmental defects, and some genes may serve a different function in adults than during development [20]. In addition, the loss of a gene during development and throughout adulthood may result in compensatory actions by other mechanisms, thereby possibly altering multiple systems and making interpretation of the phenotypes difficult. In such cases, conditional and inducible gene knockout technology can be used. Unlike the conventional knockout technology where the gene of interest is completely absent from all cells, conditional knockouts by contrast inactivate the gene of interest in a tissue-specific manner, allowing for greater control over gene expression. An inducible knockout enables the researcher to activate or inactivate the gene of interest at specific developmental time points, using antibiotic-controlled transcriptional activation, so that gene expression can be reversibly induced in the presence or absence of antibiotics such as tetracycline [20].

There are several different ways to make conditional knockout models, but the most widely used is the Cre-*lox*P recombinase system, where the target gene is modified through Cre action [21]. Cre expression can also be controlled temporally, by introducing an element into the promoter that requires a ligand, such as a drug, for induction [22]. This technique has recently been used to create a mouse model where selective genes can be deleted in sensory ganglia, which will allow researchers to investigate how specific genes in peripheral sensory neurons affect pain pathways [23], and may help identify the different components between cortical and spinal mechanisms.

REGIONAL INHIBITION OF PROTEIN FUNCTIONS

siRNA

Recently, RNA interference (RNAi) has greatly enhanced our ability to characterize gene function [24]. Through the introduction of synthesized small interfering RNA (siRNA), RNAi allows researchers to target and silence genes [24]. Small interfering RNA (siRNA), sometimes known as short interfering RNA or silencing RNA, is generally 20-25 basepairs long, and can be artificially introduced into cells by various transfection methods to bring about the specific knockdown of a gene of interest [25]. Any gene with a known sequence can be targeted based on sequence complementarity with an appropriately tailored siRNA, making it a roubst tool for gene function studies.

The efficacy of siRNA depends on multiple factors, including GC content and the thermodynamic stability of the duplex at the 5' antisense end [26]. Several types of vectors for siRNA delivery have been developed that aim to enhance cellular uptake, protect siRNAs from degradation, and to achieve site-specific delivery. While non-viral vectors have the advantages of low toxicity, ease of synthesis and low immune response [27], viral vectors have the advantage of delivering siRNAs to neurons as well as glial cells, making it the most commonly used carriers for gene transfer. Additionally, as vector assisted siRNA delivery can be micro-injected into specific brain regions, it is increasingly providing valuable observations in neuroscience; albeit targeting specific cell types still remains a challenge. The advantage of efficient, economical knockdown offered by RNAi and the large amount of data it provides makes it the current technology of choice, although overcoming the delivery obstacle for *in vivo* studies remains.

Electroporation

Much like RNAi, the ability to transfer a gene of interest to a specific brain region has also been a much sought-after technique. Several approaches have been used to target gene transfer to neurons *in vivo*, including the use of adenoviral-based vectors [28-30] and liposomal-plasmid complexes [31, 32]. Electroporation, a technique where brief electrical pulses are applied to tissue to temporarily increase cellular permeability to DNA, has been used to enhance gene delivery into cells of adult organs, as well as neurons in the brains of embryonic mice [33]. Similar to transgenic manipulations, electroporation enables selective delivery of a gene to specific regions of the central nervous system (CNS), but unlike the transgenic approach, electroporation provides a method for unilateral gene transfer, allowing overexpression of a gene within a highly specific brain region [33]. Additionally, electroporation provides a way to express a gene at any time during development or adulthood, and provides a method to assess possible rescuing effects of defects observed in genetically altered mice with fewer toxic effects when compared to other methods of gene transfer [33]. We have previously developed a microelectroporation method and selectively expressed both green fluorescent protein (GFP) and mutant calmodulin (CaM) in the ACC of adult rodents, the first instance where electroporation was used to deliver genes into the adult mammalian brain [33]. More recently, we have successfully combined siRNA and electroporation to reduce the levels of NMDA receptor subtype NR2B in the ACC of mice [34].

GENETIC MANIPULATIONS IN INVESTIGATING CHRONIC PAIN IN THE ACC

Elucidating the function of particular neurotransmitters and their receptors has traditionally relied upon pharmacological approaches using compounds that act as specific agonists or antagonists at defined receptors on the pre- and postsynaptic membranes [35]. Unfortunately, selective antagonists of ion channels are limited, and often fail to selectively inhibit specific subtypes. Moreover, studies employing lesions studies or local chemical injections aiming to dissociate behavioral responses to acute stimuli from long-term sensitization to the injury have failed, as these manipulations alone significantly affect behavioral responses to acute stimuli [8]. As a result, researchers have turned to knockout and transgenic manipulations to investigate pain pathways.

THE USE OF AC1 AND AC8 GENE KNOCKOUT MICE

The cAMP-related signaling pathway has previously been shown to contribute to behavioral responses to tissue injury and inflammation in the periphery and the spinal cord [36-39], but few studies have addressed the potential roles of forebrain cAMP in such changes. AC1 and AC8, two CaM-stimulated adenylyl cyclases (AC) in the CNS, couple NMDA receptor-induced cytosolic Ca^{2+} increases to cAMP signaling pathways [40, 41]. To examine the roles of AC1 and AC8 in behavioral sensitization within the ACC, an area mediating the emotional component of pain, we used AC1 and AC8 knockout mice as well as AC1&8 double knockout (DKO) mice to determine the roles of these ACs in chronic pain [8] (Fig. **1**).

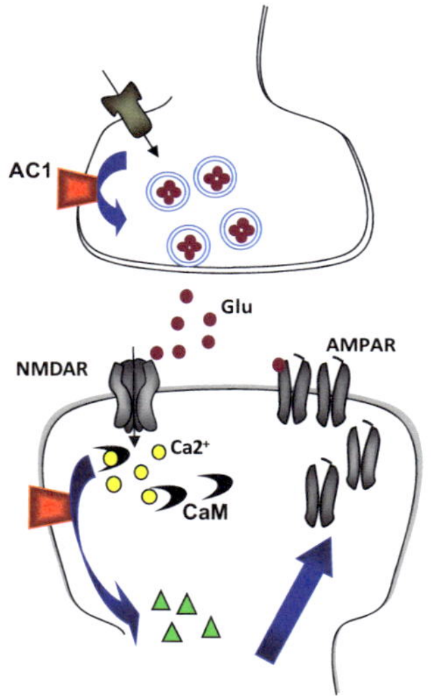

Figure 1: Identifying a plasticity-selective gene for treating chronic pain: Inflammatory or nerve injuries cause both presynaptic and postsynaptic changes in the ACC synapses. Activity triggers the release of excitatory neurotransmitter glutamate, and activation of NMDAR leads to an increase in postsynaptic Ca^{2+}. Ca^{2+} binds to CaM and leads to activation of calcium-stimulated AC1 and Ca^{2+}/CaM-dependent protein kinases, resulting in possible postsynaptic trafficking of AMPA GluR1 receptors as well as increasing AMPAR sensitivity to glutamate. Enhanced release of glutamate as well as postsynaptic changes in AMPAR-mediated responses contributes to enhanced noxious sensory information within the brain, and may be a possible cellular mechanism for chronic pain.

Behavioral nociceptive responses to subcutaneous formalin injection or nerve injury were significantly reduced in mice lacking AC1 or AC8, and more profoundly compromised in AC1&8 DKO mice. Allodynia was significantly reduced in AC1 KO mice and absent in AC1&8 DKO mice, but not in AC8 KO mice, indicating that ablation of AC8 alone is not sufficient to affect allodynia. We found that AC1 and AC8 were both expressed at high levels in two pain-related forebrain areas, the ACC and the insular cortex. These observations led to the discovery that AC1 mediated protein kinase M zeta (PKMζ) is necessary for the enhancement in post-synaptic glutamatergic transmission associated with peripheral nerve injury [42]. More recently, we reported that NB001, a potent and selective AC1 inhibitor, robustly reduces behavioral allodynia when applied systemically or specifically within the ACC [43] in mouse models of neuropathic pain. The discoveries made using the AC1 knockout mouse are thus an excellent example of how genetic mouse models can help identify pharmacological targets.

CHEMICALLY-ENGINEERED CaMKII TRANSGENIC MICE

Ca^{2+}/CaM-dependent kinase II (CaMKII) is known to be a key molecule involved in regulating tissue-injury evoked persistent pain [44]. To determine the role of forebrain CaMKII in the development of persistent pain, we used CaMKII F89G transgenic mice to spatially and temporally restrict the expression of CaMKII to the forebrain areas [45, 46]. These mice were developed using the conditional protein knockout/manipulation technique to achieve monospecific inhibition of a single kinase in one region of the mouse brain [45]. This method combined high temporal resolution of chemical inhibition with molecular and spatial specificity of genetic techniques [45]. We found that overexpressing CaMKII in the forebrain caused selective changes in synaptic LTD without causing any obvious anatomical abnormality or change in neuronal excitability. As well, enhanced CaMKII activity in the forebrain resulted in significant reduction in mechanical allodynia 1 and 3 days after CFA injection into the dorsum of a single hindpaw, as well as reduction in thermal hyperalgesia. CaMKII mice were then treated with the inhibitor 1NM-PP1 to turn off the overexpressed CaMKII activity. We found that the inhibitor treatment completely blocked the effects of CaMKII overexpression on behavioral allodynia and hyperalgesia in transgenic mice, and synaptic transmission was rescued as well. The results suggest that the enhanced forebrain CaMKII activity reduces behavioral sensitization to peripheral injury, and that the loss of LTD in CaMKII transgenic mice may be responsible for the reduction of nociceptive responses after tissue injury (Fig. **2A**).

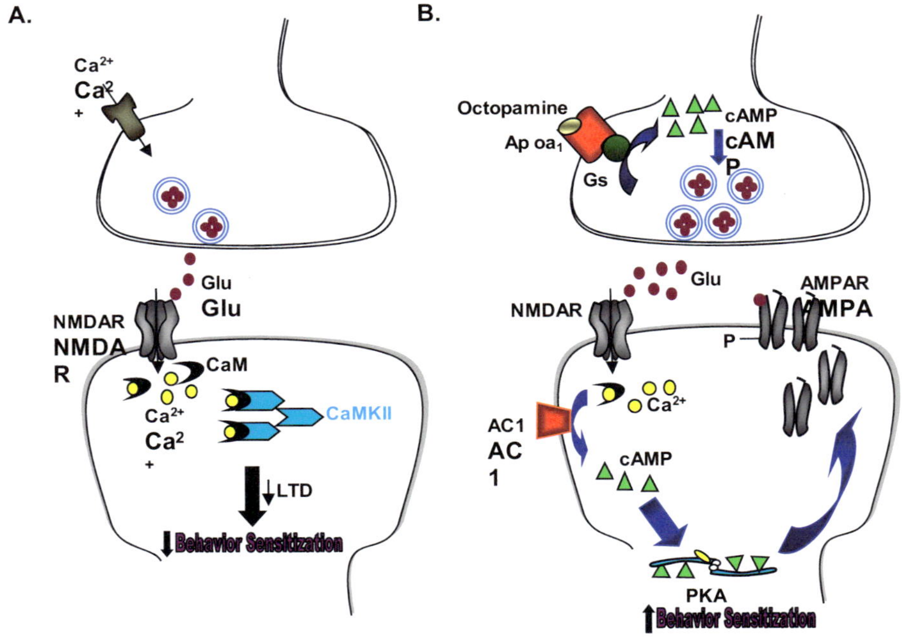

Figure 2: Two examples of combining chemical and genetic approaches: A. Inflammatory pain triggers the release of excitatory neurotransmitter glutamate, and activation of NMDAR leads to an increase in postsynaptic Ca^{2+}. Ca^{2+} binds to CaM and leads to activation Ca^{2+}/CaM-dependent protein kinases such as CaMKII. Overexpression of CaMKII leads to reduction in LTD as well as allodynia and hyperalgesia, suggesting that the observed reduction in behavioral sensitization may be due to the loss of LTD.; B. Activation of presynaptic Ap oa_1 by octopamine leads to the production of cAMP. The cAMP-related signaling pathways then facilitate the release of glutamate, which in turn acts on NMDAR and AMPAR. This triggers the downstream signaling pathways, leading to increased behavioral sensitization.

OCTOPAMINE MICE

To investigate synaptic mechanisms related to inflammatory pain, we used transgenic mice with heterologously expressed receptors impacting the cAMP pathway [47]. The mice heterologously expressed an *Aplysia* octopamine receptor (Ap oa_1) [48], a G protein-coupled receptor selectively activating the cAMP pathway after the binding of its natural ligand, octopamine [49]. There were at least two advantages in using these transgenic mice: First, as Ap oa_1 receptor expression is driven by CaMKIIα promoter, the receptor is selectively expressed in forebrain neurons, allowing the cAMP pathway coupled to Ap oa_1 receptor to be spatially activated; and second, that since the receptor is not activated endogenously, the exogenous application of octopamine is the only way of activating Ap oa_1 receptor and the associated cAMP pathway.

Octopamine is a trace amine and its endogenous level is far below those of the classical neurotransmitters. Therefore, we selectively activated cAMP pathways by microinjecting octopamine into the ACC and then assessed behavioral responses to the formalin test [47]. We found enhancement in the intermediate inflammatory phase but not in the acute phase, indicating that the enhancement is selective for late phase responses. Interestingly, activation of Ap oa_1 receptors in the ACC enhanced presynaptic glutamate release but did not modulate postsynaptic function, suggesting that these heterologous receptors are expressed in presynaptic terminals. More importantly, these observations show that increased presynaptic glutamate release in the ACC is sufficient for enhancing chronic pain (Fig. **2B**).

FOREBRAIN NR2B OVEREXPRESSED MICE

Functional NMDA receptors contain heteromeric combinations of the NR1 subunit plus one or more of NR2A-D [50] subunits, of which the NR2A and NR2B subunits are the major NR2 subtypes found in forebrain structures [51]. Using transgenic mice with forebrain-targeted overexpression of the NMDA receptor subunit NR2B, we examined their response to tissue injury and inflammation [9]. We found that the transgenic mice exhibited prominent NR2B expression and enhanced NMDA receptor-mediated synaptic responses in two pain-related forebrain areas, the ACC and insular cortex. Additionally, the mice exhibited enhanced responsiveness to peripheral injection of two inflammatory stimuli: Formalin, where the enhancement was observed in the last phase; and CFA, where at day 1 and 3 after CFA injection, mechanical allodynia was significantly enhanced in the transgenic mice. We then showed that the NR2B receptor undergoes prolonged upregulation in the ACC after tissue inflammation, and also observed enhanced NMDA receptor-mediated responses in ACC slices *in vitro* [52]. Furthermore, administration of selective NR2B antagonists (Ro 25-6981 and Ro 63-1908) systemically or specifically into the ACC inhibited inflammation-related allodynia, suggesting that NR2B subunit increases within the ACC contributes to chronic pain ACC (Fig. **3**).

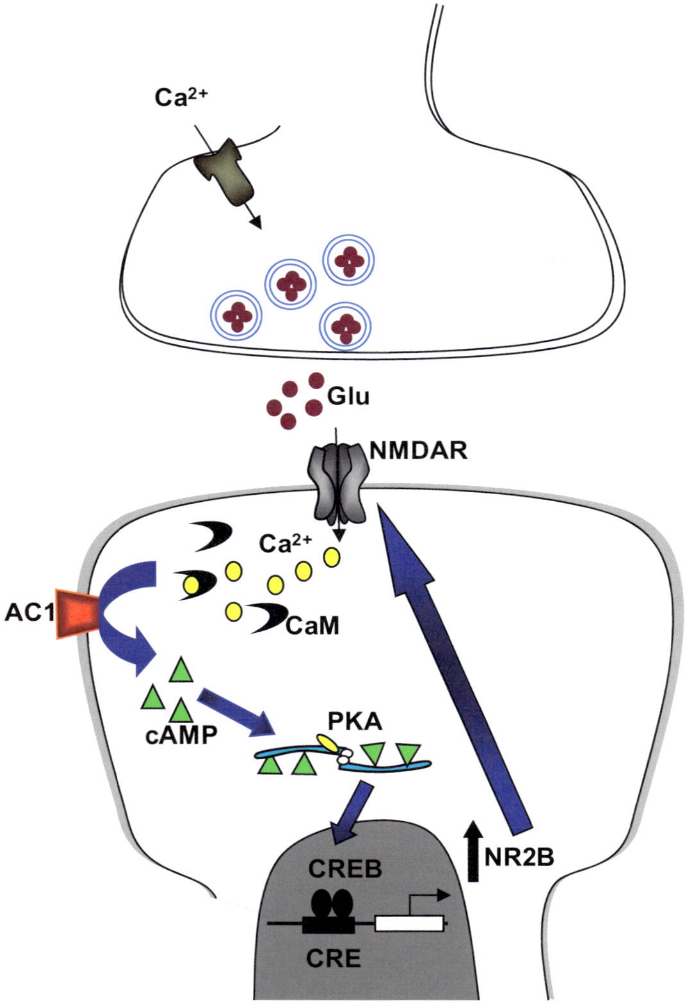

Figure 3: Identifying a cognitive gene in chronic pain: Peripheral injury such as tissue inflammation or nerve injury leads to a burst of abnormal activity, and subsequently activates postsynaptic NMDAR. This triggers Ca^{2+} influx, leading to activation of Ca^{2+}/CaM-dependent pathways, including AC1. AC1 activation leads to the generation of cAMP, which activates PKA. PKA then translocates to the nucleus and phosphorylates CREB. Postsynaptic synthesis of NMDA NR2B is then increased, and the new NR2B subunits are added to postsynaptic NMDAR. This may further enhance neuronal excitability and contribute to chronic pain.

CaMKIV GENE KNOCKOUT MICE

There has been some debate over whether increased behavioral responses seen in "smart" mice on chronic pain tests genuinely reflect the severity of pain rather than the mice being better learners. In order to address this concern, we conducted fear memory as well as pain tests on CaMKIV knockout mice. CaMKIV, like CaMKII, is a key Ca^{2+}/calmodulin-dependent kinase, and functions as a transcriptional activator [53-55]. We found that fear memory was significantly reduced in CaMKIV knockout mice, but behavioral responses to an acute noxious stimulus or prolonged injury remained comparable to wild-type mice [56]. The results indicate that CaMKIV is preferentially involved in fear memory induced by a noxious shock but not in behavioral responses to acute noxious stimuli or tissue inflammation, and that the two can be disassociated at the molecular level (Fig. **4**). Therefore, although many signaling pathways may contribute to normal physiological functions as well as to distorted pathological conditions, it is possible to identify signaling molecules that preferentially contribute to one but not the other.

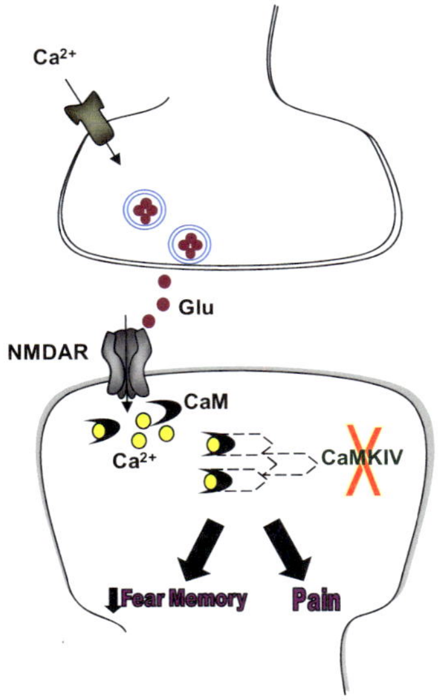

Figure 4: CaMKIV distinguishes fear memory and chronic pain: Inflammatory or nerve injuries trigger the release of excitatory neurotransmitter glutamate, and activation of NMDAR leads to an increase in postsynaptic Ca^{2+}. Ca^{2+} binds to CaM and leads to activation of Ca^{2+}/CaM-dependent protein kinases, such as CaMKIV. In the absence of CaMKIV, fear memory was significantly reduced but behavioral responses to painful stimuli remained unchanged, indicating that fear memory and pain may be dissociated at the molecular level.

SUMMARY AND FUTURE DIRECTIONS

Recent advances in genetic manipulations have transformed physiology and medicine by allowing for a direct investigation into the function of specific genes, significantly improving our knowledge of pain pathways. Studies have indicated that in the ACC, the AC1 dependent activation of the Ca^{2+}-stimulated cAMP pathway is necessary for chronic pain behaviors, and that selective elevation of cAMP is sufficient to enhance chronic pain. As well, NMDA receptors and their related signaling pathways in the forebrain have been implicated in injury-related behavioral sensitization [9, 33, 57]. Consistently, behavioral sensitization after tissue injury appears to be regulated or modulated by neuronal activity within forebrain areas. Knockout and transgenic mice provide a powerful tool for investigating the roles of genes and proteins in chronic pain. In combination with established pharmacological, behavioral, and electrophysiological methods, genetic manipulation of mice has already helped identify several molecular substrates of chronic pain, and will likely lead to novel therapeutic targets in the near future.

ACKNOWLEDGEMENTS

Supported by grants from the EJLB-CIHR Michael Smith Chair in Neurosciences and Mental Health, CIHR operating grants CIHR81086 and CIHR84256, Canada Research Chair, and WCU program to MZ. GD is funded through the E. Fidlar Ontario Graduate Scholarship.

CONFLICT OF INTEREST

The authors confirm that this chapter contents have no conflict of interest.

DISCLOSURE

The chapter submitted for Advances in Genome Science, Vol. 3 entitled **"Investigation of molecular mechanism of chronic pain in the anterior cingulate cortex using genetically engineered mice by Kim SS, Descalzi G, & Zhuo M"** is an update of our article published in *Current Genomics*. 2010 March 11(1):70-6.

REFERENCES

[1] Zhuo, M., Molecular mechanisms of pain in the anterior cingulate cortex. *J Neurosci Res* **2006,** *84* (5), 927-33.

[2] Zhuo, M., Cortical excitation and chronic pain. *Trends Neurosci* **2008,** *31* (4), 199-207.

[3] Woolf, C. J.; Salter, M. W., Neuronal plasticity: increasing the gain in pain. *Science* **2000,** *288* (5472), 1765-9.

[4] Zhuo, M., Neuronal mechanism for neuropathic pain. *Mol Pain* **2007,** *3*, 14.

[5] Porreca, F.; Ossipov, M. H.; Gebhart, G. F., Chronic pain and medullary descending facilitation. *Trends Neurosci* **2002,** *25* (6), 319-25.

[6] Waterston, R. H.; Lindblad-Toh, K.; Birney, E.; Rogers, J.; Abril, J. F.; Agarwal, P.; *et al.*, Translational investigation and treatment of neuropathic pain. *Molecular Pain* **2012,** *8* (1), 15.

[7] Xu B, Descalzi G, Ye HR, Zhuo M, Wang YW. 2012. Translational investigation and treatment of neuropathic pain. *Molecular pain* 8: 15.

[8] Wei, F.; Qiu, C. S.; Kim, S. J.; Muglia, L.; Maas, J. W.; Pineda, V. V.; *et al.*, Genetic elimination of behavioral sensitization in mice lacking calmodulin-stimulated adenylyl cyclases. *Neuron* **2002,** *36* (4), 713-26.

[9] Wei, F.; Wang, G. D.; Kerchner, G. A.; Kim, S. J.; Xu, H. M.; Chen, Z. F.; Zhuo, M., Genetic enhancement of inflammatory pain by forebrain NR2B overexpression. *Nat Neurosci* **2001,** *4* (2), 164-9.

[10] Kim, S. J.; Calejesan, A. A.; Li, P.; Wei, F.; Zhuo, M., Sex differences in late behavioral response to subcutaneous formalin injection in mice. *Brain Res* **1999,** *829* (1-2), 185-9.

[11] Wei, F.; Zhuo, M., Potentiation of sensory responses in the anterior cingulate cortex following digit amputation in the anaesthetised rat. *J Physiol* **2001,** *532* (Pt 3), 823-33.

[12] Costigan, M.; Scholz, J.; Woolf, C. J., Neuropathic pain: a maladaptive response of the nervous system to damage. *Annu Rev Neurosci* **2009,** *32*, 1-32.

[13] Attal, N.; Jazat, F.; Kayser, V.; Guilbaud, G., Further evidence for 'pain-related' behaviours in a model of unilateral peripheral mononeuropathy. *Pain* **1990,** *41* (2), 235-51.

[14] Bennett, G. J.; Xie, Y. K., A peripheral mononeuropathy in rat that produces disorders of pain sensation like those seen in man. *Pain* **1988,** *33* (1), 87-107.

[15] Kim, S. H.; Chung, J. M., An experimental model for peripheral neuropathy produced by segmental spinal nerve ligation in the rat. *Pain* **1992,** *50* (3), 355-63.

[16] Vadakkan, K. I.; Jia, Y. H.; Zhuo, M., A behavioral model of neuropathic pain induced by ligation of the common peroneal nerve in mice. *J Pain* **2005,** *6* (11), 747-56.

[17] Choi, Y.; Yoon, Y. W.; Na, H. S.; Kim, S. H.; Chung, J. M., Behavioral signs of ongoing pain and cold allodynia in a rat model of neuropathic pain. *Pain* **1994,** *59* (3), 369-76.

[18] Malmberg, A. B.; Basbaum, A. I., Partial sciatic nerve injury in the mouse as a model of neuropathic pain: behavioral and neuroanatomical correlates. *Pain* **1998,** *76* (1-2), 215-22.

[19] Seltzer, Z.; Dubner, R.; Shir, Y., A novel behavioral model of neuropathic pain disorders produced in rats by partial sciatic nerve injury. *Pain* **1990,** *43* (2), 205-18.

[20] Sauer, B., Inducible gene targeting in mice using the Cre/lox system. *Methods* **1998,** *14* (4), 381-92.

[21] Orban, P. C.; Chui, D.; Marth, J. D., Tissue- and site-specific DNA recombination in transgenic mice. *Proc Natl Acad Sci U S A* **1992,** *89* (15), 6861-5.

[22] Kuhn, R.; Schwenk, F.; Aguet, M.; Rajewsky, K., Inducible gene targeting in mice. *Science* **1995,** *269* (5229), 1427-9.

[23] Lau, J.; Minett, M.; Zhao, J.; Dennehy, U.; Wang, F.; Wood, J.; *et al.*, Temporal control of gene deletion in sensory ganglia using a tamoxifen-inducible Advillin-Cre-ERT2 recombinase mouse. *Molecular Pain* **2011,** *7* (1), 100.

[24] Sen, G. L.; Blau, H. M., A brief history of RNAi: the silence of the genes. *Faseb J* **2006,** *20* (9), 1293-9.

[25] Caplen, N. J.; Parrish, S.; Imani, F.; Fire, A.; Morgan, R. A., Specific inhibition of gene expression by small double-stranded RNAs in invertebrate and vertebrate systems. *Proc Natl Acad Sci U S A* **2001,** *98* (17), 9742-7.

[26] Khvorova, A.; Reynolds, A.; Jayasena, S. D., Functional siRNAs and miRNAs exhibit strand bias. *Cell* **2003,** *115* (2), 209-16.

[27] Zhang, S.; Zhao, B.; Jiang, H.; Wang, B.; Ma, B., Cationic lipids and polymers mediated vectors for delivery of siRNA. *J Control Release* **2007,** *123* (1), 1-10.

[28] Costantini, L. C.; Bakowska, J. C.; Breakefield, X. O.; Isacson, O., Gene therapy in the CNS. *Gene Ther* **2000,** *7* (2), 93-109.

[29] During, M. J.; Ashenden, L. M., Towards gene therapy for the central nervous system. *Mol Med Today* **1998,** *4* (11), 485-93.

[30] Luo, D.; Saltzman, W. M., Synthetic DNA delivery systems. *Nat Biotechnol* **2000,** *18* (1), 33-7.

[31] Schwarze, S. R.; Ho, A.; Vocero-Akbani, A.; Dowdy, S. F., *In vivo* protein transduction: delivery of a biologically active protein into the mouse. *Science* **1999,** *285* (5433), 1569-72.

[32] Shi, N.; Pardridge, W. M., Noninvasive gene targeting to the brain. *Proc Natl Acad Sci U S A* **2000,** *97* (13), 7567-72.

[33] Wei, F.; Xia, X. M.; Tang, J.; Ao, H.; Ko, S.; Liauw, J.; *et al.*, Calmodulin regulates synaptic plasticity in the anterior cingulate cortex and behavioral responses: a microelectroporation study in adult rodents. *J Neurosci* **2003,** *23* (23), 8402-9.

[34] Zhao, M. G.; Toyoda, H.; Lee, Y. S.; Wu, L. J.; Ko, S. W.; Zhang, *et al.*, Roles of NMDA NR2B subtype receptor in prefrontal long-term potentiation and contextual fear memory. *Neuron* **2005,** *47* (6), 859-72.

[35] Woolf, C. J.; Mannion, R. J.; Neumann, S., Null mutations lacking substance: elucidating pain mechanisms by genetic pharmacology. *Neuron* **1998,** *20* (6), 1063-6.

[36] Aley, K. O.; Levine, J. D., Role of protein kinase A in the maintenance of inflammatory pain. *J Neurosci* **1999,** *19* (6), 2181-6.

[37] Malmberg, A. B.; Brandon, E. P.; Idzerda, R. L.; Liu, H.; McKnight, G. S.; Basbaum, A. I., Diminished inflammation and nociceptive pain with preservation of neuropathic pain in mice with a targeted mutation of the type I regulatory subunit of cAMP-dependent protein kinase. *J Neurosci* **1997,** *17* (19), 7462-70.

[38] Sluka, K. A., Activation of the cAMP transduction cascade contributes to the mechanical hyperalgesia and allodynia induced by intradermal injection of capsaicin. *Br J Pharmacol* **1997,** *122* (6), 1165-73.

[39] Taiwo, Y. O.; Levine, J. D., Further confirmation of the role of adenyl cyclase and of cAMP-dependent protein kinase in primary afferent hyperalgesia. *Neuroscience* **1991,** *44* (1), 131-5.

[40] Chetkovich, D. M.; Sweatt, J. D., nMDA receptor activation increases cyclic AMP in area CA1 of the hippocampus *via* calcium/calmodulin stimulation of adenylyl cyclase. *J Neurochem* **1993,** *61* (5), 1933-42.

[41] Wong, S. T.; Athos, J.; Figueroa, X. A.; Pineda, V. V.; Schaefer, M. L.; Chavkin, C. C.; *et al.*, Calcium-stimulated adenylyl cyclase activity is critical for hippocampus-dependent long-term memory and late phase LTP. *Neuron* **1999,** *23* (4), 787-98.

[42] Li, X. Y.; Ko, H. G.; Chen, T.; Descalzi, G.; Koga, K.; Wang, H.; *et al.*l Alleviating neuropathic pain hypersensitivity by inhibiting PKMzeta in the anterior cingulate cortex. *Science* **2010,** *330* (6009), 1400-4.

[43] Wang, H.; Xu, H.; Wu, L.-J.; Kim, S. S.; Chen, T.; Koga, K.; *et al.* Identification of an Adenylyl Cyclase Inhibitor for Treating Neuropathic and Inflammatory Pain. *Science Translational Medicine* **2011,** *3* (65), 65ra3.

[44] Zeitz, K. P.; Giese, K. P.; Silva, A. J.; Basbaum, A. I., The contribution of autophosphorylated alpha-calcium-calmodulin kinase II to injury-induced persistent pain. *Neuroscience* **2004,** *128* (4), 889-98.

[45] Wang, H.; Shimizu, E.; Tang, Y. P.; Cho, M.; Kyin, M.; Zuo, W.; *et al.*, Inducible protein knockout reveals temporal requirement of CaMKII reactivation for memory consolidation in the brain. *Proc Natl Acad Sci U S A* **2003,** *100* (7), 4287-92.

[46] Wei, F.; Wang, G. D.; Zhang, C.; Shokat, K. M.; Wang, H.; Tsien, J. Z.;*et al.*, Forebrain overexpression of CaMKII abolishes cingulate long term depression and reduces mechanical allodynia and thermal hyperalgesia. *Mol Pain* **2006,** *2*, 21.

[47] Wu, L. J.; Steenland, H. W.; Kim, S. S.; Isiegas, C.; Abel, T.; Kaang, B. K.; Zhuo, M., Enhancement of presynaptic glutamate release and persistent inflammatory pain by increasing neuronal cAMP in the anterior cingulate cortex. *Mol Pain* **2008,** *4*, 40.

[48] Isiegas, C.; McDonough, C.; Huang, T.; Havekes, R.; Fabian, S.; Wu, L. J.; *et al.*, A novel conditional genetic system reveals that increasing neuronal cAMP enhances memory and retrieval. *J Neurosci* **2008,** *28* (24), 6220-30.

[49] Chang, D. J.; Li, X. C.; Lee, Y. S.; Kim, H. K.; Kim, U. S.; Cho, N. J.; *et al.*, Activation of a heterologously expressed octopamine receptor coupled only to adenylyl cyclase produces all the features of presynaptic facilitation in aplysia sensory neurons. *Proc Natl Acad Sci U S A* **2000,** *97* (4), 1829-34.

[50] Hollmann, M.; Heinemann, S., Cloned glutamate receptors. *Annu Rev Neurosci* **1994,** *17*, 31-108.

[51] Monyer, H.; Burnashev, N.; Laurie, D. J.; Sakmann, B.; Seeburg, P. H., Developmental and regional expression in the rat brain and functional properties of four NMDA receptors. *Neuron* **1994,** *12* (3), 529-40.

[52] Wu, L. J.; Toyoda, H.; Zhao, M. G.; Lee, Y. S.; Tang, J.; Ko, S. W.; *et al.*, Upregulation of forebrain NMDA NR2B receptors contributes to behavioral sensitization after inflammation. *J Neurosci* **2005,** *25* (48), 11107-16.

[53] Bito, H.; Deisseroth, K.; Tsien, R. W., Ca2+-dependent regulation in neuronal gene expression. *Curr Opin Neurobiol* **1997,** *7* (3), 419-29.

[54] Ho, N.; Liauw, J. A.; Blaeser, F.; Wei, F.; Hanissian, S.; Muglia, L. M.; *et al.*, Impaired synaptic plasticity and cAMP response element-binding protein activation in Ca2+/calmodulin-dependent protein kinase type IV/Gr-deficient mice. *J Neurosci* **2000,** *20* (17), 6459-72.

[55] Hook, S. S.; Means, A. R., Ca(2+)/CaM-dependent kinases: from activation to function. *Annu Rev Pharmacol Toxicol* **2001,** *41*, 471-505.

[56] Wei, F.; Qiu, C. S.; Liauw, J.; Robinson, D. A.; Ho, N.; Chatila, T.; Zhuo, M., Calcium calmodulin-dependent protein kinase IV is required for fear memory. *Nat Neurosci* **2002,** *5* (6), 573-9.

[57] Zhuo, M., Central plasticity in pathological pain. *Novartis Found Symp* **2004,** *261*, 132-45; discussion 145-54.

Index

A

B

C

D

E

H

I

K

L

M

R

S

Made in the USA
Las Vegas, NV
10 November 2026

57851952R00188